Deduct

Rossella Lupacchini, Giovanna Corsi (Eds.)

Deduction, Computation, Experiment

Exploring the Effectiveness of Proof

Springer

Rossella Lupacchini
Giovanna Corsi
Dipartimento di Filosofia
Università degli Studi di Bologna

Library of Congress Control Number: 2008932581

ISBN 978-88-470-0783-3 Springer Berlin Heidelberg New York
e-ISBN 978-88-470-0784-0

This work is subject to copyright. All rights are reserved, whether the whole or part of the material is concerned, specifically the rights of translation, reprinting, reuse of illustrations, recitation, broadcasting, reproduction on microfilm or in any other way, and storage in data banks. Duplication of this publication or parts thereof is permitted only under the provisions of the German Copyright Law of September 9, 1965, in its current version, and permission for use must always be obtained from Springer. Violations are liable to prosecution under the German Copyright Law.

Springer is a part of Springer Science+BusinessMedia

© Springer-Verlag Italia 2008

The use of general descriptive names, registered names, trademarks, etc. in this publication does not imply, even in the absence of a specific statement, that such names are exempt from the relevant protective laws and regulations and therefore free for general use.

Cover design: Simona Colombo, Milano
Cover figure: Edoardo Romagnoli, www.edoardoromagnoli.it
Typeset by the authors using a Springer Macro package
Printing and binding: Grafiche Porpora, Segrate, Milano

Printed on acid-free paper 57/3141/NN - 5 4 3 2 1 0

Printed in Italy

Springer-Verlag Italia Srl, Via Decembrio 28, I-20137 Milano

Preface

This volume is located in a cross-disciplinary field bringing together mathematics, logic, natural science and philosophy. Reflection on the effectiveness of proof brings out a number of questions that have always been latent in the informal understanding of the subject. What makes a symbolic construction significant? What makes an assumption reasonable? What makes a proof reliable? Gödel, Church and Turing, in different ways, achieve a deep understanding of the notion of effective calculability involved in the nature of proof. Turing's work in particular provides a "precise and unquestionably adequate" definition of the general notion of a formal system in terms of a machine with a finite number of parts. On the other hand, Eugene Wigner refers to the unreasonable effectiveness of mathematics in the natural sciences as a miracle.

Where should the boundary be traced between mathematical procedures and physical processes? What is the characteristic use of a proof as a computation, as opposed to its use as an experiment? What does natural science tell us about the effectiveness of proof? What is the role of mathematical proofs in the discovery and validation of empirical theories? The papers collected in this book are intended to search for some answers, to discuss conceptual and logical issues underlying such questions and, perhaps, to call attention to other relevant questions.

Can every 'real' proof be translated into a 'formal' proof? Although Hilbert and Gentzen's positive answer is widely shared, there are also reasons for disagreement. To deals with this matter Carlo Cellucci addresses two fundamental questions - Why proof? What is a proof? - which he settles by contrasting the notion of axiomatic proof with the notion of analytic proof.

The contribution by Andrea Cantini concentrates on the nature and role of formal proofs. It is argued that formal proofs do not target certainty or formalistic foundations. Recent results in proof theory are considered in order to illustrate the role of formal proofs in exploring ideas and clarifying foundational questions in mathematics. The question is raised to what extent are proofs for mathematics what experimental procedures are for empirical sciences?

Closely related to this topic is the question as to what role mathematics should play in certain physical theories lacking both of rigorous mathematical structures and of experimental verifications. The notion of "theoretical mathematics", as a synthesis of theoretical physics and mathematics, is elucidated by Annalisa Marzuoli. By creating "toy models", *i.e.* simplified models of complex physical systems, not only can mathematics set up a basis for testing physical theories, but mathematical proofs may become more compelling than experiments.

On the other hand, the significance and the diverse degrees of involvement of 'experimental' methods in mathematics are investigated by Gabriele Lolli. Examples are taken throughout the history of mathematics to throw doubt on the empiricist view of mathematics as a "quasi-empirical" science and to maintain the distinctive symbolic character of mathematics.

These same problems have made it necessary for mathematics to be concerned about proofs produced by machines. Dag Prawitz's paper explores conceptual questions as to the use of deductive machinery to verify the correctness of computer programs and to the running of programs on computers to produce proofs.

Giovanna Corsi presents a Gentzen-style calculus as a case study for the discussion of typical metatheoretical properties: when is that a proof-tree is closed? when is a proof-tree cut-free? when is it analytic?

The intercalation method for proof search is extended from pure first-order logic to parts of mathematics by Wilfried Sieg and Clinton Field. By interweaving general logical strategies with specific mathematical heuristic, proofs of significant theorems are found in a fully automated way. They present a solution for Gödel's incompleteness theorems.[1]

New perspectives in computational complexity theory are examined by Ugo Dal Lago and Simone Martini in the frame of the "Curry-Howard correspondence". By adopting a kind of proof-theoretical approach, the so-called "Implicit Computational Complexity" takes into account single machine-free models of computation and analyses complexity classes with respect to language constraints.

The contribution by Mario Rasetti suggests how one might benefit from quantum information tools in order to elicit wider fields than mere computation. In particular, the problem of 'combing' finite groups brings to light relevant connections between language-theoretical and algorithmic-structural issues. To open up more comprehensive fields a complex blend of notions from different theoretical regions - such as formal languages, finite groups and quantum computation - is needed.

From an 'outside' point of view, *i.e.* the detached view of the philosopher of science, Dag Westerståhl discusses the conflict between classical and intuitionistic mathematics. By focusing on proofs rather than on explanations

---

[1] This paper is reprinted from *Annals of Pure and Applied Logic* **133** (2005) pp. 319-338

of meaning, mutual understanding between classical and intuitionistic mathematicians may be significantly improved. 'Distinguishability' may help to clarify how intuitionists and classical mathematicians understand proofs by grading classical and constructive proofs.

Rossella Lupacchini's paper examines how a search for 'more effective' distinguishability leads quantum theory in complex Hilbert spaces and opens up new computational paths. Specifically the bilateral symmetry involved in the very notion of distinguishability, emerging out of quantum probability amplitudes, can provide an argument for the irreducibility of quantum structures to classical ones.

While there is a classical standard model of computability on natural numbers, *i.e.* in the style of the Turing machine, there are several nonequivalent theories of computability for real numbers. In order to illustrate different approaches, Guido Gherardi investigates the computability of the wave equation as a key example of a partial differential equation system in theoretical physics. Here the difficulty of reconciling the discrete nature of computability with the continuous character of motion equations stands out.

A reflection on the nature of incompleteness and its crucial role in the evaluation of the effectiveness of both mathematics and physics is developed by Francis Bailly and Giuseppe Longo. The phenomenon of incompleteness is analysed in the context of Gödel's logical results as well as of quantum theory. A general constructivist approach to knowledge underlies this attempt to achieve a 'unified' understanding of apparently unrelated theoretical issues.

Acknowledgements

This book is largely drawn from a Workshop hosted by the University Library of Bologna (3-4 April 2007). We are grateful to all the participants for their valuable contributions and to the Director of the Library for her assistance. Funds came from the University of Bologna, to which we express our thanks. Much is due to Giorgio Sandri for the organization of the Workshop and for his encouragment in the preparation of this volume. We are also grateful to the reviewers for their detailed comments on the manuscripts and to Guido Gherardi for his help with the revision of the texts.

Bologna,
June 2008

Giovanna Corsi
Rossella Lupacchini

Contents

Why Proof? What is a Proof?
Carlo Cellucci 1

On Formal Proofs
Andrea Cantini 29

Toy Models in Physics and the Reasonable Effectiveness of Mathematics
Annalisa Marzuoli 49

Experimental Methods in Proofs
Gabriele Lolli 65

Proofs Verifying Programs and Programs Producing Proofs: A Conceptual Analysis
Dag Prawitz 81

The Logic of the Weak Excluded Middle: A Case Study of Proof-Search
Giovanna Corsi 95

Automated Search for Gödel's Proofs
Wilfried Sieg, Clinton Field 117

Proofs as Efficient Programs
Ugo Dal Lago, Simone Martini 141

Quantum Combing
Mario Rasetti 159

Proofs instead of Meaning Explanations: Understanding Classical *vs* Intuitionistic Mathematics from the Outside
Dag Westerståhl 175

Proof as a Path of Light
Rossella Lupacchini 195

Computability and Incomputability of Differential Equations
Guido Gherardi 223

Phenomenology of Incompleteness: From Formal Deductions to Mathematics and Physics
Francis Bailly and Giuseppe Longo 243

Index 273

List of Contributors

Francis Bailly
Physique, CNRS, Meudon
9, Rue Jules Hetzel
92190 Meudon (France)
francis.bailly@noos.fr

Andrea Cantini
Università di Firenze
Dipartimento di Filosofia
Via Bolognese 52
50139 Firenze (Italy)
cantini@philos.unifi.it

Carlo Cellucci
Università di Roma 'La Sapienza'
Dipartimento di Studi Filosofici ed
Epistemologici
Via Carlo Fea 2
00161 Roma (Italy)
carlo.cellucci@uniroma1.it

Giovanna Corsi
Università di Bologna
Dipartimento di Filosofia
Via Zamboni 38
40126 Bologna (Italy)
giovanna.corsi@unibo.it

Ugo Dal Lago
Università di Bologna
Dipartimento di Scienze
dell'Informazione
Mura Anteo Zamboni 7
40127 Bologna (Italy)
dallago@cs.unibo.it

Clinton Field
Carnegie Mellon University
Department of Philosophy
Baker Hall 135
Pittsburgh PA 15213-3890 (US)

Guido Gherardi
Università di Bologna
Dipartimento di Filosofia
Via Zamboni 38
40126 Bologna (Italy)
guido.gherardi@unibo.it

Gabriele Lolli
Università di Torino
Dipartimento di Matematica
Via Carlo Alberto 10
10123 Torino (Italy)
gabriele.lolli@unito.it

Giuseppe Longo
Ecole Normale Supérieure
CNRS et Département
d'Informatique
45, rue d'Ulm
75230 Paris Cedex 05 (France)
Giuseppe.Longo@ens.fr

Rossella Lupacchini
Università di Bologna
Dipartimento di Filosofia
Via Zamboni 38
40126 Bologna (Italy)
rossella.lupacchini@unibo.it

Simone Martini
Università di Bologna
Dipartimento di Scienze
dell'Informazione
Mura Anteo Zamboni 7
40127 Bologna (Italy)
martini@cs.unibo.it

Annalisa Marzuoli
Università di Pavia
Dipartimento di Fisica Nucleare e
Teorica
Via Bassi 6
27100 Pavia (Italy)
Annalisa.Marzuoli@pv.infn.it

Dag Prawitz
Stockholms Universitet
Filosofiska institutionen
106 91 Stockholm (Sweden)
dag.prawitz@philosophy.su.se

Mario Rasetti
Politecnico di Torino
Dipartimento di Fisica
C.so Duca degli Abruzzi 24
10129 Torino (Italy)
mario.rasetti@polito.it
Fondazione ISI
Via Settimio Severo 65
10133 Torino (Italy)
rasetti@isi.it

Wilfried Sieg
Carnegie Mellon University
Department of Philosophy
Baker Hall 135
Pittsburgh PA 15213-3890 (US)
sieg@cmu.edu

Dag Westerståhl
Göteborgs Universitet
Filosofiska institutionen
Box 200, SE 405 30 Göteborg
(Sweden)
dag.westerstahl@phil.gu.se

Why Proof? What is a Proof?

Carlo Cellucci

Dipartimento di Studi Filosofici ed Epistemologici
Università di Roma 'La Sapienza' (Italy)
carlo.cellucci@uniroma1.it

1 The Hilbert-Gentzen Thesis

This paper is concerned with real proofs as opposed to formal proofs, and specifically with the ultimate reason of real proofs ('Why Proof?') and with the notion of real proof ('What is a Proof?').

Several people believed and still believe that real proofs can be represented by formal proofs. A recent example is provided by Macintyre who claims that "one could go on to translate" all "classical informal proofs into formal proofs of some accepted formal system", where such translations "do map informal proofs to formal proofs" [44, p. 2420].

This view is to a certain extent implicit in Frege – to a certain extent only, because for Frege in a sense "every inference is non-formal in that the premises as well as the conclusions have their thought-contents which occur in this particular manner of connection only in that inference" [18, p. 318].

Anyway, the view that real proofs can be represented by formal proofs is explicitly stated by Hilbert and Gentzen.

For Hilbert claims that formal proofs are "carried out according to certain definite rules, in which the technique of our thinking is expressed" [34, p. 475]. These are "the rules according to which our thinking actually proceeds". They "form a closed system that can be discovered and definitively stated".

Similarly, Gentzen claims that formal proofs in his natural deduction systems have "a close affinity to actual reasoning" [22, p. 80]. They reflect "as accurately as possible the actual logical reasoning involved in mathematical proofs" [22, p. 74].

Thus one may state the following:

Hilbert-Gentzen Thesis. Every real proof can be represented by a formal proof.

Although the Hilbert-Gentzen thesis is widely held, there are various reasons for thinking that it is inadequate. To discuss this matter we must answer the questions: Why proof? What is a proof?

2 What is a proof?

As already mentioned, 'Why proof?' is a question about the ultimate reason of real proofs. Such question is strictly connected to the question 'What is a proof?', for the ultimate reason of real proofs depends on what real proofs are, so one can expect that an answer to the question 'What is a proof?' will yield an answer to the question 'Why proof?'.

Of course, this holds only if 'Why proof?' is meant as a question about the ultimate reason of real proofs, not as a question about their possible uses, which are multifarious. (Thirty-nine such uses are listed in Lolli [42]). Here 'Why proof?' will be meant in that sense.

There are two distinct answers to the question 'What is a proof?', which yield two essentially different and indeed alternative notions of real proof. All known apparently different notions of proof can be reduced to such two notions.

A) *The notion of axiomatic proof.* Proofs are deductive derivations of propositions from primitive premisses that are true in some sense of 'true'. They start from given primitive premisses and go down to the proposition to be proved. Their aim is to give a foundation and justification of the proposition.

B) *The notion of analytic proof.* Proofs are non-deductive derivations of plausible hypotheses from problems, in some sense of 'plausible'. They start from a given problem and go up to plausible hypotheses. Their aim is to discover plausible hypotheses capable of giving a solution to the problem.

The notion of axiomatic proof was first stated by Aristotle in *Posterior Analytics* and then modified by Pascal, Pieri, Hilbert, Padoa in that order (see Cellucci [4, Chs. 4-5]). It is a very familiar one and so does not seem to require further explanation.

The notion of analytic proof was first stated by Plato in *Meno* and *Phaedo* (see [4, pp. 270-308]). It is less familiar and so requires some explanation. The main points seem to be the following.

1) A problem is any open question.
2) A hypothesis is any means that can be used to solve a problem.
3) A hypothesis is said to be plausible if and only if it is compatible with the existing data – that is, all mathematical notions and results available at that moment – in the sense that, comparing the arguments for and the arguments against the hypothesis on the basis of the existing data, the arguments for prevail over those against. (A typical example is provided by the discussions

concerning the plausibility of the axiom, or rather hypothesis, of choice at the beginning of the twentieth century).

4) The process by which problems are solved is an application of the analytic method, which can be described as follows. One looks for some hypothesis that is a sufficient condition for solving the problem. The hypothesis is obtained from the problem, and possibly other data, by some non-deductive inference: inductive, analogical, diagrammatic, metaphorical, metonymical, by generalization, by specialization, by variation of the data, and so on. (On different kinds of non-deductive inferences for finding hypotheses, see Cellucci [6, pp. 235-295]. The hypothesis must not only be a sufficient condition for solving the problem, but must also be plausible, that is, compatible with the existing data. However the hypothesis, in turn, is a problem that must be solved, and will be solved in the same way. That is, one will look for another hypothesis that is a sufficient condition for solving the problem posed by the former hypothesis, it is obtained from it, and possibly other data, by some non-deductive inference, and must be plausible. And so on, *ad infinitum*. Thus the solution of a problem is a potentially infinite process.

In the course of this process the statement of the problem may be modified to a certain extent to make it more precise, or may even be radically changed as new data emerge. Thus the development of the statement of the problem and the development of the solution of the problem may proceed in parallel.

The analytic method is both a method of discovery and a method of justification. It involves two distinct processes, the formulation of candidates for hypothesis by means of non-deductive inferences, and the choice among such candidates on the grounds of their plausibility. Such choice is necessary because non-deductive inferences can yield different conclusions from the very same premisses. To choose among such conclusions one must carefully assess the arguments for and the arguments against each of them on the basis of the existing data. Such assessment is a process of justification, so justification is part of discovery. (For more on the analytic method, see [4], [6], [7], [11]).

The axiomatic method is what results from the analytic method when the hypotheses stated at a certain stage are considered as an absolute starting point, for which no justification is given. Thus the axiomatic method is an unjustified truncation of the analytic method.

One of the oldest examples of analytic proof concerns the problem of the duplication of the cube a^3. Hippocrates of Chios solved it by showing that the hypothesis 'One can find two mean proportionals x and y in continued proportion between a and $2a$' is a sufficient condition for its solution. Then Menaechmus solved the problem posed by such hypothesis by showing that a certain other hypothesis is a sufficient condition for its solution. And so on.

A recent example of analytic proof concerns Fermat's Problem. Ribet solved it by showing that the Taniyama-Shimura conjecture – or hypothesis – is a sufficient condition for its solution. For "let E be an elliptic curve over $\mathbf{Q}$. The Taniyama-Shimura Conjecture (also known as the Weil-Taniyama Conjecture) states that E is modular" [55, p. 123]. Ribet showed: "Conjecture of

Taniyama-Shimura $\Longrightarrow$ Fermat's Last Theorem" [55, p. 127]. Thus, as stated above, Ribet solved Fermat's problem by showing that the Taniyama-Shimura hypothesis is a sufficient condition for its solution. Then Wiles and Taylor solved the problem posed by the Taniyama-Shimura hypothesis by showing that certain other hypotheses are a sufficient condition for its solution. And so on.

But already one can hear the objection: Surely Ribet did not solve Fermat's Problem, for his alleged solution depended on a hypothesis that at the time had not been proved yet! (One could hear a similar objection about Hippocrates of Chios's solution of the problem of the duplication of the cube).

Now, if Ribet did not solve Fermat's Problem because his solution depended on a hypothesis, the Taniyama-Shimura hypothesis, that at the time had not been proved yet, then Wiles and Taylor have not solved Fermat's problem because their solution depends on a hypothesis, the axioms of set theory, that to this very day has not been proved yet. (Similarly as regards Hippocrates of Chios and Menaechmus. Ribet stands to Hippocrates of Chios as Wiles and Taylor stand to Menaechmus).

One can also hear the objection: A so-called 'analytic proof' is not a proof. Admittedly, working backwards to find the needed ingredients to prove a potential theorem is a standard method in the work of mathematicians. But this cannot justifiably be referred to as proof, for what is customarily understood by 'proof' is a sequence of arguments to justify an assertion.

Now, at each stage in the development of an analytic proof, only a finite piece of the proof is given and, reading it top down rather than bottom up, one has a sequence of arguments that justifies an assertion. The sequence justifies it because hypotheses must be plausible. Reading the sequence top down rather than bottom up is inessential, it is just a matter of convention.

3 Analytic and axiomatic proof

The point of analytic proof can be seen in terms of Gödel's first incompleteness theorem.

Suppose that you want to solve a problem, say, of elementary number-theory and want to find a hypothesis to solve it. By Gödel's result there is no guarantee that the hypothesis can be derived from the axioms of Peano Arithmetic, so you must be prepared to look for hypotheses of any kind, concerning objects of any mathematical field. Think, for instance, of Fermat's problem, a problem concerning natural numbers that, as we have already mentioned, Ribet solved using a hypothesis, the Taniyama-Shimura, concerning elliptic curves over $\mathbf{Q}$ – the set of rational numbers.

Moreover, again by Gödel's result, there is no guarantee that the hypothesis can be derived from any known axioms. Think, for instance, of Gödel's suggestion that we might need new infinity axioms to solve number-theoretic problems (see [23, p. 269]). Thus solving a problem generally consists in looking for hypotheses in an open, that is, not predetermined space.

The point of analytic proof can be also seen, perhaps more vividly, in terms of Hamming's statement: "If the Pythagorean theorem were found to not follow from postulates, we would again search for a way to alter the postulates until it was true. Euclid's postulates came from the Pythagorean theorem, not the other way" [26, p. 87]. In mathematics "you start with some of the things you want and you try to find postulates to support them" [27, p. 645]. The idea that you simply lay down some arbitrary postulates and then make deductions from them "does not correspond to simple observation" [26, p. 87].

In addition to giving alternative answers to the question 'What is a Proof?', the notions of axiomatic proof and analytic proof give alternative answers to the question 'Why Proof?'. For the ultimate reason of axiomatic proof is to give a foundation and justification of a proposition, and the ultimate reason of analytic proof is to discover plausible hypotheses capable of giving a solution to a problem.

Since their ultimate reasons are different, the notions of axiomatic and analytic proof play different roles in the development of mathematics.

Axiomatic proof, being meant to give a foundation and justification of an already acquired proposition, is not intrinsically fruitful for the creation of new mathematics.

On the contrary, analytic proof has a great heuristic value, not only because it is meant to discover plausible hypotheses capable of giving a solution to a problem, but also because such hypotheses may belong to areas of mathematics different from the one to which the problem belongs. Thus they may establish connections between the problem and concepts and results of other areas of mathematics. This may reveal unexpected relations between different areas, which may suggest new perspectives and new problems and so may be very fruitful for the development of mathematics. As Grosholz says, these new perspectives and problems may allow one "to explore the analogies among disparate things, a practice which in the formal sciences tends to generate new intelligible things" [25, p. 49].

The notions of axiomatic and analytic proof yield alternative notions of mathematical theory.

In terms of axiomatic proof, a mathematical theory is a closed set of primitive premisses and propositions obtained from them by deductive inferences – a closed set, because primitive premisses are predetermined and given once for all, and the propositions belonging to the set in question are entirely determined by the primitive premisses. Briefly, a mathematical theory is a closed system.

In terms of analytic proof, a mathematical theory is an open set of problems and hypotheses for their solution obtained from the problems by non-deductive inferences – an open set, because hypotheses are not predetermined or given once for all and the solution of a problem may generate new problems.

Briefly, a mathematical theory is an open system. (For more on these notions of closed and open system, see [4, pp. 309-347], [5], [6, Chs. 7 and 26]).

4 Analytic and analytic-synthetic method

The analytic method must not be confused with the analytic-synthetic method. While the analytic method is a method for finding hypotheses to solve given problems, the analytic-synthetic method is a method for finding deductions of given propositions from given primitive premisses (axioms, rules, definitions), thus it is only a heuristic pattern within axiomatized mathematics.

In the analytic-synthetic method, to find a deduction of a given proposition from given primitive premisses, one looks for premisses from which that proposition will follow, then one looks for premisses from which those premisses will follow, and so on until one arrives at some primitive premisses among the given ones. If this process is successful, then inverting the path direction – that is, repeating the steps in inverse order – one gets a deduction of the given proposition from the given primitive premisses, as desired.

While in the analytic method finding a solution of a given problem is a potentially infinite process, in the analytic-synthetic method finding a deduction of a given proposition from given primitive premisses is a finite process.

Actually, there are two versions of the analytic-synthetic method, originally described by Aristotle and Pappus, respectively, which differ as to the direction of the analysis. In Aristotle's version the direction is upward, in Pappus's version it is downward (see [4, pp. 289-299], [11, Ch. 15]). Here we need only consider Aristotle's version, that is the one described above. (For more on the analytic-synthetic method, see Hintikka-Remes [37], Knorr [40], Mäenpää [45], Timmermans [59]).

5 Frege's Thesis

Supporters of the notion of axiomatic proof assume that all proofs come under that notion. This is due to the influence of Frege, who sharply separates the context of discovery from the context of justification, limiting logic to the latter and confining the context of discovery to individual psychology (see [19, p. 5], [17, p. 3]). Contemporary supporters of the notion of axiomatic proof follow Frege. Thus one may state the following:

Frege's Thesis. Every real proof is an axiomatic proof.

The Hilbert-Gentzen Thesis can be viewed as an extreme form of Frege's Thesis, for it implies that every real proof not only is an axiomatic proof but is also, 'up to representation', a formal proof. Azzouni states such implication by saying that "ordinary mathematical proofs indicate (one or another) mechanically checkable derivation of theorems from the assumptions those ordinary mathematical proofs presuppose" [2, p. 105]. So "it's derivations, derivations in one or another algorithmic system, which underlie what's characteristic of mathematical practice" [2, p. 83]. (For a critical appraisal of Azzouni's views, see Rav [54]).

6 Proofs as means of discovery or justification

While Hilbert and Gentzen build on Frege's Thesis, Aristotle who, as we have already mentioned, first stated the notion of axiomatic proof, would have rejected it. For, although Aristotle sharply distinguishes between the procedure by which new propositions are obtained and the procedure by which propositions already obtained are organized and presented, he considers both such procedures as belonging to logic. Aristotle views the former as the procedure of the working mathematician, the latter as the procedure for teaching and learning propositions already obtained, and attributes only the latter to the notion of axiomatic proof.

For Aristotle states that the procedure by which new propositions are obtained consists in a method that will tell us "how we may always find a deduction to solve any given problem, and by what way we may reach the primitive premisses adequate to each problem" (Aristotle, *Analytica Priora*, A 27, 43a 20-22). The method "is useful with respect to the first elements in each science" (Aristotle, *Topica*, A 2, 101a 36-37). For, "being used in the investigation, it directs to the primitive premisses of all sciences" (*ibid.*, A 2, 101b 3-4). On the other hand, the procedure by which propositions already obtained are organized and presented consists in the axiomatic method, for "we know things through demonstrations", where demonstration is "scientific deduction" (Aristotle, *Analytica Posteriora*, A 2, 71b 17-19). That is, it is a deduction which proceeds "from premisses that are true and primitive" (*ibid.*, A 2, 71b 20-21). The importance of demonstration depends on the fact that "all teaching and intellectual learning" is obtained by means of it, in particular "the mathematical sciences are acquired in this way" (*ibid.*, A 1, 71a 1-4).

Aristotle's description of the procedure by which propositions already obtained are organized and presented corresponds to the notion of axiomatic proof. On the other hand, his description of the procedure by which new propositions are obtained does not correspond to that notion. It does not correspond to the notion of analytic proof either because, for Aristotle, the process by which new propositions are obtained is finite. Rather, it corresponds to the procedure by which deductions of given propositions from given primitive premisses are obtained in – Aristotle's version of – the analytic-synthetic method.

That, for Aristotle, the procedure by which new propositions are obtained is finite depends on his argument that infinite regress is inadmissible, otherwise one could prove everything, including falsehood. To stop infinite regress Aristotle assumes that there must be some primitive premisses that are true, and must also be known to be true, otherwise one would be unable to tell whether something is a demonstration.

For Aristotle claims that, if the series of premisses "did not terminate and there was always something above whatever premiss has been taken, then there would be demonstrations of all things" (Aristotle, *Analytica Posteriora*, A 22, 84a 1-2.). Thus there must be premisses that must be "primitive and indemonstrable, because otherwise there would be no scientific knowledge", and moreover "must be true, because it is impossible to know what is not the case" (*ibid.*, A 2, 71b 25-27). In addition to being true, primitive premisses must also be known to be true, for "if it is impossible to know the primitive premisses, then it is impossible to have scientific knowledge of what proceeds from them absolutely and properly" (*ibid.*, A 2, 72b 13-14). And to know the primitive premisses amounts to knowing that they are true, for "grasping and stating" them "is truth" (Aristotle, *Metaphysica*, Θ 10, 1051b 24).

Moreover, Aristotle claims that we know that primitive premisses are true by intuition. For since "there cannot be scientific knowledge of the primitive premisses, and since nothing except intuition can be truer than scientific knowledge, it will be intuition that apprehends the primitive premisses" (Aristotle, *Analytica Posteriora*, B 19, 100b 10-12). So "it is intuition that grasps the unchangeable and first terms in the order of proofs" (Aristotle, *Ethica Nicomachea*, Z 11, 1143b 1-2).

However reasonable such Aristotle's claims may appear, nevertheless they are untenable.

Aristotle's claim that, since nothing except intuition can be truer than scientific knowledge, it will be intuition that apprehends the primitive premisses, is untenable because intuition is an unreliable source of knowledge. Kripke states: "I think" that intuition "is very heavy evidence in favor of anything, myself. I really don't know, in a way, what more conclusive evidence one can have about anything, ultimately speaking" [41, p. 42]. Actually just the opposite is true. Being completely subjective and arbitrary, intuition cannot be used as evidence for anything. One really doesn't know what less conclusive evidence one could have about anything, ultimately speaking. For instance, Frege considered his paradoxical Basic Law V completely intuitive since, in his opinion, it "is what people have in mind, for example, where they speak of the extensions of concepts" [18, p. 4]. But Russell's paradox showed that Frege's intuition was wrong. On the other hand, completely counterintuitive propositions, the so-called 'monsters', have been proved in various parts of mathematics. (On intuition and 'monsters', see [6, Ch. 12] and [11, Ch. 8]).

Moreover, Aristotle's claim that, if the series of premisses did not terminate, then there would be demonstrations of all things, is untenable because in the analytic method, which involves a potentially infinite regress, premisses – that is, hypotheses – must be plausible, that is, compatible with the existing data, so there can only be demonstrations of things using plausible premisses.

Of course, a price has to be paid for that. Since plausible premisses are not certain, the things proved by demonstrations are not certain, so mathematics is not certain. But, in view of the unreliability of intuition, there is no alternative to that. As Xenophanes said, "as for certain truth, no man has known it, nor will he know it" for "all is but a woven web of guesses" [16, 21 B 34]. And yet knowledge, uncertain knowledge, is possible, for "with due time, through seeking, men may learn and know things better" [16, 21 B 18].

7 The status of the Hilbert-Gentzen Thesis

In the light of what has been stated above, the status of the Hilbert-Gentzen Thesis can be assessed as follows.

1) If by 'proof' one means 'analytic proof', then the Hilbert-Gentzen Thesis is obviously inadequate because formal proofs don't represent analytic proofs.

2) If by 'proof' one means 'axiomatic proof', then the Hilbert-Gentzen Thesis is inadequate because, for instance, even the very first proof in Hilbert's *Grundlagen der Geometrie* cannot be represented by a formal proof since it makes an essential use of properties of a figure (see [9] and [11, Ch. 9]). This belies Hilbert's claim that "a theorem is only proved when the proof is completely independent of the figure" [36, p. 75]. Admittedly, one can give a purely formal proof of the same result, but this involves replacing the use of the figure by the use of additional primitive premisses (see Meikle-Fleuriot [46]). Then the resulting formal proof is essentially different from, and hence cannot be considered a representation of, Hilbert's proof. Generally the use of figures is crucial in mathematics. As Grosholz says, "number and figure are the Adam and Eve of mathematics" [25, p. 47].

3) If by 'proof' one means 'axiomatic proof', then the Hilbert-Gentzen Thesis is inadequate also for the the more basic reason that the notion of axiomatic proof itself is inadequate. It is widely believed that the axiomatic method "guarantees the truth of a mathematical assertion" [56, p. 135]. This belief depends on the assumption that proofs are deductive derivations of propositions from primitive premisses that are true, in some sense of 'true'. Now, as we will presently see, generally there is no rational way of knowing whether primitive premisses are true. Thus either primitive premisses are false, so the proof is invalid, or primitive premisses are true but there is no rational way of knowing that they are true, then one will be unable to see whether something is a proof, and hence will be unable to distinguish proofs from nonproofs. In both cases, the claim that the axiomatic method guarantees the truth of a mathematical assertion is untenable.

8 The truth of primitive premisses

We have claimed that generally there is no rational way of knowing whether primitive premisses are true. This can be seen as follows.

That primitive premisses are true can be meant in several distinct senses. The main ones are the following: 1) truth as possession of a model; 2) truth as consistency; 3) truth as convention.

8.1 Truth as possession of a model

Primitive premisses are true in the sense that they have a model, that is, there is a domain of objects in which they are true.

For instance, Tarski says that we "arrive at a definition of truth and falsehood simply by saying that a sentence is true" in a given domain "if it is satisfied by all objects" in that domain, "and false otherwise" [58, p. 353]. Then a sentence is true if and only if there is a domain of objects in which it is true.

But, if primitive premisses are true in the sense that they have a model, then to know that they are true one must be able to prove that they have a model. However, by Gödel's second incompleteness theorem, the sentence 'Primitive premisses have a model' will not be provable from such primitive premisses but only from a proper extension of them, whose primitive premisses have a model. However, by Gödel's second incompleteness theorem, the sentence 'The primitive premisses of the proper extension have a model' will not be provable from such primitive premisses but only from a proper extension of them, whose primitive premisses have a model. And so on, *ad infinitum*.

Thus there is no rational way of knowing whether primitive premisses are true in the sense that they have a model.

8.2 Truth as consistency

Primitive premisses are true in the sense that they are consistent, that is, no contradiction is provable from them.

For instance, Hilbert says that, "if arbitrarily given axioms do not contradict one another with all their consequences, then they are true" [35, p. 39]. Thus "'non-contradictory' is the same as 'true' " [33, p. 122].

But, if primitive premisses are true in the sense that they are consistent, then to know that they are true one must be able to prove that they are consistent. However, by Gödel's second incompleteness theorem, the sentence 'The primitive premisses are consistent' will not be provable from such primitive premisses but only from a proper extension of them, whose primitive premisses are consistent. However, by Gödel's second incompleteness theorem, the sentence 'The primitive premisses of the proper extension are consistent' will not be provable from such primitive premisses but only from a proper extension of them, whose primitive premisses are consistent. An so on, *ad infinitum*.

Thus there is no rational way of knowing whether primitive premisses are true in the sense that they are consistent.

8.3 Truth as convention

Primitive premisses are true in the sense that they are conventions, that is, they may be chosen arbitrarily, subject to no condition whatsoever.

For instance, Carnap says that "it is not our business to set up prohibitions, but to arrive at conventions" [3, p. 51]. Primitive premisses "may be chosen quite arbitrarily" and this "choice, whatever it may be, will determine what meaning is to be assigned to the fundamental logical symbols" [3, p. xv]. Thus "no question of justification arises at all, but only the question of the syntactical consequences to which one or the other of the choices leads". A sentence is said to be 'determinate' if its truth or falsity is settled by the syntactical consequence relation alone, which thus provides "a complete criterion of validity for mathematics" [3, p. 100]. Sentences "may be divided into logical and descriptive, i.e. those which have a purely logical, or mathematical, meaning" and those which express "something extralogical – such as empirical" facts, "properties, and so forth" [3, p. 177]. Then "every logical sentence is determinate; every indeterminate sentence is descriptive" [3, p. 179].

But, if primitive premisses are true in the sense that they are conventions, then to know that they are true one must know that they are true with respect to the meaning their choice assigns to the fundamental logical symbols. However, by Gödel's first incompleteness theorem, there are sentences of Peano Arithmetic, say, that are indeterminate and hence descriptive, so for Carnap they express something extralogical. This means that the primitive premisses of Peano Arithmetic don't fully determine the meaning of the fundamental logical symbols, which will then be partly extralogical. Thus, to know that the primitive premisses of Peano Arithmetic are true involves considering something extralogical, say, some empirical facts.

To overcome this problem Carnap considers the possibility of expanding the primitive premisses of Peano Arithmetic by adding an inference rule with infinitely many premisses, the ω-rule, which allows one to infer $\forall x A(x)$ from $A(0), A(1), A(2), \ldots$ and makes all sentences of Peano Arithmetic determinate. Carnap claims that "there is nothing to prevent the practical application of such a rule" [3, p. 173]. But the syntactical consequence relation resulting from this addition is not recursively enumerable, and hence a fortiori, in Carnap's parlance, it is indefinite. For, according to Carnap, "every definite" relation "can be calculated", whereas in this case there exists no "definite method by means of which this calculation" can "be achieved" [3, p. 46]. So the ω-rule yields "a method of deduction which depends upon indefinite individual steps" [3, p. 100].

Thus any choice of primitive premisses for Peano Arithmetic either will not fully determine the meaning of the fundamental logical symbols – which will then be partly extralogical – or will yield an indefinite syntactical consequence relation.

Moreover, by Gödel's second incompleteness theorem, one cannot know whether primitive premisses are consistent. This is problematic for, if primitive premisses were inconsistent, then it would be worthless to know that a sentence is a syntactical consequence of them. Since one cannot know whether primitive premisses are consistent, the only ground one would have to believe that their syntactical consequences include no contradiction would be inductive, that is, it would consist in the fact that until then no contradiction has been drawn from them. But then induction, not convention, would be the basis of the choice of primitive premisses.

Thus we may conclude that there is no rational way of knowing whether primitive premisses are true in the sense that they are conventions.

This is the substance of Plato's criticism of the axiomatic method. Plato asks: "When a man does not know the principle, and when the conclusion and intermediate steps are also constructed out of what he does not know, how can he imagine that such a fabric of convention can ever become science?" (Plato, *Republic*, VII 533 c 4-5). Carnap has no answer to this question. (For more on Plato's criticism of the axiomatic method, see [4, pp. 286-291]).

9 Decline and fall of axiomatic proof

As we have already said, to stop infinite regress Aristotle assumes that there must be primitive premisses that are true and are also known to be true. By what we have just seen, however, such primitive premisses cannot exist, for there is generally no rational way of knowing whether primitive premisses are true, in any sense of 'true'.

Thus the very foundation on which Aristotle and his modern followers wanted to build an alternative to analytic proof breaks down. Axiomatic proof is no viable alternative to analytic proof since it is inadequate. One is not justified in using it for generally there is no rational way of knowing whether the starting points of axiomatic proofs are true, in any sense of 'true'.

Axiomatic proof is inadequate also because there is no non-circular way of proving that deduction from primitive premisses is truth-preserving, that is, such that, if primitive premisses are true, then the propositions deduced from them are also true (see [8] and [11, Ch. 26]).

In addition to implying that axiomatic proof is inadequate, the fact that generally there is no rational way of knowing whether primitive premisses are true has another important consequence. It entails that primitive premisses of axiomatic proofs are simply 'accepted opinions', *endoxa* in Aristotle's parlance, or rather plausible propositions in the sense explained above. Thus they have the same status as hypotheses in analytic proofs. Then the notion of axiomatic proof collapses into that of analytic proof.

Even some supporters of the axiomatic method acknowledge that. For instance, Pólya states that analogy and other non-deductive inferences "not only help to shape the demonstrative argument and to render it more understandable, but also add to our confidence to it. And so we are led to suspect that a good part of our reliance on demonstrative reasoning may come from plausible reasoning" [52, p. 168].

Thus Frege's Thesis depends on a misunderstanding. An instance of such misunderstanding is the claim that Wiles and Taylor solved Fermat's Problem. What they actually solved is the problem posed by the Taniyama-Shimura hypothesis.

Admittedly, in the last century most mathematicians have thought themselves to be pursuing axiomatic proof. But, as the case of Fermat's problem shows, they weren't. Their belief to be pursuing axiomatic proof has been a matter of trend and fashion, so essentially a sociological fact: a result of the predominance of the ideology of the Göttingen School and the Bourbaki School over the mathematicians of the last century (see [4, Ch. 5]).

Mathematicians who think themselves to be pursuing axiomatic proof don't seem to be generally aware that Frege's Thesis together with Hilbert-Gentzen Thesis would make mathematics trivial. For then there would be an algorithm that in principle could generate all possible proofs from given axioms in systematic manner, checking each time if the final proposition is the proposition to be proved. Thus theorem proving would become an activity requiring no intelligence.

Some supporters of axiomatic proof seem however to be aware of that, at least to a certain extent. For instance Rota, while maintaining that the axiomatic method guarantees the truth of a mathematical assertion, says that the "identification of mathematics with the axiomatic method has led to a widespread prejudice among scientists that mathematics is nothing but a pedantic grammar, suitable only for belaboring the obvious" [56, p. 142].

10 Proving and re-proving

Even if a rational way of knowing whether primitive premisses are true generally existed, the notion of axiomatic proof would have other basic defects.

For instance, in terms of that notion one cannot explain why, once a proof of a proposition has been found, mathematicians look for alternative proofs.

Several research papers in mathematics are concerned not with proving but with re-proving. For instance, well over four hundred distinct proofs of the Pythagorean Theorem have been given, a Fields Medal has been awarded to Selberg for producing a new proof of a theorem, the prime-number theorem, for which a proof was already known, and so on.

Now, if proofs were meant to provide a foundation and justification of a proposition, once a proof has been found and hence a foundation and justification has been given, what would be the point of looking for other proofs, even hundreds of them? No adequate answer to this question can be given in terms of the notion of axiomatic proof.

A suitable answer can be given only in terms of the notion of analytic proof, by which, to solve a problem, one may use several distinct hypotheses.

For a problem may have several sides, so one may look at it from several distinct perspectives, each of which may suggest a distinct hypothesis, thus a different proof and hence a different explanation. (On the notion of mathematical explanation involved here, see [10]). As we have already pointed out, this may have a great heuristic value, so it may be very fruitful for the development of mathematics. (For other approaches to the question of re-proving, see [56, Ch. XI], Avigad [1], Dawson [13]).

11 Mathematics and intuition

Since generally there is no rational way of knowing whether primitive premisses are true, supporters of axiomatic proof may only resort to assuming that there is an irrational faculty, intuition, by which one can grasp mathematical concepts and see that primitive premisses are true of them – an irrational faculty, because intuition is a faculty of which no account can be given.

This is the solution that, as we have seen, Aristotle suggested and most supporters of axiomatic proof have since adopted.

For instance, Gödel claims that ultimately for the "axioms there exists no other" foundation except that they "can directly be perceived to be true" by means of "an intuition of the objects falling under them" [24, pp. 346-347].

However, appealing to intuition not only bases mathematical knowledge on an irrational – and completely unreliable – faculty, but reduces proofs to rhetorical flourishes.

This is made quite clear by Hardy, who states that a mathematician is "in the first instance an observer, a man who gazes at a distant range of mountains and notes down his observations" [28, p. 18]. If "he sees a peak" and "wishes someone else to see it, he points to it, either directly or through the chain of summits which led him to recognize it himself". When "his pupil also sees it, the research, the argument, the proof is finished". Seeing a peak corresponds to having an intuition of certain mathematical objects. That mathematical activity consists in seeing peaks and pointing to them entails – Hardy argues – that "there is, strictly, no such thing as mathematical proof; that we can, in the last analysis, do nothing but point"; that proofs are merely "gas, rhetorical flourishes designed to affect psychology".

That appealing to intuition bases mathematical knowledge on an irrational and completely unreliable faculty and reduces proofs to rhetorical flourishes, conflicts with the intended aim of axiomatic proof to give a foundation and justification of a proposition.

Moreover, appealing to intuition is inconclusive. For suppose that you have an intuition of the concept of set S which tells you that your axioms of set theory T are true of S. By Gödel's first incompleteness theorem there is a sentence A of T which is true of S but is unprovable in T. Then the theory $T \cup \{\neg A\}$ is consistent, so it has a model, say S'. Thus $\neg A$ is true of S', and hence A is false of S'. Then S and S' are both models of T, so they are both concepts of set, but A is true of S and false of S'. Therefore S and S' cannot be isomorphic, so S and S' are essentially different.

Now suppose that, by reflecting on the way S' has been obtained, you get an intuition of the concept of set S' which tells you that the axioms of T are true of S'. Then you have two distinct intuitions, one ensuring that S is the genuine concept of set, the other one ensuring that S' is the genuine concept of set. Since S and S' are essentially different, this raises the question: Which of S and S' is the genuine concept of set? Intuition gives no answer.

This confirms that axiomatic proof is inadequate. As Hersh says, "the view that mathematics is in essence derivations from axioms is backward. In fact, it's wrong" [32, p. 6]. Only analytic proof is adequate, so axiomatic proof is not on a par with it.

12 Mathematics and evolution

Axiomatic proof is not on a par with analytic proof also in another respect. While axiomatic proof is simply a way of organizing and presenting results already obtained and so, as Aristotle says, is essentially aimed at teaching and learning, analytic proof goes deeply into the nature of organisms for it reflects the way in which they mainly solve their problems, starting from the most basic one: survival.

All organisms survive by making hypotheses on the environment by a process that is essentially an application of the analytic method. Thus analytic proof is based on the procedure by which organisms provide for their most basic needs. As our hunting ancestors solved their survival problem by making hypotheses about the location of preys on the basis of hints – crushed or bent grass and vegetation, bent or broken branches or twigs, mud displaced from streams, and so on – provided by them, mathematicians solve mathematical problems by making hypotheses for the solution of problems on the basis of hints provided by them.

Some of the hypotheses on the environment are chosen by natural selection and are embodied in the biological structure of organisms, and some of them concern mathematical properties of the environment. As a result, all organisms have at least some of the following innate capabilities: space sense, number sense, size sense, shape sense, order sense. Such capabilities are mathematical in kind. They have a biological function and are a result of biological evolution that has selected and embodied them in organisms.

Mathematical capabilities embodied in organisms, also non-human ones, can even be rather sophisticated.

For instance, if standing on a beach with a dog at the water's edge you throw a tennis ball into the waves diagonally, the dog will not plunge into the water immediately swimming all the way to the ball. It will run part of the way along the water's edge, and only then will plunge into the water and swim out to the ball. For, since the dog's running speed is greater than the dog's swimming speed, the dog will choose to plunge into the water at a point that will minimize the time of travel to the target. Such point can be determined by calculus, and the point actually chosen by the dog broadly agrees with the one given by calculus (see Pennings [51]). Does that mean that dogs know calculus? Of course not. They are capable of choosing an optimal point thanks to natural selection, which gives a definite survival advantage to organisms that exhibit better judgment. Thus the calculation required to determine an optimal point is not made by the dog but has been made by nature through natural selection. It is thanks to natural selection that dogs are able to solve this calculus problem. (For further examples of mathematical capabilities embodied in non-human organisms, see Devlin [15]).

Natural selection has hardwired organisms to perform certain mathematical operations building mathematics in several features of their biological structure, such as locomotion and vision, which require some sophisticated embodied mathematics. Such mathematical operations are essential to escape from danger, to search for food, to seek out a mate.

One may then distinguish a 'natural mathematics', that is, the mathematics embodied in organisms as a result of natural selection, from 'artificial mathematics', that is, mathematics as discipline. (Devlin calls artificial mathematics 'abstract mathematics', but 'artificial mathematics' seems more suitable here since it expresses that it is a mathematics that is not a natural product, being not a direct result of biological evolution but rather a human creation [15, p. 249]).

Natural mathematics, however, is necessarily limited since biological evolution is slow. On the contrary, artificial mathematics has developed relatively fast in the past five thousand years or so since it is a result of cultural evolution, which is relatively fast. This raises serious doubts about Cooper's claim that artificial "mathematics must itself be evolutionarily reducible" [12, p. 135]. It seems more reasonable to conclude that artificial mathematics cannot be reduced to natural mathematics.

The fact that natural mathematics is a result of biological evolution, whereas artificial mathematics is a result of cultural evolution, leads to a program of interpreting mathematics in terms of biological and cultural evolution. Although, of course, such program cannot be carried out here, some of its preconditions can be briefly discussed.

13 Mathematics and logic

Natural mathematics is based on natural logic, which is that natural capability to solve problems that all organisms have and is a result of biological evolution. On the other hand, artificial mathematics is based on artificial logic, which is a set of techniques invented by organisms to solve problems and is a result of cultural evolution.

Unlike the distinction between natural mathematics and artificial mathematics, the distinction between natural logic and artificial logic is not a new one. A similar distinction was made in the sixteenth century, for instance, by Ramus, and was still alive two centuries later when Kant used it in his logic lectures (see [39, pp. 252, 434, 532]). At that time, however, artificial logic was restricted to deductive inferences. But the notion of analytic proof requires that artificial logic include non-deductive inferences. Natural logic too requires non-deductive inferences, since the process by which all organisms provide for their most basic needs is essentially an application of the analytic method. However, natural logic requires not only non-deductive inferences but also non-propositional unconscious inferences, for the latter are essential, for instance, in vision. (On the role of non-propositional unconscious inference in vision, and generally on the characters of natural and artificial logic as intended here, see [11, Chs. 16-17]).

Since natural and artificial logic are based on two different forms of evolution, biological and cultural evolution, they are distinct. That, however, does not mean that they are opposed. For artificial logic ultimately depends on capabilities of organisms that are a result of biological evolution. Moreover, both natural and artificial logic depend on the very same basic procedure: the analytic method. The latter then provides a link between natural and artificial logic, and hence between natural mathematics and artificial mathematics.

14 Logic and reason

The main aim of natural logic is to find hypotheses on the environment to the end of survival. This implies that there is a strict connection between logic and the search of means for survival and that, since generally all organisms seek survival, natural logic does not belong to humans only but to all organisms.

On the contrary, logic has been traditionally viewed as the organ of reason meant as a higher faculty belonging to humans only, which allows them to overcome the limitations of their biological constitution, limitations within which animals and plants are instead constrained. In particular, such higher faculty has been supposed to be capable of intuitively and directly apprehending certain primitive truths, and specifically certain primitive premisses which are the necessary basis of any demonstrative reasoning.

But reason is not such a higher faculty, it is rather the capability of choosing means adequate to a given end. As Russell says, 'reason' "signifies the choice of the right means to an end that you wish to achieve" [57, p. 8]. Thus, in conformity with the original meaning of 'ratio', reason is a relation between means and ends. Then nothing is rational in itself but only relative to a given end. Now, since the primary end of all organisms is survival, the choice of means adequate to that end can be viewed as an expression of the faculty of reason, which then does not belong to humans only.

One might think that the concept of reason could be made less relative by stating that 'rational' – that is, 'compliant with reason' – is what is compliant with human nature. That, however, would not solve the problem of explaining what reason is but would simply refer it back to the problem of explaining what human nature is.

Now human nature is the result of two factors, biological evolution and cultural evolution. In explaining what human nature is biological evolution plays an important role, for our biological structure has a basic importance in determining what we are.

This view is fiercely opposed by those who, like Heidegger, deny that "the essence of man consists in being an animal organism", claiming that "the aberration of biologism" consists in considering the body of man as that of "an animal organism", and that the fact "that the physiology and biochemistry of man as an organism can be investigated in a natural scientific way is no proof that the essence of man lies in this organicity, that is, in the scientifically explained body" [26, p. 324].

But these claims are unjustified, because our biological structure really plays an essential role in determining what we are. For instance, monozygotic twins, when separated at birth and grown up in distinct environments with no possibility of mutual communication, have similar personalities, their behaviours resemble under several respects, they even take similar positions on the most disparate questions.

Those who deny that our biological structure has a basic importance in determining what we are, claim that the behaviour of humans is not largely governed by biological functions shared by all humans. There is no biological basis of our most important behaviours, the latter are a result of cultural evolution.

But the claim that our most important behaviours are a result of cultural evolution is not in conflict with the claim that our biological structure has a basic importance in determining what we are, for cultural evolution develops on the basis of biological evolution. Culture is not an ethereal substance independent of our biological structure. It depends on the neural networks with which biological evolution has provided us, for it is a product of our biological structure and so is bound to it. To separate cultural from biological evolution is to neglect what the subject of cultural evolution is: a biological organism which is an outcome of biological evolution.

Since biological and cultural evolution are what determines human nature, they are the relative terms with which we must commensurate rationality. Of course, only relative terms, for there is nothing necessary in biological evolution or in cultural evolution. In particular, biological evolution does not work by design: it has gone that way but could have gone otherwise. Thus, if 'rational' is what is compliant with human nature, there is nothing absolute in rationality. 'Rational' is a term relative to the contingent character of human nature, which is a contingent result of biological evolution and cultural evolution.

To view logic as the organ of reason, meant as a higher faculty belonging to humans only, is to misjudge the nature of reason. Logic can be said to be the organ of reason, though of a reason intended not as a higher faculty but as the capability of choosing means adequate to a given end, starting from survival, and hence as belonging to all organisms. Natural logic is the organ of reason for it provides all organisms with means adequate to their ends.

Here 'organisms' are supposed to include not only animals but also plants. Some of them, when attacked by herbivores, implement sophisticated defense strategies. They produce complex polymers that reduce plant digestibility, or toxins that repel or even kill the herbivores. They use other insects against the herbivores, emitting volatile organic compounds that attract other carnivorous insects which kill the attacking herbivores. These volatile organic compounds may be also perceived by neighboring yet-undamaged plants to adjust their defensive phenotype according to the present risk of attack, thus they function as external signal for within-plant communication (see Heil and Silva Bueno [31]).

15 Logic and evolution

That natural logic belongs to all organisms does not mean that non-human organisms choose means adequate to their ends on the basis of learned logical cognitions. But several humans too do not choose means adequate to their ends on the basis of learned logical cognitions. They use logical means such as induction, the cause-effect relation, the identity principle, and generally make inferences, without having attended to any logic course. They are capable of using logical means because biological evolution has designed them to do so.

Not only biological evolution has designed humans to use logical means, but natural logic, in addition to being a means for survival, is itself a result of natural selection. The natural logic system we have inherited is such that, on average, it increases the possibility of surviving and reproducing in the environment in which our most ancient ancestors evolved. Thus the first and deepest origin of reason and logic is natural selection, which has provided humans with those capabilities that have allowed them to survive.

The importance of reason and logic stems from the fact that the world changes continually and irregularly, so organisms are confronted all the time with the need to adapt to new situations. To deal with them they need logic, which helps them to cope with new situations, thus increasing their overall adaptive value.

The logic useful to this end is not only natural logic but also artificial logic, though an artificial logic including not only deductive propositional inferences, but also non-deductive and non-propositional inferences.

Biological evolution has embodied a series of informations in organisms concerning their evolutionary past, and also suitable kinds of behaviour by which they are able to cope with situations similar to those that already occurred in their evolutionary past. Moreover, they are able to cope with them automatically, that is, with no need for the single organism to reinvent the means to cope with them. To that end, natural logic is sufficient.

But, since the world changes continually and irregularly, it presents situations dissimilar from those that already occurred in the evolutionary past of organisms and, to cope with them, the means embodied in organisms by biological evolution are generally insufficient, new means are necessary. Providing them is the task of artificial logic, a logic which, like natural logic, includes non-deductive and non-propositional inferences, but goes essentially beyond natural logic because it includes stronger kinds of inference. This raises serious doubts about Cooper's claim that "logic is reducible to evolutionary theory" [12, p. 2]. On the other hand, just because artificial logic goes essentially beyond natural logic, it can supplement the work of biological evolution.

16 Mathematics and human activities

Like logic mathematics, being based on logic – both natural and artificial – is an organ of reason and so is bound to our biological structure.

Hart claims that "not only are there infinitely many primes, but also, since Euclid's proof" of the infinity of primes "makes no reference to living creatures, there would have been infinitely many primes even if life had never evolved. So the objects required by the truth of his theorem cannot be mental" [29, p. 3].

But Hart's claim depends on the assumption that Euclid's proof makes no reference to living creatures, which seems unwarranted for Euclid's proof uses concepts that are man-made and hence 'mental'. In particular, humans introduced the concept of prime number in pre-Greek mathematics in connection with such concrete human activities as dividing rations among workers. Thus, if life had never evolved, the concepts Euclid uses in his proof would not have been formed, in particular there would have been no concept of prime number.

Mathematics is strictly related to several human activities, and most of its concepts arise – directly or indirectly – from them. As van Benthem says, "mathematics is not some isolated faculty of the human mind which needs to be approached with special reverence", on the contrary, there is "a fluid transition all the way from common sense reasoning to mathematical proof, and from knowledge structures in daily life to mathematical theories" [60, p. 41].

It could be objected that stating that, if life had never evolved, there would have been no concept of prime number, says nothing about the relation between the human concept of prime number and prime numbers. The question at issue is the relation between human concepts and the mathematical objects involved in those concepts. Such relation is one of the most important topics in the philosophy of matematics and needs special investigation.

Now, if one interprets mathematics in terms of biological and cultural evolution, then mathematical objects are not independently existing entities but rather cultural products. Specifically, they are hypotheses introduced by humans to solve mathematical problems. For instance, a prime number is the hypothesis of an integer greater than 1 whose only positive divisors are 1 and itself. (On mathematical objects as hypotheses – not to be confused with fictions – see [6, pp. 300-303]. On the distinction between hypotheses and fictions, see Vaihinger [61, pp. 147-148, 152, 606], [6, pp. 303-307]).

This broadly agrees with Plato's claim that "practitioners of geometry, arithmetic and similar sciences hypothesize the odd, and the even, the geometrical figures, the three kinds of angle, and any other things of that sort which are relevant to each subject" (Plato, *Republic*, VI 510 c 2-5). Such hypotheses are in turn a problem that must be solved, and will be solved by introducing other hypotheses, contrary to practitioners of the axiomatic method who "don't feel any further need to give an account of them either to themselves or to anyone else" but simply "make them their starting-points and draw conclusions from them" (*ibid.*, VI 510 c 6-d 2).

Of course, considering the odd, the even, the geometrical figures, and so on, as hypotheses entails that one must distinguish two different kinds of hypotheses: hypotheses consisting of jugments, such as the Taniyama-Shimura hypothesis, and hypotheses consisting of objects, such as prime numbers. Thus hypotheses can be either jugments or objects.

17 Proof and evolution

Unlike the notion of axiomatic proof which, as we have seen, can be maintained, if ever, only at the cost of falling into irrationalism, unreliability and rhetoric, the notion of analytic proof is completely rational since it is based on logic – both natural and artificial logic.

Analytic proof is not a device aimed at the rather futile end of providing a justification of a proposition based on primitive premisses for which no absolute justification can be given anyway, and can only be shown to be plausible. It is rather a continuation of strategies resulting from natural selection by which organisms solve their problems, starting from survival.

Mach says that, although "science apparently grew out of biological and cultural development as its most superfluous offshoot", today "we can hardly doubt that it has developed into the factor that is biologically and culturally the most beneficial. Science has taken over the task of replacing tentative and unconscious adaptation by a faster variety that is fully conscious and methodical" [43, p. 361]. Thus modern science is a cultural artifact with a biological role.

Then artificial mathematics too is a cultural artifact with a biological role, for modern science is intrinsically mathematical. It originated from a philosophical turn, Galilei's decision to replace Aristotle's view that science must "seek to penetrate the true and intrinsic essence of natural substances", by the view that we must "content ourselves with a knowledge of some of their properties" such as "location, motion, shape, size" [21, V, pp. 187-188]. Such properties are quantitative and hence mathematical in kind, unlike essences that are the object of Aristotle's science, which are non-mathematical in kind. For that reason modern science is intrinsically mathematical whereas Aristotle's science is intrinsically non-mathematical. For that very same reason artificial mathematics is 'unreasonably effective' (see [6, Ch. 42] and [11, Ch. 13]).

Since modern science is intrinsically mathematical and is a cultural artifact with a biological role, artificial mathematics too, being an inherent constituent of modern science, is a cultural artifact with a biological role. Thus artificial mathematics has a biological role like natural mathematics, though less directly.

Artificial mathematics is a cultural artifact with a biological role roughly in the same sense in which animal-made tools are cultural artifacts with a biological role. As New Caledonian crows, say, make a wide variety of tools by means of which they develop techniques that help them to solve their survival problem (see Hunt-Gray [38]), humans make proofs by means of which they develop techniques that help them to solve their survival problem. Although there are obvious differences between proofs and animal-made tools, viewing proofs as having a biological role helps to make sense of the phenomenon of proof.

Such phenomenon is hardly comprehensible if 'proof' is intended as 'axiomatic proof', that is, as a means to justify propositions by deducing them from primitive premisses for which, as we have already stressed, no absolute justification can be given. It is comprehensible only if 'proof' is intended as 'analytic proof', that is, as a means to discover plausible hypotheses capable of giving solutions to problems that meet needs, even basic needs, of humans.

Rota states that, "of all escapes from reality, mathematics is the most successful ever", all other escapes, "sex, drugs, hobbies, whatever", being "ephemeral by comparison", and speaks of "the mathematician's feeling of triumph as he forces the world to obey the laws his imagination has freely created" [56, p. 70].

But things stand otherwise. Mathematics is no escape from reality for it is an answer to needs, even basic needs, of humans. Mathematicians don't force the world to obey the laws their imagination has created, for such laws are just the way mathematicians make the world understandable to themselves, and the working of the world does not depend on them. Moreover, their creations are not completely free, for they are a product of the mathematicians's biological structure and so are bound to it.

18 Proof, teaching and learning

That the notion of axiomatic proof is inadequate does not mean that it is of no use. Its main use remains the one originally stated by Aristotle, that is, teaching and learning.

This has been the main use of axiomatic proof from the very beginning, as it appears from Euclid's *Elements* that were intended to be a textbook "for elementary teaching", in which Euclid "did not bring in everything he could have collected, but only what could serve as elements" [53, 69.6-9].

Euclid's use of axiomatic proof in the *Elements* does not mean that this was the notion of proof he used as a working mathematician. Writing textbooks is one thing, doing mathematical research is another thing. As Knorr says, "the writing of textbooks is the end of mathematical research only in the sense that death is the end of life" [40, p. 7]. Euclid's notion of proof as a working mathematician is to be found not in the *Elements* but in his research work, where he "proceeds by analysis and synthesis" [50, 634.10-11].

Although the main use of axiomatic proof remains teaching and learning, some reservations can be made even about that use. Descartes claims that the notion of axiomatic proof makes proofs appear "discovered more through chance than through method", so by using it "we get out of the habit of using our reason" [14, X, p. 375]. Therefore the notion of axiomatic proof "does not completely satisfy the minds of those who are eager to learn" [14, VII, p. 156]. Descartes's claim seems justified since axiomatic proofs are often unnatural and unmemorable. Presenting propositions in a way different from the one in which they were obtained, they conceal the real process, thus contributing to make mathematics hard.

Using axiomatic proof for teaching and learning is more a matter of trend and fashion than of effectiveness. For instance, Descartes did not use it in presenting his *Geometry*. In the seventeenth century this practice was so widespread that Newton wrote: "The Mathematicians of the last age have very much improved Analysis but stop there & think they have solved a Problem when they have only resolved it" – that is, solved it by the method of Analysis – "& by this means the method of Synthesis", that is, the axiomatic method, "is almost laid aside" [49, p. 294]. To them this "synthetic style of writing is less pleasing, whether because it may seem too prolix and too akin to the method of the ancients, or because it is less revealing of the manner of discovery" [48, VIII, p. 451].

Newton also makes quite clear that using axiomatic proof for teaching and learning is a matter of trend and fashion: "The Propositions in the following book were invented by Analysis" [49, p. 294]. And "certainly I could have written analytically what I had found out analytically with less effort than it took me to compose it" [48, VIII, p. 451], that is, to write it axiomatically. But "considering that the Ancients (so far as I can find) admitted nothing into Geometry before it was demonstrated by Composition", that is, axiomatically, "I composed what I invented by Analysis to make it Geometrically authentic

& fit for the publick" [49, p. 294]. But "this makes it now difficult for unskilful Men to see the Analysis by which those Propositions were found out" [47, p. 206].

Thus Newton agrees with Descartes that axiomatic proof makes it difficult for the learner to see how propositions were found out. This makes it advisable to reconsider the use of axiomatic proof even for teaching and learning, in sharp contrasts with our time when "the axiomatic method of presentation has reached a pinnacle of fanaticism" [56, p. 142]. Admittedly, the axiomatic method can sometimes supply more compact proofs, but it conceals how they were discovered, and this may negatively affect teaching and learning.

Acknowledgements

I would like to thank Jeremy Avigad, Riccardo Chiaradonna, Cesare Cozzo, Emily Grosholz, Reuben Hersh, Robert Thomas, Johan van Benthem and two anonymous referees for useful comments on earlier drafts of this paper.

References

1. J. Avigad: Mathematical method and proof. *Synthese*, **153** (2006) pp 105-159
2. J. Azzouni: The derivation-indicator view of mathematical practice. *Philosophia Mathematica*, vol. 12 (2004) pp 81-105
3. R. Carnap: *The logical syntax of language* (Humanities Press, New York 1951)
4. C. Cellucci: *Le ragioni della logica* (Laterza, Rome 1998; fifth ed 2008)
5. C. Cellucci: The growth of mathematical knowledge: an open world view. In: *The growth of mathematical knowledge*, ed by E. Grosholz and H. Breger (Kluwer, Dordrecht 2000) pp 153-176
6. C. Cellucci: *Filosofia e matematica* (Laterza, Rome 2002) English translation: Introduction to *18 unconventional essays on the nature of mathematics*, ed by R. Hersh (Springer, Berlin 2006) pp 17-36
7. C. Cellucci: Mathematical discourse *vs* mathematical intuition. In: *Mathematical reasoning and heuristics*, ed by C. Cellucci and D. Gillies (College Publications, London 2005) pp 137-165
8. C. Cellucci: The question Hume didn't ask: why should we accept deductive inferences? In: *Demonstrative and non-demonstrative reasoning in mathematics and natural science*, ed by C. Cellucci and P. Pecere (Edizioni dell'Università, Cassino 2006) pp 207-235
9. C. Cellucci: Gödel aveva qualcosa da dire sulla natura del ragionamento? In: *La complessità di Gödel*, ed by G. Lolli and U. Pagallo (Giappichelli, Turin 2008) pp 31-64
10. C. Cellucci: The nature of mathematical explanation. *Studies in History and Philosophy of Science* (2008) To appear
11. C. Cellucci: *Perché ancora la filosofia* (Laterza, Rome 2008) To appear
12. W. S. Cooper: *The evolution of reason. Logic as a branch of biology* (Cambridge University Press, Cambridge 2001)

13. J. W. Dawson: Why do mathematicians re-prove theorems? *Philosophia Mathematica* **14** (2006) pp 269-286
14. R. Descartes: *Oeuvres*, ed by C. Adam and P. Tannery (Vrin, Paris 1996)
15. K. Devlin: *The math instinct. Why you're a mathematical genius (along with lobsters, birds, cats, and dogs)* (Thunder's Mouth Press, New York 2005)
16. H. Diels: *Die Fragmente der Vorsokratiker* ed by W. Krantz (Weidmann, Berlin 1934)
17. G. Frege: *The foundations of arithmetic. A logico-mathematical enquiry into the concept of number* (Blackwell, Oxford 1953)
18. G. Frege: *The basic laws of arithmetic*, ed by M. Furth (University of California Press, Berkeley and Los Angeles 1964)
19. G. Frege: Begriffsschrift, a formula language, modeled upon that of arithmetic, for pure thought. In: *From Frege to Gödel. A source book in mathematical logic, 1879-1931*, ed by J. van Heijenoort (Harvard University Press, Cambridge MA 1967) pp 5-82
20. G. Frege: On the foundations of geometry: second series. In: *Collected papers on mathematics, logic, and philosophy* ed by B. McGuinness (Blackwell, Oxford 1984) pp 293-340
21. G. Galilei: *Opere*, ed by A. Favaro (Barbera, Florence 1968)
22. G. Gentzen: Investigations into logical deduction. In: *The collected papers of Gerhard Gentzen*, ed by M. E. Szabo (North-Holland, Amsterdam 1969) pp 68-131
23. K. Gödel: What is Cantor's continuum problem? – (1964). In: *Collected works*, vol. I, ed by S. Feferman et al. (Oxford University Press, Oxford 1990) pp 254-270
24. K. Gödel: Is mathematics syntax of language? – Version III. In: *Collected works*, vol. III, ed by S. Feferman et al. (Oxford University Press, Oxford 1995) pp 334-356
25. E. Grosholz: *Representation and productive ambiguity in mathematics and the sciences* (Oxford University Press, Oxford 2007)
26. R. W. Hamming: The unreasonable effectiveness of mathematics. *The American Mathematical Monthly* **87** (1980) pp 81-90
27. R. W. Hamming: Mathematics on a distant planet. *The American Mathematical Monthly* **105** (1998) pp 640-650
28. G. H. Hardy: Mathematical proof. *Mind* **38** (1929) pp 1-25
29. W. D. Hart: Introduction to *The philosophy of mathematics*, ed by W. D. Hart (Oxford University Press, Oxford 1996) pp 1-13
30. M. Heidegger: Brief über den Humanismus. In: *Gesamtausgabe*, vol. 9: *Wegmarken* (Klostermann, Frankfurt am Mein 1976) pp 313-364
31. M. Heil and J. C. Silva Bueno: Within-plant signaling by volatiles leads to induction and priming of an indirect plant defense in nature. *Proceedings of the National Academy of Sciences USA* **104** (2007) pp 5467-5472
32. R. Hersh: *What is mathematics, really?* (Oxford University Press, Oxford 1997)
33. D. Hilbert: Beweis des tertium non datur. *Nachrichten von der Gesellschaft der Wissenschaften zu Göttingen*, Math.-Phys. Klasse (1931) pp 120-125
34. D. Hilbert: The foundations of mathematics. In: *From Frege to Gödel. A source book in mathematical logic, 1879-1931*, ed by J. van Heijenoort (Harvard University Press, Cambridge MA 1967) pp 464-479
35. D. Hilbert: Letter to Frege 29.12.1899. In: *Philosophical and mathematical correspondence* (The University of Chicago Press, Chicago 1980) pp 38-41

36. D. Hilbert: Die Grundlagen der Geometrie. In: *David Hilbert's lectures on the foundations of geometry (1891-1902)*, ed by M. Hallett and U. Majer (Springer, Berlin 2004) pp 72-81
37. J. Hintikka and U. Remes: *The method of analysis: its geometrical origin and its general significance* (Reidel, Dordrecht 1974)
38. G. R. Hunt and R. D. Gray: Tool manufacture by New Caledonian crows: chipping away at human uniqueness. *Acta Zoologica Sinica* **52** (2006) Supplement, pp 622-625
39. I. Kant: *Lectures on logic*, ed. J. M. Young (Cambridge University Press, Cambridge 1992)
40. W. R. Knorr: *The ancient tradition of geometric problems* (Dover, Mineola NY 1993)
41. S. Kripke: *Naming and necessity* (Blackwell, Oxford 1980)
42. G. Lolli: *QED. Fenomenologia della dimostrazione* (Bollati Boringhieri, Turin 2005)
43. E. Mach: *Knowledge and error. Sketches on the psychology of enquiry* (Reidel, Dordrecht 1976)
44. A. Macintyre: The mathematical significance of proof theory. *Philosophical Transactions of the Royal Society* **363** (2005) pp 2419-2435
45. P. Mäenpää: *The art of analysis. Logic and history of problem solving*. Dissertation (University of Helsinki, Helsinki 1993)
46. L. I. Meikle and J. D. Fleuriot: Formalizing Hilbert's *Grundlagen* in Isabelle/Isar. In: *Theorem proving in higher order logics*, LNCS 2758, ed by D. Basin and B. Wolff (Springer, Berlin 2003) pp 319-334
47. I. Newton: An account of the book entitled *Commercium Epistolicum Collinii et aliorum, de analysi promota*. *Philosophical Transactions*, vol. 29 (1715) pp 173-224 The quoted passage is also reprinted in I. B. Cohen, *Introduction to Newton's 'Principia'* (Cambridge University Press, Cambridge 1971) p. 295
48. I. Newton: *The mathematical papers*, ed by D. T. Whiteside (Cambridge University Press, Cambridge 1967-81)
49. I. Newton: MS Add. 3968, f. 101. In: I. B. Cohen, *Introduction to Newton's 'Principia'* (Cambridge University Press, Cambridge 1971) pp 293-294
50. Pappus of Alexandria: *Book 7 of the Collection*, ed by A. Jones (Springer, Berlin 1986)
51. T. J. Pennings: Do dogs know calculus? *College Mathematics Journal* **34** (2003) pp 178-182
52. G. Pólya: *Mathematics and plausible reasoning* (Princeton University Press, Princeton 1954)
53. Proclus: *In primum Euclidis Elementorum librum commentarii*, ed by G. Friedlein (Olms, Hildesheim 1992)
54. Y. Rav: A critique of a formalist-mechanist version of the justification of arguments in mathematicians' proof practices. *Philosophia Mathematica* **15** (2007) pp 291-320
55. K. Ribet: From the Taniyama-Shimura conjecture to Fermat's Last Theorem. *Annales de la Faculté des Sciences de Toulouse – Mathématiques*, vol. 11 (1990) pp 116-139
56. G. C. Rota: *Indiscrete thoughts*, ed by F. Palombi (Birkhäuser, Boston 1997)
57. B. Russell: *Human society in ethics and politics* (Allen & Unwin, London 1954)
58. A. Tarski: The semantic conception of truth and the foundations of semantics. *Philosophy and Phenomenological Research*, vol. 4 (1944) pp 341-376

59. B. Timmermans: *La résolution des problèmes de Descartes à Kant* (Presses Universitaires de France, Paris 1995)
60. J. van Benthem: Interview. In: *Philosophy of mathematics: 5 questions*, ed by V. F. Hendricks and H. Leitgeb (Automatic Press / VIP Press, New York 2007)
61. H. Vaihinger: *Die Philosophie des Als Ob* (Felix Meiner, Leipzig 1927)

On Formal Proofs

Andrea Cantini

Dipartimento di Filosofia, Università di Firenze (Italy)
cantini@unifi.it

1 Introduction

We address two questions:
- what is the use of formal proofs?
- how do we proceed from a formal proof to a computation?

Indeed, the latter is subsidiary to the former and is briefly dealt with by means of concrete examples in the final section. We shall argue that formal proofs yield in many cases genuine methods for ideal experiments with problematic principles and rules, and for gaining semantical and computational information. The points are briefly illustrated by cases taken from proof theory; we emphasize the fact that even within proof theory, the variety (*vs* purity) of methods is common practice and it is justified by theoretical results.

The main motivation is to contrast the claim that there would be a standard thesis among logically oriented philosophers of mathematics, tending to reduce the essence of usual proofs of mathematical statements to their formal counterparts within a given axiomatic theory. This tenet is not justified today; actual genuine uses of formal proofs are peculiar and technical, and they do not aim at certainty or formalistic foundations.

On the other hand, formal proofs have a role with respect to a numbers of issues, ranging from structural analysis (if one uses sensible calculi), to proof mining (e.g. extracting constructive information, typically the rate of growth of provably total functions). Above all, logical analysis of proofs is essential in order to reveal internal symmetries (sequent calculi, cut elimination and beyond) and to make the denotational semantics of proofs possible. Think of Gentzen's natural deduction[1] and to the unveiling of the structure of proofs since the fundamental investigations of D. Prawitz [33].

---

[1] G. Rota [36] claims that Gentzen's natural deduction is "an instance of a beautiful theory which has never been matched in beauty of presentation". The semantics of proofs requires that proofs are regarded as definite mathematical objects,

The final point is that the separation between *formal* and *informal* is not so sharp as it could be thought of. This can be seen by dealing with advanced results in proof theory and its applications to ordinary mathematics, where the interaction between formal logical methods and concrete mathematics is apparent. Also, insofar as the computational aspects are stressed in present mathematical research, the impact of formal techniques is bound to increase.

As to the contents, we first deal (§ 2) with recent criticism of formal proofs, i.e. with the thesis that formals proofs are certainly not enough to represent concrete proofs for a number of reasons. We consider a stimulating paper of Y. Rav [35] and certain ideas therein. In particular, he holds that a plain philosophical understanding of proofs has been neglected. According to Rav's claim, proofs are for the mathematicians what experimental procedures are for the experimental scientist. In § 3 we reconsider this thesis in the light of ideas of Paul Bernays and Bertrand Russell, which point to the experimental (in a suitable sense) and conjectural nature of the so-called foundational investigations.

The second part of the paper (§ 4) is devoted to illustrate the role of formal proofs and methods with examples taken from recent investigations: we survey a few results taken from analytic combinatorics, proof mining, the theory of self-applicable operations and constructive set theory. The cases should convey a succinct but significant idea of the positive role of formal proofs.

2 Why do we prove theorems?

There is a misunderstanding, concerning the role of formal proofs and logic in mathematics, which derives from assuming that logic is by its very nature (doomed to be) infected by foundational bias or dogmas, or nowadays sterile and old-fashioned views about mathematical thought. In order to clarify our view, we first briefly consider the discussion in Rav [35]) emphasizing the priority of informal genuine proofs against formal proofs. We share some positive remarks with [35] (what a genuine mathematical proof is), but we do not share the implicit representation of what formal proofs are, and we try to complete the picture with a few remarks on the positive role thereof.

2.1 (Informal) proofs

First of all, a crucial thesis in [35] is that proofs are central in mathematics, not the truth of the mathematical statements. The mere fact that a statement A is true is negligeable in comparison to the questions: how do we get to know that A holds and *why*. A genuine proof of A has to tell us *why* it

---

namely think of the Curry-Howard correspondence. That proof theory is a well-established mathematical discipline is explicitly acknowledged by Rav [35, p. 12].

is so. This insistence on the reasons for A being true recalls the classical distinction between *demonstratio propter quid* (the reason for the truth of A) and *demonstratio quia* (the fact that A is true).[2]

A genuine proof of an interesting statement A has a *conceptual nature*: it is usually the starting point for elaborating a novel theory which ought *to explain the reason*, the grounds for accepting the truth of A. In this sense *mathematical explanation* is tightly linked to proofs.[3] A real proof is *not* a simple *verification*, but it creates new notions and often in the long run it gives rise to a new theory. Indeed, if we look at the history of mathematics, we see that it is important not quite *the truth value of a statement*, but the proof generated methods and notions.[4]

This thesis is corroborated by cases taken from the history of modern and contemporary mathematics. For instance, by considering the failed attacks to number theoretic conjectures like Goldbach's and the twin primes conjecture, one comes to know that new powerful methods have been developed.[5]

Similar remarks apply to problems coming from logic and the foundations of mathematics, e.g. the Continuum Hypothesis CH. At the outset, topological methods yield partial results, in the sense that, if one restricts the notion of *arbitrary set of reals* with topological conditions, one does obtain positive theorems (Cantor-Bendixson, CH for *closed* sets; Hausdorff's theorem, CH for *Borel* sets; Suslin's theorem, CH for *analytic* sets). Later on, in the twenties of the last century, Hilbert's attack on CH unveiled new ideas leading to Gödel's invention of the constructible universe. But a powerful method had to be invented by Cohen in the sixties that has been applied in many fields of mathematical logic, also outside set theory (from recursion theory to proof theory and constructive mathematics).

2.2 On the opposition *proofs vs derivations*

A further elucidation of the conceptual role of proofs can be gained by a closer comparison with formal derivations. According to [35], ordinary proofs are semantical objects and cannot be exhausted by formal derivations, which depend *upon a fixed formal system* and can be codified as purely syntactical

---

[2] A thorough discussion on the philosophical side of this topic is to be found in Casari [12].

[3] We cannot go into the important issue of mathematical explanation here; for recent contributions, we send the reader to Mancosu [31], [20] and Cellucci [13], who proposes his own variant of the so-called Aristotle-Pólya tradition, based on a version of the analytic method and a criticism of the axiomatic method.

[4] E. Giusti [18] in his book on the existence of mathematical objects remarks that several important mathematical notions are proof-generated, see his discussion of the notion of group. A similar view of proofs is also presented by J. Avigad in [4].

[5] [35] mentions sieve methods in number theory and a famous theorem proved by Schnirelmann in 1930: for some positive c, every natural number bigger than 1 can be written as the sum of at most c primes.

finite objects. An ordinary proof better corresponds to *a potentially infinite object* and, with reference to Kreisel [26], the incompleteness results can be regarded as justification of such views.[6] In order to clarify the opposition *formal/informal*, one might be tempted to recall by analogy the pair *effectively computable function (in the intuitive sense)/partial recursive function* and Church's Thesis CT, according to which effectively computable functions are partial recursive. At this point one might invoke a parallel *Hilbert's thesis* claiming that *every proof can be turned into a formal proof or derivation*. But according to Rav, this is not so good an analogy; by contrast with the situation in computability the connection is only one-way: *there is no means to restore the ordinary proof content, once we have the formal derivation.*[7]

This is right up to a certain extent, but we wish to add a few remarks. First of all, in some cases, there is a standard (perhaps trivial) means to restore ordinary proof content, that is, to apply a *soundness theorem* with respect to a class of models or to a given intended semantics, which is often available for a given axiomatic theory.

More important, metatheorems and theorems, even in axiomatic contexts or in formal logic, are given in the same *informal high level style* as the one adopted in mathematical texts or journals. This common style is grounded upon theoretical reasons. Let us give a concrete example: a modification of Selberg's elementary proof of the prime number theorem PNT has been given in [14] for the system $I\Delta_0 + exp$. The authors do not literally exhibit a *formal proof*, say, in Hilbert's style or natural deduction, but: (i) they show that (a suitable version of) PNT can be carefully formalized in the language of $I\Delta_0 + exp$; (ii) they work out in full details the basic steps and computations in an *arbitrary model M* of $I\Delta_0 + exp$, dealing with rationals and integers of M, and pointing out from time to time if the given bounds are standard elements of M. *By completeness*, it follows that there exists a formal proof of PNT in $I\Delta_0 + exp$.[8] Of course, the formalization of PNT in $I\Delta_0 + exp$ is highly non-trivial, since one has to show that the results involving the logarithm function work under a corresponding approximate function and the steps involving infinite sums and products can be carried out, applying *only*

---

[6] Another argument for holding the infinitary nature of proofs, which is reminiscent of a well-known puzzle by Carroll [11], refers to the process of understanding a proof: in order to see that $A \to B$, we may need $A \to C$ and $C \to B$, for some interpolant C. In turn these steps might require additional interpolations (see [35], p. 15), which correspond to subsequent lemmata. This could lead to a sort of potential infinite regress; at each step, there could be some further premise to accept, before we are compelled to accept the conclusion that $A \to B$.

[7] An additional reason one ought to stress is that there is no informal conceptual analysis of the notion of proof comparable to the notion of effective procedure, say, in the style of Turing.

[8] PNT as a case study for the program of formalizing important results from mathematical practice is discussed by Avigad [3]. Quite recently, [5] yields a formally verified version of Selberg's proof via the Isabelle proof assistant.

induction on bounded formulas. This in turn has byproducts, since one can potentially apply proof theory and estimate the rate of growth of the Skolem functions implicit in the proofs, and so on.

When we work within an axiomatic system and produce formal derivations, the *essential difference* is that we try to carry out *constructions by limited means*, like solving a geometric problem by ruler and compass: we need to invent the successful combination of axioms, lemmata, definition and (logical and mathematical) rules, but we also need a careful representation of the concepts involved. The main reason why we impose such restrictions is not foundational in a narrow, traditional sense (say, because we like to hold a formalistic philosophy of mathematics or we force our arguments to meet the canons of an ideal absolute rigor). Most likely, it is because we are after the computational content of a statement, or we search for more general models (constructive, recursive), or because we are investigating new principles we do not understand well and we wish to experiment with (see below 3.2).

Therefore, the opposition *formal/informal* is not so neat and, up to a point, is a matter of degree. Grasping the proof content of a deep and complex ordinary proof may be very difficult for a number of reasons. For instance, we might gain a local understanding, that is, we might follow the single transformations, steps and lemmata, and verify at a certain level that a given argument is flawless and sound. Yet we might fail to grasp the leading ideas (why it is so on general grounds). This might require further hard work and we might be puzzled why certain technical choices have been made and why they do their job (this is often the case in computational and combinatorial arguments).[9] Deep proofs, both as global and local constructions, usually remain objects of reflection over several years until a new conceptual approach is discovered, which *transforms a theorem with a difficult proof in a straightforward corollary of more general theoretical facts, that explain (in some sense) it*. Logicians can offer a typical instance of this situation: think of the many attempts to clarify the deep structure of cut elimination, from its combinatorial procedural assessment to its understanding via models deriving from functional analysis.

Thus the problem of restoring the proper semantical content is still there even with an ordinary proof, which has a lot of *implicit knowledge* to be made explicit, both as a whole (the strategy behind, the main ideas) and locally (the invention of the right formulas, etc.). This part is important both conceptually and technically and it is not given by the informal proof as such.

By their very nature, real proofs have to be concise and streamlined insofar as *they are codified in a finite text*, which usually tends to conceal or simply hint the motivations, the false starts and the attempts that justify in the very

---

[9] A related urgent issue we neglect here is that of *highly complex proofs*; we send the reader to the interesting discussion by Aschbacher [2] of the proof of the classification theorem of finite simple groups.

end the final product. So the opposition already at a cognitive level should be somewhat softened.

At this stage, it is perhaps worth recalling S. Mac Lane's ideas about proofs. He is well-known for holding a peculiar version of structuralism and also for contrasting the view that identifies mathematics with set theory, hence as a strong opponent of the view that the natural philosophy of mathematics is set theoretic realism. A mathematician, like a painter or a poet, is concerned with creating *general patterns or forms*[10] and it is the abstract nature of mathematics that requires a peculiar form of verification, i.e. mathematical proof. According to Mac Lane, mathematics is that branch of science in which the concepts are *protean*: each concept applies not to one aspect of reality, but to many. Each mathematical idea is protean, thus deals with different realities, so does not have a proper ontology. But this leads to forms of arguments that have to be uniform and universal in the different interpretations of the main predicates. This means that we need verifications involving the forms themselves and a real proof is formal in an essential natural way [30, p. 150]:

> Proof (and not experiment or speculation) is what is required in all of that part of science which is mathematics, and this requirement is there because of the very nature of mathematics. This is the case in all the branches of mathematics. Thus, group theory is the study of symmetry wherever it appears. The same axioms describe a group, whether it be a symmetry group of a crystal, of a Moorish ornament, or a physical system ... Any theorem about groups is intended to apply to all of these cases. Such a theorem may be suggested by the circumstances of one of these applications, but *the theorem itself is not about any one use and so must be established by a formal proof from the definitions. Thus, the protean character of mathematics as a part of science also explains why proofs are essential to mathematics.*

3 Experiments?

Rav claims that proofs are for the mathematicians what experimental procedures are for the experimental scientist. To a certain extent, this is acceptable. My point is that the construction of formal systems and proofs sometimes has a peculiar role, which is often forgotten. This leads to reconsider the term *experience*; the idea is that a relevant part of the foundational work can be regarded as experimental at least in the sense in which *Gedankenexperimente* are used in science.[11] This is forcefully illustrated by Paul Bernays. Furthermore, there is a peculiar interpretation of the foundational work given by

---

[10] This is reminiscent of Hardy's dictum: *a mathematician, like a painter or a poet is a maker of patterns*, see [21, p. 84].

[11] This part summarizes a lengthier paper *Logical analysis and the philosophy of mathematics. Formal systems as Gedankenexperimente*, presented at the Workshop *Philosophy of Mathematics Today*, Centro di Ricerca Matematica "Ennio De Giorgi", Scuola Normale Superiore, Pisa, January 23-28, 2006.

Russell which considers the foundational work as mainly inductive. And we believe that both views ought to be reconsidered in the proper light.

3.1 *Geistiges Experimentieren*

Bernays [7], commenting on Gonseth's philosophy and arguing for *the dual nature of mathematical knowledge*, which combines rational and empirical factors, just like natural sciences, says:

> For the abstract fields of mathematics and logic this means specifically that *thought-formations are not purely a priori, but grow out of a kind of intellectual experimentation* (geistiges Experimentieren). This view is confirmed when we consider the foundational research in mathematics. Indeed, it becomes apparent here that one is forced to adapt the methodological framework to the requirements of the task by trial and error (*durch Probieren den Erfordernissen der Aufgabe anzupassen*). Such experimentation, which must be judged as *an expression of failure* according to the traditional view, seems entirely appropriate from the viewpoint of intellectual experience. In particular, from this standpoint experiments that turned out to be unfeasible cannot *eo ipso* be considered methodological mistakes. Instead, they can be appreciated as stages in intellectual experimentation (if they are set up sensibly and are carried out consistently). Seen in this light, the variety of competing foundational undertakings is not objectionable, but appears in analogy with the multiplicity of competing theories encountered in several stages of developments of research in the natural sciences...

We believe this is a masterful synthesis of both the aims of foundational investigations and the role of formal methods; this could be hardly better expressed.

The view of Bernays is just at variance with the standard thesis concerning mathematical knowledge, that is, the view [35, p. 15] that the function of proofs is simply to validate by deductive means theorems from axioms, and that the notion of logical consequence does capture in full the essence of mathematical proof. The methodological frame and the consequent formal tools for mathematical investigations are the outcome of a dynamical process, and the whole enterprise might be better assimilated to a sort of ideal experimental work. Bernays proposes a nice parallel: "just as in physics the theoretical language and the theoretical attitude is complemented by the attitude and the language of the experimentalist, the theoretical attitude in mathematics is also complemented by a manner of reflection that is directed toward *the procedural aspect of mathematical activity*" [7]. This means expressions, operations, definitions, methods of finding solutions, etc.

3.2 Back to Russell: the regressive method

Bernays's paper stresses the role of ideal experiments in proper mathematical activity, and it shows the recognition of a peculiar trial-and-error methodology by a mathematician and philosopher deeply involved in Hilbert's program.

If we go back to the early years of the XX century and to an unpublished paper [37] of Bertrand Russell on the so-called *regressive method*, we see that foundational work was interpreted, at least at a certain stage, in a similar vein.

> In mathematics, except in the earliest parts, the propositions from which a given proposition is deduced generally give the reason why we believe the given proposition. But *in dealing with the principles of mathematics, this relation is reversed*. Our propositions are too simple to be easy, and thus their consequences are generally easier than they are. Hence *we tend to believe the premises because we can see that their consequences are true*, instead of believing consequences because we know the premises to be true. But inferring the premises from consequences is the essence of induction; *thus the method of investigating the principles of mathematics is really an inductive method, and is substantially the same as the method of discovering general laws in any other science... Our reasons for believing logic and pure mathematics are, in part, only inductive and probable* [37, pp. 273–274].

We accept axioms in the foundations of mathematics not because they are evident or certain or even true in some sense, but because they yield the "right" consequences.[12] Similar statement in the Preface to *Principia Mathematica* (1910):

> The justification for this and the chief reason in favour of any theory of the principles of mathematics must always be inductive, i.e. it must lie in the fact that the theory in question enables us to deduce ordinary mathematics. In mathematics, the greatest degree of self-evidence is usually not to be found quite at the beginning, but at some later point; hence the early deductions, until they reach this point, give reason rather for believing the premisses because true consequences follow from them, than for believing the consequences because they follow from the premisses...

Research in the foundations is parallel to theoretical physics. Principles are often chosen not because they are evident, but because they have nice consequences and unifying power, and we have inductive evidence for them. This is in strong opposition to the Euclidean model; it is important to devise bold hypotheses with a strong explanatory power.

Thus both Bernays and Russell are also well aware that foundations are tentative, conjectural, aim at guessing unifying principles, not true statements, but points to be used in order to progress in our understanding.

A more indirect conclusion is that the opposition between formal and informal methods is rather thin, because formal methods – insofar they are used

---

[12] This general attitude is already in Mill's *Utilitarianism* (1863): "The truths which are ultimately accepted as the first principles of a science, are really the last results of metaphysical analysis, practised on the elementary notions with which the science is conversant; and their relation to the science is not that of foundations to an edifice, but of roots to a tree, which may perform their office equally well though they be never dug down to and exposed to light."

for foundational investigations – have similar features to informal methods. If they are not used for foundational methods, then they have plenty of applications, which go well beyond the level of mere soundness verification and logical certification in a narrow sense.

4 Why do we prove formal theorems?

In this sections we survey results, which concern:

(i) applications of formal methods and proofs to combinatorial problems (Weiermann), and to various parts of analysis (Kohlenbach);
(ii) principles concerning self-applicable operations (combinatory logic and lambda calculus), and alternative foundations of the continuum in the light of predicative constructive foundations.

4.1 A threshold phenomenon related to formal provability

Usually the notion of *phase transition* belongs to physics. The essential feature of a phase transition is the sudden change of a physical property (e.g. passing from liquid to gaseous state) of a physical system in consequence of a small change of some parameter (e.g. temperature), marked by a threshold point. Recently, a discovery by A. Weiermann [39], further investigated by G. Lee [27], shows that there are phase transition phenomena involving combinatorial statements and formal systems of interest for logicians. Thus an idea originally stemming from physics is unexpectedly lifted to the domain of logical analysis and its applications to combinatorics.[13] Let us briefly sketch the main notions and results.

By a *finite tree* T we understand a finite partial ordering $T := \langle T, \preceq \rangle$, with a minimum such that, for $b \in T$, every set $\{a \in T \mid a \preceq b\}$ is totally ordered. We write $T_1 \hookrightarrow T_2$ as an abbreviation for: *there is an injective map of the first tree into the second preserving* inf. $|T|$ is the number of nodes of the tree T.

A celebrated theorem of Kruskal (KT in short) then states:

Theorem 1. *For every infinite sequence $\{T_i \mid i \in \omega\}$ of finite trees, there are indices $i < j$ such that $T_i \hookrightarrow T_j$.*

Apart from important applications, the conceptual interest of KT lies in the fact observed by H. Friedman that KT is independent of ATR_0, an important fragment of second order arithmetic,[14] whose arithmetical content is coexten-

---

[13] Threshold results are known since 1960 in connection with random graphs (Erdös and Rényi; see [17]); for results separating satisfiability from unsatisfiability, see [19].
[14] For the basic definitions and results, also concerning ATR_0, see Simpson's book [38], to which we refer for the definitions of other standard systems involved in the present discussion, like Peano arithmetic PA, Primitive recursive arithmetic PRA and arithmetical analysis ACA_0.

sive with the set of predicatively acceptable arithmetical truths in the sense of Feferman-Schütte.

Now, by a suitable miniaturization of KT, one obtains a Π_2^0-true arithmetical statement, which is true but predicatively unprovable, and this is considered a step for arguing that *impredicative methods are necessary for ordinary mathematics*.[15] Let us give more details. Let $\mathrm{SWQ}(f)$ be the statement, depending on a given function parameter f:

> For every K there exists M so big such that, for every M-sequence of finite trees $T_1 \ldots T_M$, satisfying $|T_i| < K + f(i)$, then there exist i, j with $i < j \leq M$ and $T_i \hookrightarrow T_j$.

$\mathrm{SWQ}(f)$ is a finite miniaturization of KT with a *slowness* condition on the potential descending sequences controlled by f, and it can be proved by a compactness argument. It is known:

Theorem 2 (Friedman). *Let $f(i) = i$. Then $\mathrm{ATR}_0 \nvdash \mathrm{SWQ}(f)$.*

Let $|i| =$ the length of the binary representation of i.

Theorem 3 (Löbl-Matousek).

(i) Let $f(i) = 1/2.|i|$. Then $\mathrm{PRA} \vdash \mathrm{SWQ}(f)$.
(ii) Let $f(i) = 4|i|$. Then $\mathrm{PA} \nvdash \mathrm{SWQ}(f)$.

Given the previous results, it is natural to ask whether there exists a real number separating provability from unprovability of $\mathrm{SWQ}(f)$. This threshold problem is solved by the subsequent

Theorem 4 (Weiermann). *There exists a real number c such that, if r is a primitive recursive real number, and $f_r(i) = r|i|$, then*

1. *if $r > c$, then $\mathrm{ACA}_0 + \Pi_2^1 - \mathrm{TI} \nvdash \mathrm{SWQ}(f_r)$;*
2. *if $r \leq c$, then $\mathrm{PRA} \vdash \mathrm{SWQ}(f_r)$.*

(c is related to the so-called Otter's constant α[16]*).*

It is worth mentioning that the proof methods involve concrete mathematics, and in particular complex analysis (for details see [27], chapter 6, where the method of generating functions, the Pringsheim lemma and the Weierstrass preparation theorem are applied).

---

[15] All this would deserve separate discussion, but we skip it since the issue is irrelevant to the point we like to clarify.

[16] $c = 1/\lg(\alpha)$, where α is a real number given by

$$\alpha = \lim_{n \to \infty} T_n/T_{n-1}$$

and T_n is the number of non-isomorphic finite rooted trees with n nodes. $\mathrm{ACA}_0 + \Pi_2^1 - \mathrm{TI}$ is the fragment of second order arithmetic based on arithmetical comprehension and transfinite induction along arbitrary wellorderings for Π_2^1-predicates; see [38]. The system proves the consistency of ATR_0 (in fact more, i.e. the existence of ω-models for ATR_0).

4.2 Proof-Mining: mixing formal and informal

The general idea is well-established: use proof theoretic techniques in order to extract *effective bounds* from *ineffective* proofs (Kreisel's unwinding of proofs). In the last fifteen years, due mainly to the work of U. Kohlenbach and his students, applications of carefully designed systems and functional interpretations, based on Gödel primitive recursive functionals of finite type and variants thereof, have produced a number of theorems in analysis (approximation theory, metric fixed point theory; see [24], [25]). Below we present a brief (and rough) account of typical results.

The chosen framework $\mathcal{A}^\omega$ is usually rather liberal and is represented by a classical version of arithmetics in all finite types where choice is restricted to quantifier free conditions, while dependent choice is applied to arbitrary formulas. The formalization makes the language and the principles to be applied definite, and it clarifies the logical structure of the theorem to be proved. The formal restrictions are essential in order to apply suitable extraction techniques at large, and they allow to put upper bounds on the computational complexity.

This part is, so to speak, *a priori*. However, the general frame is suitably tailored for a special class of theorems one is interested in. Then $\mathcal{A}^\omega$ is uniformly expanded with data describing *concrete mathematical structures*, for instance a metric space (X, d).

Moreover, the emphasis is put on theorems whose logical form is actually found in ordinary mathematical texts. A typical instance (given by [24]) has the form[17]

$$(\forall x \in X)(\forall y \in K)(f(x,y) \stackrel{\mathbb{R}}{=} 0 \rightarrow g(x,y) \stackrel{\mathbb{R}}{=} 0), \qquad (1)$$

where X, K are Polish space, K is also compact,[18] $f, g : X \times K \rightarrow \mathbb{R}$ are continuous. The information to be extracted here is: how close to zero $f(x,y)$ must be in order to make sure that $g(x,y)$ is ϵ-close to zero (for any given $\epsilon > 0$). In other words, one would like to find a functional Φ – a modulus functional – such that, if $|f(x,y)| \leq \Phi(x,y,\epsilon)$, then $|g(x,y)| \leq \epsilon$. In general the compactness of the space K guarantees that such a Φ is *independent* of y. It turns out that in many cases the missing information can be extracted by purely logical analysis out of prima-facie ineffective proofs of the theorem.

Similar results can be given in the form of powerful metatheorems providing majorization functionals, having the following form:

Let X be a Polish space), K be a compact Polish space. Then we can extract from ineffective proofs of theorems of the form

$$\forall x \in X \forall y \in K \exists z \in \mathbb{N}. A(x,y,z),$$

---

[17] For examples, the reader is sent to § 4 of [24].
[18] Polish space = complete separable metric space, e.g. the Baire space $\mathbb{N}^\mathbb{N}$; a typical compact Polish space is the Cantor space $2^\mathbb{N}$.

where A *is existential, effective uniform (in K) bounds* $\Phi(x)$ *such that*

$$\forall x \in X \forall y \in K \exists z \leq \Phi(x).A(x,y,z)$$

Φ *is usually* subrecursive *(i.e. it belongs to a lower complexity class)*.

Concerning the proof interpretations used, these are purely syntactical transformations and hence, given a formalized proof, the extraction of information can be carried out in principle automatically via a computer. But, as the authors remark, "the difficult part of proof mining would then consist in fully formalizing a mathematical proof originally given in ordinary mathematical terms. That can be in general very tiresome and intricate." So one usually needs only partial formalization and the extraction can be carried out 'by hand'. Interestingly, the typical theorems involve steps, where formal proofs and formal manipulations are intertwined with informal ordinary mathematical arguments.

This shows *in re* that the opposition between proofs and derivations is not that sharp: formal proofs and methods are after all generated by mathematical questions, and there are not good grounds to reject them. If one looks for applications, it is likely that one needs explicit computational content and computational methods. If one likes to apply computational methods, it is not unlikely that one is forced to interpolate formal pieces of works. This, once more, has nothing to do with joining a philosophy of mathematics based on identifying derivations with the essence of actual proofs.

4.3 Problems about self-applicable operations

Making sense of self-application is not entirely trivial either in mathematics or in logic. On the other hand, it is certainly a mature topics, having been dealt by logicians and mathematicians since the twenties (Schönfinkel, Curry, Church and, as earlier as 1905, by Hilbert, see Kahle [23]).

We consider two problems illustrating how formal methods can be used to understand better the interplay of suitable principles, computational assumptions and analysis of formal proofs. After all, the argument should be attuned with the main topics of the workshop, like computation and effectiveness of proof.

Implicit operations and choice

Given a structure $\mathcal{M}$ whose elements represent "computable operations" in abstract sense, we like to investigate the closure properties of $\mathcal{M}$, e.g. under elementary definability. In order to make the question definite, we rephrase it by means of a well-known natural notion of *representable function*.

The representability problem for functions

Fix a universe M and a binary function $Ap : M \times M \longrightarrow M$.
$F : M \longrightarrow M$ is *Ap-representable* iff, for some $c \in M$,
$$(\forall a \in M)(Ap(c, a) = F(a))$$
(Ap is the *application operation*; as usual, we henceforth use ab as an abbreviation for $Ap(a, b)$. For unexplained notions, see [6]).
This has a natural reading: c represents the rule computing F; if M is a combinatory algebra, the idea is that *the representable functions are the computable functions*. Of course, no maximal solution (i.e. the whole space M^M) is available, for cardinality reasons. But *sensible solutions do exist* for *logic-free* definable objects. Indeed, it is well-known at least since the thirties:

Theorem 5 (Church–Rosser). *There exist non-trivial applicative structures $\mathcal{M}$ satisfying the condition*
$$\mathcal{M} \models (\exists f)(\forall \mathbf{x})(f\mathbf{x} = t)) \qquad (2)$$

(t is arbitrary application term over $\mathcal{M}$, $\mathbf{x}$ stands for a list of variables including the variables occurring free in t; 'non-trivial' means $card(\mathcal{M}) > 1$).

(2) is known as *combinatory completeness* and, in words, it states that application is universal for *algebraic functions*;[19] $\mathcal{M}$ is usually presented as a *combinatory algebra* (in short we write $\mathcal{M} \in CA$), i.e. a structure satisfying, for fixed K, S, the equations $Kxy = x$ and $Sxyz = xz(yz)$.

Problem 1. Do $\mathcal{M} \in CA$ exist, where every *implicit function is representable*, i.e. $\mathcal{M}$ satisfying (c fixed in M)
$$(\forall x)(\exists! y)(F(x,y) = c) \rightarrow (\exists f)(\forall x)(F(x,fx) = c) \quad ? \qquad (3)$$

Indeed, we can require more, i.e. whether there exist $\mathcal{M} \in CA$, satisfying AC_V:
$$(\forall x)(\exists y)\, \varphi(x,y) \rightarrow (\exists f)(\forall x)\varphi(x,fx) \qquad (4)$$

Note that AC_V states that *every (elementarily definable) total binary relation is uniformized by a representable function.*

From Hilbert's Paradox to latest developments

The negative answer to the stronger version of the previous problem is surprisingly present already in unpublished 1905 lecture notes of Hilbert, as detailed by Kahle [23]. Let $QF - AC!$ be the schema (4), restricted to *quantifier free-formulas* and where the first occurrence of $\exists$ is replaced by the existential-uniqueness quantification $\exists!$. Clearly, $QF - AC!$ trivially follows from AC_V.

---
[19] Once you naturally identify *"algebraic"* with "definable by a term of the equational language of the structure".

We can easily obtain a contradiction understanding negation classically and using self-application. Indeed, by classical logic, if one assumes that there exist two distinct objects a and b, then

$$(\forall x)(\exists! y)\left((y = a \wedge xx \neq a) \vee (y = b \wedge xx = a)\right)$$

By $QF - AC!$ there is an operation f such that $(\forall x)(fx = a \leftrightarrow \neg xx = a)$, which implies a contradiction, with $x := f$. To sum up:

Fact (Hilbert). $QF - AC! + (\exists x)(\exists y)(x \neq y) \vdash \bot$

Constructive move

Is the problem settled once and for all? Not at all: the question is, so to speak, opened again by D. Scott in the early seventies. The intuition is that under the microscope of constructive logic, more operations are tame: Scott conjectures that AC_V might be consistent with an intuitionistic theory of self-applicable operations, in the version, say, of combinatory logic CL.[20] The conjecture was then proved in 1973. Let CL_i be the intuitionistic theory with equality whose non-logical axioms are the equations of combinatory logic. Let Ext be the extensionality axiom for operations:

$$(\forall f)(\forall g)((\forall x)(fx = gx) \to f = g).$$

Theorem 6 (Barendregt). $CL_i + Ext + AC_V$ *proves the same equations as combinatory logic with extensionality* $CL + Ext$. *Hence* $CL_i + Ext + AC_V$ *is consistent.*

The proof is semantical in nature; it involves an abstract realizability model, where realizers are elements of a combinatory algebra.

Indeed, what about *classical* systems? Note that the original question involves after all a version (3) of the principle AC_V, where negation is not involved. Thus this suggests to consider a weakening $Pos - AC_V$, where the defining conditions are positive, i.e. $\neg$, $\to$-free, but within CL_c, i.e. combinatory logic embedded in classical logic. We obtain a positive answer in [9]:

Theorem 7. *A primitive recursive procedure ψ exists such that, if $\mathcal{D}$ is a derivation of $t = s$ in $CL_c + Ext + Pos - AC_V$, then $\psi(\mathcal{D})$ is an equational derivation of $t = s$ in combinatory logic with extensionality.*

The result is enough strong to grant existence for operations representing implicitly defined functions; it also implies that *every surjective representable function has a representable right inverse*.

The proof actually is rather involved and requires a *tour de force*, through a chain of *non-classical interpretations*.

---

[20] In this subsection we assume the reader is moderately familiar with the basics of combinatory logic and lambda calculus; see [6].

The *input* is a classical derivation of $t = s$ in $\text{CL}_c + \text{Ext} + \text{Pos} - \text{AC}_V$. The *output* of the procedure ψ is an equational derivation of $t = s$, that is, a *computation*, and the result shows how to transform a proof using logical notions and suitable choice principles in a purely computational construction.

Without giving any details, we summarize the different steps:

1. validating choice requires a constructive reading; this costs the embedding of classical logic into an *almost constructive* environment via the $\neg\neg$-translation; *almost-constructive* means that there remains a classical residue around (a special positive instance of the double negation law which is a sort of generalized Markov principle, named after truth stability);
2. validating TS requires an internal forcing model, where forcing conditions are *arbitrary positive conditions*;
3. the treatment of arbitrary positive conditions is carried out by extending the whole frame with a self-referential truth predicate; the soundness result under forcing requires the law CD of constant domains $(\forall x)(A \vee B(x)) \to A \vee (\forall x)B$ (x not free in A) in the case of universal quantification;
4. CD is validated via a suitable realizability model which is trivial on quantifiers;
5. positive choice is then explained away by a realizability model à la Barendregt, but we are still left with the truth predicate around;
6. eliminating truth is achieved by means of cut elimination and asymmetric interpretation, which allows to get rid of the truth predicate T by separating positive and negative occurrences of T and replacing T by its finite approximations;
7. the final step applies a witnessing technique for extracting equational derivations from derivations using intuitionistic logic; a basic difficulty in the construction is that we have to cope with undecidable atomic formulas (as = in combinatory logic is essentially undecidable).

Each step roughly corresponds to proving that certain notions and principles are superfluous, but each reduction is obtained by producing a suitable internal model. The result has, of course, a traditional reading in the style of reductive proof theory: *eliminating ideal elements* by finitary means, and *restoring purity of methods*. The interesting point is however that purity is recovered by using semantical ideas; it is crucial that logical methods allow to *change the meaning* of the basic logical operations (step from classical to constructive logic, asymmetric interpretation), and this is really essential for the realizability arguments lying at the heart of the proof.

Another interesting point is again computational, but we cannot go into details: what about the computation given by ψ? Apparently it is much longer in comparison to the original proof, and one might ask for a precise speed-up theorem in this direction.

A final byproduct of the argument is that one can even refine Barendregt's result. For instance, one obtains conservation (for a suitable class of formulas) over $\text{CL}_i + \text{Ext}$, even if $\text{CL}_i + \text{Ext} + \text{AC}_V$ is extended with the (classically inconsistent) *Unzerlegbarkeit* principle UZ:

$$(\forall x)(A(x) \vee B(x)) \to (\forall x)A \vee (\forall x)B$$

and hence with the constructively unsound law of constant domains

$$(\forall x)(A \vee B(x)) \to A \vee (\forall x)B.$$

Combining extensionality, CT and enumeration

The second problem concerns the notion of computation and computable operation in the abstract: to which extent can we consistently assume that the ground operations are extensional, the universe is enumerated (it is the surjective image of the natural numbers) and Church's thesis CT holds (in the sense that the space N^N of number theoretic functions contains only computable number-theoretic operations)?

We believe that this problem is typical of the *geistiges Experimentieren* Bernays talked about in his paper.

First of all, the problem can be rephrased as a question about a formal system. Let TON be the first-order theory including (i) CL; (ii) axioms on the type N of natural numbers, conditional on N, successor, predecessor, induction. The *enumeration axiom* EA has the form $(\exists f)(\forall x)(\exists n)(fn = x)$; Church's thesis CT is the statement $(\forall f : N \to N)(\exists n)(\forall m)(fm = \mathcal{E}nm)$,[21] where $\mathcal{E}$ is a fixed combinatory term defining an enumeration of the recursive functions. Now the problem becomes whether there is a model of TON + Ext + EA + CT. Clearly, due to CT, the model must be *effective* in some sense. This leads to consider *term models* i.e. suitable effective congruences on the set of lambda terms. However, in order to verify EA, one would like to deal only with *closed term models*, for which an enumerator can be defined. But then we have serious difficulties in validating extensionality because of the well-known Plotkin's counterexample[22]. Of course, one would like to kill the counterexample by closing under the so-called ω-rule. To make this precise, if T is a set of equations in the language of the lambda calculus, define T *consistent* iff there is at least one closed equation $t = s$, such that $T \not\vdash t = s$; if T is consistent, T is called a *lambda-theory*. The ω-rule is the inference (ω):

$$ts = rs, \text{ for each closed term } s \Rightarrow t = r \qquad (5)$$

T is *recursively enumerable* (r.e. in short) if the corresponding derivability relation $T \vdash t = s$ is r.e. The crucial (simple) observation is then:

Lemma 1. *If there exists a λ-theory T which is recursively enumerable and closed under the ω-rule, then the induced closed term model $\mathcal{M}_0(T)$ can be made into a model of* TON + Ext + EA + CT.

Problem 2. Does a λ-theory T, which is r.e. and closed under ω-rule exist?

---

[21] Of course, n,m range over natural numbers.
[22] There are closed terms M and N behaving in the same way on closed terms, which are not convertible (in the sense of $\lambda\eta$-conversion, see [6]).

We *cannot* use the usual lambda theories, because the usual term models falsify either Ext or CT or else EA. However, by a clever extension of Böhm trees, [22] proves:

Theorem 8 (Intrigila-Statman). *There exists a theory* T_Ω *which is closed under ω-rule and recursively enumerable.*

Corollary 1 (Cantini [10]). TON + Ext + CT + EA *is consistent.*

Of course, it remains to be seen if there other interesting models and applications thereof that make the "thought experiment" resulting in the formal system above of any real use.

4.4 On the Dedekind continuum, constructively

Formal methods can be *relevant for traditional philosophical and foundational issues*. This is evident if we consider the debate about the notion of set. Impredicative notions and classical logic notwithstanding, Aczel, following previous ideas of Myhill, succeeded in 1977 in designing a constructive version CZF of ZF which has a distinctive feature: CZF can be shown to be sound under an interpretation into Martin-Löf's type theory. This shows that (predicative) constructive set theory has a semantics of proofs, i.e. it can be based upon a theory of meaning grounded upon an informal notion of proof (or construction).

In this context, one can profitably analyze certain basic notions of classical mathematics, for instance that of *real number*. The least upper bound(= l.u.b.) principle for the reals [23] has traditionally been classified as typically impredicative, and it is known that Dedekind reals can be handled in fragments of second order arithmetics, which are usually regarded as *impredicative* (e.g. the system based upon the Π_1^1-comprehension schema, allowing full universal quantification on sets of natural numbers). However, Aczel and Rathjen [1] could rather smoothly define a version of the Dedekind continuum in CZF by identifying reals with (suitable) Dedekind *cuts*.[24] In particular, the Dedekind reals form a set, by invoking the principle of fullness,[25] a powerful generalization of exponentiation, still sound under the constructive type-theoretic interpretation. By construction, the Dedekind reals satisfy an appropriate form of completeness corresponding to the l.u.b. principle.

---

[23] Each set of reals having an upper bound has a least upper bound.
[24] Let X be a set with a binary relation $<$. A pair (L, U) of subsets of X is a D-cut iff (i) L, U are disjoint; (ii) L, U are both inhabited (constructively non-empty); (iii) L, U are located, i.e. $a \in L$ or $b \in U$, whenever $a < b$; (iv) L, U are open, i.e. for every $a \in L$, there is $b \in L$ and $a < b$ and for every $b \in U$ exists $a \in U$ with $a < b$.
[25] Fullness states that, given sets A, B, there exists a set C whose elements are multivalued functions from A to B such that, if R is a multivalued function from A to B, then there exists $S \in C$ such that $S \subseteq R$.

Crosilla, Ishihara, Schuster [15] recently sharpened the theorem, proving that the Dedekind cuts in an ordered set form a set in the sense of CZF set theory. They deduce the statement from a new principle of refinement, which, together with exponentiation, yields fullness.[26] This can be further generalised: the completion of a separable metric space turns out to be a set, *even if the original space is a proper class*; in particular, *every complete separable metric space automatically is a set*. There is a nice conceptual point behind all this: constructively, *sethood* is not only a matter of size, but also of cohesion.

We believe that the moral of the previous fact is that the classical characterization of predicativism, stemming from Poincaré and Russell and proof-theoretically characterized by Feferman and Schütte in the sixties, must be somewhat updated. This could lead to see the famous Pólya-Weyl wager[27] in a new light. Consider the two propositions:

- each bounded set of reals has a least upper bound;
- each infinite set of reals has a countable subset

According to Weyl, the notions of *real, set, countable* would have been regarded as *vague* in twenty years, and the two statements above would have been considered with respect to their own truth status on a par with the *Hauptsätze* of Hegel's *Naturphilosophie*; according to Pólya, quite the opposite.

Now, at least if one accepts CZF and its understanding within the frame of predicative constructive type theory, the notion of real number à la Dedekind and the l.u.b. principle could have been accepted even by Weyl (and possibly Russell and Poincaré).[28] This shows that *informal content* can be sharpened by formal methods and logical analysis.

4.5 Conclusion

The previous results show that the separation between formal and informal work is difficult to trace. Often formal systems and proofs act as powerful microscopes, whose aims go far beyond pure ideological (foundational) aims, and the level of mere soundness verification and logical certification in a narrow sense.

Formal methods have plenty of applications and they provide us with fine tools if properly understood. We believe they are essential for clarifying *the procedural aspects* of mathematics and abstract reasonings, and evaluating the epistemological nature of proofs.

---

[26] Quite recently, Lubarski and Rathjen have shown that exponentiation is not sufficient to prove sethood of Dedekind reals, see [29].

[27] The wager dated back to February 9, 1918; see G. Pólya [32] and S. Feferman [16, p. 57].

[28] Of course, we only mean that Pólya could have used CZF as a route to render the l.u.b. principle palatable to Weyl, even if Pólya himself would have probably rejected the constructive approach behind CZF.

On the other hand, investigating formal methods and proofs requires semantical ideas and new constructions as ever in mathematics and abstract sciences.

References

1. P. Aczel and M. Rathjen: *Notes on Constructive Set Theory*, Draft available from the Internet, http://www.cs.man.ac.uk/~petera/
2. M. Aschbacher: Highly complex proofs and implications of such proofs. *Philosophical Transactions of the Royal Society A* **363** (2005) pp 2401-2406
3. J. Avigad: Number theory and elementary arithmetic. *Philosophia Mathematica* **11** (2003) pp 257-284
4. J. Avigad: Mathematical Method and Proof. *Synthèse* **153** (2006) pp 105-159
5. J. Avigad, K. Donnelly, D. Gray and P. Raff: A formally verified proof of the prime number theorem. *ACM Transactions on Computational Logic,* **9**, 1:2 (2007)
6. H. Barendregt: *The Lambda Calculus. Its Syntax and Semantics*(Elsevier, Amsterdam, 1984)
7. P. Bernays: Mathematische Existenz und Widerspruchfreiheit. In: *Études de Philosophie des Sciences en hommage à Ferdinand Gonseth*, (Neuchatel, 1950) pp 11–25 (English Translation by W. Sieg, R. Zach and S. Goodman, available as Text No.19, Bernays Project)
8. P. Bernays: Die Schematische Korrespondenz und die idealieserte Strukturen. *Dialectica* **24** (1970) pp 53–66 (English Translation by E. Reck and C.Parsons available as Text No.27, Bernays Project)
9. A. Cantini: The axiom of choice and combinatory logic. *Journal of Symbolic Logic* **68** (2003) pp 1091–1108
10. A. Cantini: Remarks on applicative theories. *Annals of Pure and Applied Logic* **136** (2005) pp 91-115
11. L. Carroll: What the Tortoise said to Achilles. *Mind* **4** (1895) pp 278–280
12. E. Casari: Matematica e Verità. *Rivista di Filosofia* **78** (1987) pp 329–350
13. C. Cellucci: The nature of mathematical explanation, this volume (2008)
14. C. Cornaros and C. Dimitracopoulos: The prime number theorem and fragments of PA. *Arch. Math. Logic* **33** (1994) pp 265-281
15. L. Crosilla, H. Ishihara, P. Schuster: On constructing completions. *Journal of Symbolic Logic* **70** (2005) pp 969–978
16. S. Feferman: *In the Light of Logic* (Oxford University Press, Oxford 1998)
17. E. Friedgut (with an appendix by J. Bourgain): Sharp thresholds of graph properties and the k-Sat problem. *Journal of the American Mathematical Society* **12** (1999) pp 1017–1054
18. E. Giusti: *Ipotesi alla base dell'esistenza matematica* (Boringhieri, Torino 1999)
19. A. Goerdt: A Threshold for Unsatisfiability. *Journal of Computer and System Sciences* **53** (1996) pp 469–486
20. J. Hafner and P. Mancosu: The Varieties of Mathematical Explanation. In: *Visualization, Explanation, and Reasoning Styles in Mathematics*, ed by K. Jørgensen et al. (Kluwer, Dordrecht 2005).
21. G. Hardy: *A mathematician's apology* (Cambridge University Press, Cambridge, 1941)

22. B. Intrigila and R. Statman: Some results on extensionality in lambda calculus. *Annals of Pure and Applied Logic* **132** (2005) pp 109-125
23. R. Kahle: David Hilbert and functional self-application, preprint (2006)
24. U. Kohlenbach and P. Oliva: Proof mining: a systematic way of analysing proofs in mathematics. *Proc. Steklov Inst. Math.* **242** (2003) pp 136-164
25. U. Kohlenbach: Some Logical Metatheorems with Applications in Functional Analysis. *Transactions of the American Mathematical Society* **357** (2005) pp 89-128
26. G. Kreisel: Principles of proofs and ordinals implicit in given concepts. In: *Intuitionism and Proof Theory*, ed by J. Myhill, A. Kino, J. Myhill and R. E. Vesley (North Holland, Amsterdam 1970) pp 489-516
27. G. Lee: *Phase Transitions in Axiomatic Thought*, Ph. Thesis, Westfalische Wilhelms-Universität Münster (2005)
28. G. Lolli: *Capire una dimostrazione* (Il Mulino, Bologna 1988)
29. R. S. Lubarski and M. Rathjen: On the constructive Dedekind reals. To appear in *Logic and Analysis*, published online: 18 February (2008) pp 1–22
30. S. Mac Lane: Despite physicists, proof is essential in mathematics. *Synthèse* **111** (1997) pp 147–154
31. P. Mancosu: Mathematical Explanation: Problems and Prospects. *Topoi* **20** (2001) pp 97–117
32. G. Pólya: Eine Erinnerung an Hermann Weyl. *Mathematische Zeitschrift* **126** (1972) pp 296-298
33. D. Prawitz: *Natural Deduction: a Proof–theoretical Study* (Almqvist and Wiksell, Stockholm 1965)
34. D. Prawitz: Philosophical Aspects of proof theory. In: *Contemporary philosophy. A new survey, vol.I* (Nijhoff, The Hague 1981) pp 235–277
35. Y. Rav: Why do we prove theorems. *Philosophia Mathematica* **7** (1999) pp 7-41
36. G. Rota: The phenomenology of mathematical beauty. *Synthèse* **111** (1997) pp 171-182
37. B. Russell: The regressive method of discovering the premises of mathematics. In: B. Russell: *Essays on Analysis* (London 1973) pp 272–283
38. S.G. Simpson: *Subsystems of Second Order Arithmetic* (Springer, Berlin Heidelberg New York 1999)
39. A. Weiermann: An application of graphical enumeration to PA. *J. Symbolic Logic* **68** (2003) pp 5–16
40. A. Weiermann: Analytic combinatorics, proof-theoretic ordinals, and phase transition for independence results. *Ann. Pure Appl. Logic* **136** (2005) pp 189-218

Toy Models in Physics and the Reasonable Effectiveness of Mathematics

Annalisa Marzuoli

Dipartimento di Fisica Nucleare e Teorica, Università di Pavia & Sez. INFN (Italy)
annalisa.marzuoli@pv.infn.it

Toy models in theoretical physics are invented to make simpler the modelling of complex physical systems while preserving at least a few key features of the originals. Sometimes toy models get a life of their own and have the chance of emerging as paradigms. Such an upgraded role, on the one hand, makes these models likely to be considered for validation through (possibly new) experimental tests. On the other, the role played by mathematical proof – evoked in Wigner's "unreasonable effectiveness of Mathematics in the Natural Science"– could be so enhanced as to become in a sense more compelling than experiments. 'Theoretical Mathematics', a new synthesis of mathematics and theoretical physics proposed by Jaffe and Quinn in the 1990's [12], looks at pure mathematics as a sort of experimental testing ground for certain physical theories (and in particular for associated toy models).

As a case study I shall illustrate some basic features of Chern–Simons topological quantum field theory and its connections with the mathematical theory of knots. The issue of effective computability of the basic functionals and observables of such model will be briefly addressed as well.

1 Roles of mathematics

Modern mathematics is nearly characterized by the use of rigorous proofs, but this practice sometimes runs the risk of becoming 'compulsive' about details of arguments. Starting from the mid-1970's there has been a flurry of mathematical–type activities –driven or inspired to a large extent by theoretical physics– which has enhanced the role of 'intuitive reasoning' and speculation (actually the original, old pattern of history of mathematics). In the paper titled *"Theoretical Mathematics: toward a cultural synthesis of Mathematics and Theoretical Physics"* Arthur Jaffe and Frank Quinn proposed to develop a constructive context for these new trends, changing somehow the way mathematics is organized [12].

1.1 Mathematics as a natural science

Within the current practice, the discovery of mathematical structures is achieved in two stages, namely through

A. intuitive insight, conjectures based on 'reasonable' assumptions, speculation *beyond established knowledge*

B. proof–oriented phase, where conjectures are proved (validated) by
  a) adapting (or improving) known techniques within a same field;
  b) inventing a new synthesis of standard arguments from different fields;
  c) inventing a completely new strategy for proving
  or, possibly, a mixing of them.

According to Jaffe and Quinn, **A** is likely to be refereed to as 'theoretically–oriented' mathematics while **B** embodies 'rigorous' mathematical reasoning.[1] As far as proceeding from **A** to **B** requires corrections, refinement and validation, the role of rigorous mathematics is functionally analogous to the role of experiment in the natural sciences.[2] Keeping on the analogy, unexpected features coming from proofs may feed back the theoretical phase with new speculative material:

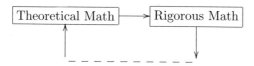

As an illustration of these phases, consider two conjectures in differential geometry, the validation of which has been recently proved.

The Poincaré conjecture.

This lonstanding problem can be stated as

> *Every closed, oriented and simply–connected 3–space is (topologically equivalent to) the 3–sphere S^3,*

where S^3 is the analog in dimension 3 of the usual spherical surface embedded in ordinary space and denoted by S^2. Such conjecture, open for more than a century, was positively answered by Perelman in the early 2000's [17] and his proof was validate by the mathematical community in 2005.

---

[1] **b)** relies at bottom on the effective procedure of *recognition* of identical fundamental entities such as for instance isomorphic underlying algebraic structures.

[2] Note that 'experimental' mathematics, as opposite to theoretical, is to be meant as computer–aided activity including both numerical calculations and computer simulations, while the interesting issue concerning 'automated proofs' is addressed in other contributions to this volume.

The geometrization programme

This conjecture, better framed as a programme, was stated by Thurston in the 1970's [18] and for more than 30 years a number of people have been working on several segments of the proof. To give an idea of its content, recall that in dimension $D = 2$ there exist only three 'standard' types of geometric structures, namely the Euclidean, flat plane $\mathbb{R}^2$, the sphere S^2 and the hyperbolic plane $\mathbb{H}^2$ (modelled for instance as a Poincaré disk). To be more precise, the uniformization theorem asserts that each (connected portion of) 2–space is equivalent to one of the above Riemannian manifolds whose canonical metrics have constant curvatures $k = 0, 1$ and -1, respectively.[3]

The geometrization programme deals with the generalization to $D = 3$ of the uniformization theorem and, leaving aside its technical formulation, its content can be summarized in

> *Every 3–space can be decomposed into a finite number of 'elementary pieces' or model 3–geometries, and only eight models exist (apart from Euclidean 3–space).*

The proof of the first conjecture has a crucial impact also on this second conjecture, previously validated only up to Poincaré conjecture.

The above examples were aimed to give the flavor of what the theoretical side of mathematics should embody. Poincaré conjecture, once explicitly formulated, has been of course addressed with methods of the proof–oriented phase **B** (and Perelman's work can be framed within the validation scheme **b)** above). On the other hand, even the precise statement of the geometrization conjecture has required quite a long phase of elaboration based on intuitive insight and analogy–based reasoning. The possible way of looking at the body of logical deductions leading to its validadion –and preceding the proof of Poincaré conjecture– is to some extent a matter of taste. Such process might fall into rigorous mathematics –but then Thurston's theorem stated above should have been emended by 'if Poincaré conjecture would be positively answered'– or rather considered as purely speculative. It is worth noting that Jaffe and Quinn do not take stand in favour of either opinion.[4]

---

[3] Two topological spaces X and Y of a same dimension D are topologically equivalent if there exists a map $f : X \to Y$ which is one–to–one and continuous together with its inverse f^{-1}. A (real) D–manifold M^D is a topological Hausdorff space locally modelled on the the standard D–dimensional linear space $\mathbb{R}^D$. A Riemannian metric on M^D is the assignment of a bilinear, positive definite quadratic form ds^2 in (the tangent space associated with) each point of the manifold. Two Riemannian manifolds are equivalent if they are isometric, namely have the same global topology and local metric propeties. Riemannian manifolds represent the proper frame for addressing D–dimensional 'non–Euclidean' geometry.

[4] Indeed they were quite troubled about the acceptance policy of mathematical journals and suggested that 'sound' conjectural papers should deserve publication, going beyond the current practice of editorial boards.

1.2 Mathematics & physics

A physical system examined in the light of

* guiding principles
* phenomenological data

provides a theory or a 'modellig scheme' for reality ('model' will be used in connection with mathematical physics, see below).

The role of experimental work can be characterized as to giving rise to

- validation of the theory;
- correction, modification of the model;
- unexpected observations, and thus new input data for improving the model.

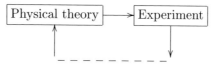

Then the theoretical side of physics mirrors the speculative phase in mathematics: here 'intuitive insights' are driven by guiding principles that have their counterparts in 'basic axioms' about fundamental mathematical entities (but of course phenomenological input data do not possess strict mathematical counterparts, unless one considers as possible input also the result of numerical 'experiments' performed on instances of assumptions).

The traditional role of mathematics in physics is that played by *mathematical physics*. Since any physical theory is expressed in a mathematical language, every phenomenon is (approximated by) a mathematical model. Thus any such model can be validated with tools proper of rigorous mathematics, namely no ambiguity exists about definitions of fundamental entities, formulation of claims, proofs of theorems. As a matter of fact, this type of 'proof' is considered less effective than any experimental verification.[5]

Indeed, the typical attitude of theoretical physicists is nicely expressed in the following quotations (see [12] for the original references).

> It is quite satisfactory to test mathematical statements by verifying a few well–chosen cases (Richard Feynman).

> No attempt is made at mathematical rigor in the treatment, since it is anyhow illusory in theoretical physics (Lev Landau).

Even with a more positive attitude for mathematics, it could happen that either

---

[5] Quantum field theory, combining quantum mechanics and special relativity, is a paradigmatic example in the present context, since it admits an 'axiomatic' formulation only in very few (quite trivial) cases, while a number of predictions have been completely validated by experiments, see the discussion in § 2.

i) the approximation of the phenomenon by means of a mathematical model is too rough, so that the model is useless in physics,

or

ii) the physical entities are not modelled exactly on mathematical structures

and/or

iii) deductions and calculation procedures are heuristic and cannot be casted into rigorous proving protocols.

As will be illustrated through the case study of § 4, both **i)** and **ii)** may provide material to feed the speculative side of mathematics.

2 Predictive & speculative physics: genesis of toy models

The 'predictive' role of a physical theory emerges when a physical phenomenon is examined in the light of

* guiding principles
* known and presumed phenomenological data.

The predictions are subjected to (present or forthcoming) experiments whose outcomes may provide either

CV) validation of the whole theory (**C**omplete **V**alidation), or
PV) validation of a few predictions (**P**artial **V**alidation), and/or
NO) observation of unexpected new effects (**N**ew **O**utcome).

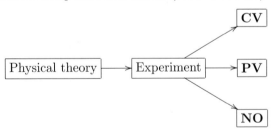

An example of case **CV)** is Einstein's General Relativity, while the Standard Model of particle physics complies with **PV)** at present, and has the chance to fall into **NO)** when the new high–energy experimental devices at CERN will be completed.

In the presence of **PV)** or **NO)** an improved, more effective theoretical framework is needed. It will be based on

* guiding principles improved with respect to the previously accepted ones
* upgraded phenomenological input from experiments

and, usually, it will be required that the correct predictions of the original theory **I** are recovered also in the new setting **II**.

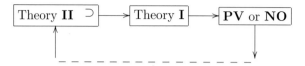

In the search for more and more 'complete' theoretical schemes including all fundamental interactions (possibly up to the 'theory of everything') hierarchies of theories can be generated. At the top, guiding principles and phenomenologcal input may shade into

* guiding principles that mimic those of 'analog' predictive theories
* input data that are only reminiscent of reality, if not physically 'unreasonable'.

Theories that reach such a stage are to be considered as *speculative* since they have not yet matured to provide observable predictions or predictions are beyond the potentialities of present and forthcoming experiments. It is the lacking of experimentally testable predictions that makes entities and procedures of such theories likely to be subjected to mathematical speculation, as discussed below.

On the other hand, also in the presence of a truly predictive theory, it can be useful to address 'unrealistic' versions of it in order to get a better insight into particular aspects or mechanisms. In both cases the resulting theories represent *toy models* since they display only analogs of physical phenomena, even if they are not necessarily much more 'simpler' than the parent theories. With respect to a same predictive reference theory, toy models can thus emerge through bottom–up and top–down processes, schematically

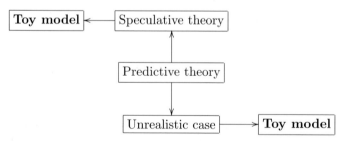

Examples of the bottom–up genesis from the Standard Model (viewed as the reference theory) are given by string theories, while General Relativity in non–physical spacetime dimensions (namely $D \neq 4$) gives rise to toy models of the top–down type (see § 3 and § 4 for a more detailed analysis).

Being the access to experiments forbidden, it may happen that specific toy models framed within strongly formalized settings –reflecting the presence of underlying algebraic or geometric symmetries– actually fall into theoretical mathematics looking for their validation. According to the discussion on the roles of mathematics in § 1.1, the following diagram should summarizes the point

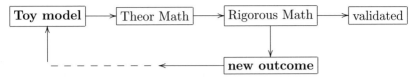

More stringently, a few toy models do not even require an intermediate phase of mathematical speculation, but find a direct access to rigorous mathematics. Such goal is achieved by means of *recognition*, namely the fundamental entities of the model are identified with known mathematical structures and the deductive operational procedures are theorems.

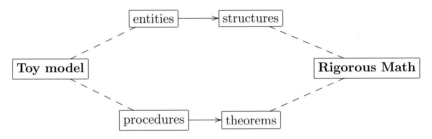

Note finally that the above correspondence could hold true only for entities, as it is going to happen for Chern–Simons topological quantum field theory –a toy model arising as a special type of gauge theory– presented in § 4. In such cases the possibility of recognizing effectively deduction procedures as mathematical proofs (the counterpart of the sequence of experimental operations performed by physical devices) would represent a major breakthrough in theoretical physics.

3 Yang–Mills gauge theories

Yang–Mills (YM) field theories, and in particular their quantized versions, represent a paradigmatic framework for addressing predictive/speculative hierarchies of theories as well as bottom–up and top–down toy models.

The treatment here is aimed to sketch the main features of such theories in connection with the remarks of § 2, leaving aside most tecnical details that can be found for instance in [5], [4], [10].

3.1 Classical field theory

A classical YM field theory is based on the action functional

$$\int_{M^D} d^D x \, \mathcal{L} \doteq S, \qquad (1)$$

where $d^D x$ represents the D–dimensional volume element and

$$\mathcal{L} = \tfrac{1}{2} \mathrm{Tr}_{\mathcal{G}} \, F^{\mu\nu} F_{\mu\nu} \qquad (2)$$

is the YM Lagrangian that describes a classical field F ($F_{\mu\nu}$ is called the 'field strenght' and $\mu, \nu = 1, 2, \ldots, D$) living in an ambient spacetime M^D of dimension D. Such theory is endowed with a (Lie) group of symmetries, the *gauge group* $\mathcal{G}$, so that the field F carries a further 'internal' label, $F^{\mu\nu} \to F^{\mu\nu(a)}$, with $a = 1, 2, \ldots, d \equiv \mathrm{rank}\,(\mathcal{G})$.

The trace operation in (2) is explicitly given by $\mathrm{Tr}_{\mathcal{G}} \, F^{\mu\nu(b)} F^{(a)}_{\mu\nu} = F^{\mu\nu(a)} F^{(a)}_{\mu\nu}$, where Einstein's convention for summation over repeated indices is assumed. The fundamental dynamical variable of the theory is the 'vector potential' A^a_μ (mathematically defined as a connection on the principal $\mathcal{G}$–bundle whose basis is the manifold M^D) and its explicit relation with the field strenght (representing the curvature associated with the connection A) reads

$$F_{\mu\nu} \equiv F^{(a)}_{\mu\nu} = \partial_\mu A^a_\nu - \partial_\nu A^a_\mu + \mathbf{g}\,(A^a_\mu A^a_\nu - A^a_\nu A^a_\mu) \qquad (3)$$

where $\mathbf{g}$ is the YM *coupling constant*.

The YM action recasted in terms of the vector potential $A^a_\mu \equiv A$

$$\mathbf{g} \int_{M^D} d^D x \, \mathcal{L}(A) \doteq S(A; \mathbf{g}) \qquad (4)$$

is required to be stationary against arbitrary variations of A according to Hamilton variational principle. Formally

$$\delta S(A; \mathbf{g}) = 0$$

which provides the Euler–Lagrange equations of the field strenght $F^{\mu\nu}$ (analog of equations of motion for a discrete physical system)

$$\mathcal{D}_\mu F^{\mu\nu} = 0 \qquad (5)$$

where $\mathcal{D}_\mu = \partial_\mu + \mathbf{g}[A_\mu,\ ]$ and $[\,,\,]$ is the commutator, *i.e.* the antisymmetrized product of its entries.

YM field equations (5) (and associated Bianchi identities) are highly non linear and have been solved exactly only in a few cases.

Example. The dynamics of the classical electromagnetic field is described by an Abelian YM theory formulated in Minkowski spacetime M^4 with the Abelian group $U(1)$ (unimodular complex numbers) as gauge group. The corresponding Euler–Lagrange equations (supplemented by Bianchi identities) are the Maxwell equations. Here the *phenomenological data* are represented by $F_{\mu\nu}$, the tensLagrangeor encoding both the electrical and the magnetic fields, while $\mathbf{g} \equiv \alpha$ is the 'fine structure' constant.

Looking in some more details at the features of classical YM field theories with respect to the categories of guiding principles and phenomenological input addressed in § 2, the main issues can be summarized as follows.

Guiding principles

- The Lagrangian is a 'scalar' –namely invariant with respect to the group of spacetime symmetries– in order to ensure that field equations are generally covariant. Moreover $\mathcal{L}(A)$ must be 'gauge–invariant', namely invariant under the action of the gauge group $\mathcal{G}$. Indeed the explicit expressions given in (1), (2) and (4) complies with these requirements.
- The *observables* of the theory –namely those quantities that can be measured experimentally in the presence of a predictive theory– cannot depend on the reference (coordinate) system used to write down field equations. Moreover they must be gauge invariant, reflecting the presence of the internal symmetry given by the gauge group $\mathcal{G}$. It can be shown that 'good' observables are of the form

$$\mathbf{W}_\gamma = \text{Tr}_\mathcal{G} \exp \left\{ \mathbf{g}\, i \oint_\gamma dz^\mu A_\mu(z) \right\} \tag{6}$$

where γ is a closed contour embedded into the spacetime M^D and z^μ represent the coordinates of points lying on the curve (mathematically these quantities are holonomies associated with the connection A).

The main characteristic of $\mathbf{W}_\gamma$ is to be *non local*, namely it depends on the curve γ and on the global topology of the ambient space M^D. However, the calculation of $\mathbf{W}_\gamma$ for arbitrary contours makes it possible, at least in principle, to reconstruct the vector potential A_μ and the associated field strenght $F_{\mu\nu}$ all over M^D.

Phenomenological input *(predictive theory)*

Since YM action describes a pure vector potential A (associated with the 'massless' field F) able to interact with other 'matter' fields, a truly predictive theory should be based on a more general action of the type

$$S(A; \mathbf{g}) + \mathcal{S}(\{\phi\}, A; \{\mathbf{k}\}, \mathbf{g}) \tag{7}$$

where $\{\phi\}$ denote collectively matter fields and $\{\mathbf{k}\}$ the set of associated coupling constants. Note that the second part of the action contains mixed terms taking into account physical interactions among A and matter fields and among matter fields themselves.

- The first requirement is that $M^D = M^4$, namely the theory is formulated in Minkowski spacetime, naturally endowed with the Lorentz (Poincaré) group of symmetries.
- The physical input data are represented by the specification of A, $\mathbf{g}$ together with the collection of the (scalar, vector and/or tensor) fields $\{\phi\}$.

3.2 Quantization

Guiding principles here must comply also with the basic principles of quantum mechanics. On the basis of the Lagrangian setting adopted for the classical case, the 'natural' quantization scheme turns out to be the Feynman's path integral prescription. The basic quantum functional is built on the basis of a democratic assessment: within the space of potentials $\mathcal{A} = \{A\}$ each configuration $A(z)$ is weighted by a dynamical phase given by the value of the corresponding classical action $S(A; \mathbf{g})$. Formally

$$\int_{\mathcal{A}} [DA]\, e^{i S(A; \mathbf{g})/\hbar} = \mathbf{Z}[\mathbf{g}; \hbar]\,, \qquad (8)$$

where $\hbar$ is the Planck constant and $[DA]$ represents a 'measure' over the configuration space, up to gauge transformations.

The *quantum observables* are suitably defined counterparts of the classical observables $\mathbf{W}_\gamma$ introduced in (6). The expectation values[6] of such gauge invariant quantum 'Wilson loop' operators are formally defined according to

$$\frac{\int_{\mathcal{A}} [DA]\, \hat{\mathbf{W}}_\gamma\, e^{i S(A; \mathbf{g})}}{\mathbf{Z}[\mathbf{g}]} = \langle \hat{\mathbf{W}}_\gamma[\mathbf{g}] \rangle\,, \qquad (9)$$

where $\mathbf{Z}[\mathbf{g}]$ corresponds to the expression given in (8) with the convention $\hbar = 1$.

Leaving aside the technical treatment of quantized interacting matter fields, the main problems rised by such quantization procedure can be summarized as follows.[7]

- The path integral prescription is heuristic because it is worked out explicitly only in very few special cases and extended by analogy to other situations. Thus, generally speaking, quantum functionals as those given in (8) and (9) do not correspond to well defined mathematical structures.
- The relevant equations [expressions] cannot be solved [calculated] *exactly* in most cases.
- As a consequence of the previous remark it becomes necessary to introduce *approximation schemes* aimed to extract numerical predictions to be compared with experimental data. The most effective approach relies on the perturbative expansion in terms of powers the relevant coupling constant(s). Referring for simplicity to the Wilson loop operator in (9) (and dropping the brackets $<>$), the form of such an expansion would look like

---

[6] In quantum field theory the expectation value of an operator (the quantum transition amplitude in ordinary quantum mechanics) is the quantity the square modulus of which gives the quantum probability of occurence.

[7] Actually the path integral prescription is not the unique possible scheme, but other approaches for addressing quantum fields –such as the second quantization method– must face foundational obstacles that ultimately can be brought back to those listed here.

$$\hat{\mathbf{W}}_\gamma[\mathbf{g}] \sim \hat{\mathbf{W}}^{(0)} + \mathbf{g}\,\hat{\mathbf{W}}^{(1)} + \mathbf{g}^2\,\hat{\mathbf{W}}^{(2)} + \ldots . \qquad (10)$$

Such kind of expansions[8] must be further improved by regularization and renormalization–group techniques aimed to resum the series to get the desired finite quantities. Quantum field theories of this kind are technically referred to as *perturbatively renormalizable*.

Notwithstanding the heuristic, not rigorous, approximate nature of (basically all) quantization procedures, they are considered *effective* in theoretical physics because well validated in a number of situations as the following examples witness.

Examples
• Quantum electrodynamics, the quantum counterpart of Maxwell theory, describing charged quantum particles interacting with photons.
• The Weinberg–Salam theory (1967-68) unifying electomagnetic and weak interactions ($\mathcal{G} = SU(2) \times U(1)$).
• Quantum cromodynamics ('t Hooft, Gross, Wilczek 1972-73) describing strong interactions and quarks ($\mathcal{G} = SU(3)$).
• The Standard Model unifying all previous theories, the complete validation of which is expected in the next future from LHC experiments at CERN.

Once validated in such a strong sense, both guiding principles and techniques can be borrowed in addressing *speculative* Yang–Mills–type quantum field theories[9] as well as 'unrealistic' toy models, thus providing a concrete application of the general observations presented in § 2.

4 Chern–Simons theory & topological invariants of knots

The case study addressed in this section belong to the class of 'topological' quantum field theories of Schwarz type [15], namely those YM gauge theories that can be consistently axiomatized [2] and share the feature of possessing observables that do not depend on local metric properties but only on globally defined, topological quantities associated with the underlying geometric structures.

On the basis of the discussion in § 2 and taking into account the definitions given in § 3, Chern–Simons theory can be characterized as follows.

- Guiding principles mimic those of YM predictive theories. In particular: the underlying ambient 'spacetime' M is 3–dimensional (and thus physically unrealistic); the YM–type classical action is selected on the basis of purely 'theoretical' considerations;

---

[8] Note that these techniques actually requires that $|\mathbf{g}| \ll 1$, otherwise other (numerical) approximation methods or non–perturbative analysis are needed.
[9] The genesis of string theories in connection with quantum gauge theories is nicely addressed at a non–technical level in [8].

the gauge group $\mathcal{G}$ can be chosen at will, namely does not reflect the existence of a physical symmetry (typically one works with the special unitary group $SU(2)$, the simplest non–Abelian compact Lie group).

- Input phenomenological data do not have a counterpart in reality, namely the vector potential A and the coupling constant **k** are not related to any physical interaction.

Chern–Simons (CS) classical action in term of the vector potential A – technically, a 1–form with values in the Lie algebra of the gauge group $\mathcal{G} = SU(2)$– reads

$$S^{(CS)}(A; \mathbf{k}) = \mathbf{k} \int_{M^3} \mathrm{Tr}_{\mathcal{G}} \left(A\, dA + \tfrac{2}{3} A \wedge A \wedge A \right), \tag{11}$$

where M^3 is a closed Riemannian 3–manifold and **k** is CS coupling constant (d denotes the differential and $\wedge$ the wedge product on differential forms).

The quantum expectation values of a Wilson loop operator is formally defined as in the YM setting (*cfr.* (9)) and is given by

$$\frac{\int_A [DA]\, \hat{\mathbf{W}}_\gamma\, e^{i S^{(CS)}(A;\mathbf{k})}}{\mathbf{Z}[M^3; \mathbf{k}]} = \langle\, \hat{\mathbf{W}}_\gamma [\mathbf{k}]\, \rangle. \tag{12}$$

The basic facts about the quantum theory, as developed by Witten [46], are summarized as follows.

1. The quantum theory is *solvable*, namely it is not only perturbatively renormalizable, but actually finite for each fixed value of the coupling constant **k** (constrained to be a positive integer by the quantization prescription). Accordingly, up to suitable normalization, the outcomes of the theory are *finite quantities*.
2. The quantum functionals $\mathbf{Z}[M^3, \mathbf{k}]$ and $< \hat{\mathbf{W}}_\gamma [\mathbf{k}] >$ share a topological nature, namely they depend only on the global topology of the ambient manifold M^3 and on intrinsic properties of the curve γ (see below).
3. Note preliminarily that the contour γ in (12) is actually a 'knotted' closed curve, namely represents a *mathematical knot*.[10] Then the expectation value of the Wilson loop operator equals a well known topological invariant of knots, the Jones polynomial [33]

$$\langle\, \hat{\mathbf{W}}_\gamma [\mathbf{k}]\, \rangle = \mathcal{J}(\gamma; t), \tag{13}$$

---

[10] Slightly more generally, a mathematical knot is a collection of circles embedded in a 3–dimensional ambient space. Its topological properties are caught by the notion of 'ambient isotopy', in the sense that two knots are equivalent (ambient isotopic) if and only if they can be superposed by continuously stretching them without cutting. Actually knots can live only in dimension 3 because in any higher dimensional ambient space all such curves can be made 'unknotted', while in a 2–dimensional space knots do not exist at all.

where the variable t of the polynomial is related to CS coupling constant by $t = \exp(2\pi i/\mathbf{k})$ (recall that $\mathbf{k}$ is a positive integer, so that t is a complex $\mathbf{k}$–th root of unity).
4. The complete solvability of the model as stated in **1.** does not rules out questions concerning *effective computablity* of such topological invariants. Indeed it can be shown that the Jones polynomial of any knot can be calculated in practice by means of a (combinatorial) recursive procedure based on 'skein relations' [14], and the same kind of relations are recovered within the field–theoretic setting [9].

The issue of *efficient* computability is more subtle and will be briefly addressed at the end of this section and in the concluding remarks.

The discussion in § 2 about toy models and the roles played by mathematics can be now revisited in the light of the above list of remarks. The present toy model, whose observables do not have any physical counterpart, is properly located within theoretical mathematics, and thus subjected to rigorous proving in view of a complete validation.

In particular, the equality stated in (13) between a quantum functional and a topological invariant represents an instance of the process referred to as *recognition* (of an entity as a well–defined mathematical structure). The validation of the heuristic procedures associated with quantization in terms of effective proving techniques is still an open issue.

Computational complexity of knot invariants

Over the years mathematicians have proposed a number of 'knot invariants' aimed to classify systematically all possible (equivalence classes of) knots. The most effective invariants turn out to be Laurent polynomials in one or two formal variables with coefficients in some ring, such as the 1-variable Jones polynomial already quoted above [33] and the 2-variable HOMFLY polynomial [16]. The reason why Jones' case is so crucial also in the computational context –besides CS theory and topology– is due to the fact that a 'simpler' link invariant, the Alexander–Conway polynomial, can be computed efficiently, while the problem of computing 2–variable polynomials, such as the HOMFLY invariant, is **NP**–hard (see [3] for an account on knot theory–based computational questions and original references).

The issue of computational complexity of the Jones polynomial is summarized in the question:

How hard is it to determine the Jones polynomial of a knot γ?

The most relevant parameter encoding the 'size' of this computational problem is given by the number of crossings of the knot, so that the time required to performed the calculation as a function of the input size is the quantity to be evaluated.

An exhaustive answer to the question within the framework of classical complexity theory was provided in [11], where the evaluation of the Jones polynomial at a root of unity t was shown to be generically #**P**–hard, namely computationally intractable in a very strong sense.[11]

The computational intractability of the above problem does not rules out the possibility of *approximating efficiently* Jones' invariant either in a classical or in a quantum computing context. Loosely speaking, the approximation in question is a number X such that, for any choice of a small $\delta > 0$, the numerical value of $\mathcal{J}(\gamma; t)$, when one substitutes in its expression the previously selected value of t, differs from X by an amount ranging between $-\delta$ and $+\delta$. In a probabilistic setting (either classical or quantum) it is required that the value X is accepted as an approximation of the value of the polynomial if

$$\text{Prob } \{|\,\mathcal{J}(\gamma; t) - X\,| \leq \delta\} \geq \frac{3}{4}. \tag{14}$$

It was proved quite recently that there actually exist explicit and efficient (polynomial time) *quantum algorithms* [12] for approximating both the Jones' and more general knot polynomials arising in the context of CS quantum field theory (see [26] and references therein).

5 Concluding remarks

The issues of solvability and computability presented within the framework of quantum Chern–Simons theory in the previous section deserve a closer analysis. Unlike perturbatively renormalizable quantum field theories (*cfr.* the end of § 3) where the only (necessarily finite) physically measurable quantities are obtained as limits of infinite series as in (10), quantum CS theory is 'solvable' in the sense specified in remark **1** of § 4. More precisely, the Wilson observable defined in (12) is the sum of a *finite number* of terms for each fixed value of the coupling constant **k**. Actually such finiteness property reflects the existence of a deeper algebraic symmetry stemming from braid group representations and associated Yang–Baxter equation, see [9, 3, 27] and references therein. This notion of solvability may be viewed as the quantum analog of the property of 'complete integrability' in classical mechanics. Recall that (Liouville) integrable systems admit a sufficient number of conserved quantities that make it

---

[11] #**P** complexity class can be defined as the class of enumeration problems in which the structures that must be counted are recognizable in polynomial time. A problem π in #**P** is said #**P**–complete if, for any other problem π' in #**P**, π' is polynomial–time reducible to π; if a polynomial time algorithm were found for any such problem, it would follow that #**P** $\subseteq$ **P**. A problem is #**P**–hard if some #**P**–complete problem is polynomial–time reducible to it. Instances of #**P**–complete problems are the counting of Hamiltonian paths in a graph and the most intractable enumerative problems in combinatorics.

[12] More precisely, such algorithms belong to **BQP**, the computational complexity class of problems which can be solved in polynomial time by a quantum computer with a probability of success at least $\frac{1}{2}$ for some fixed (bounded) error.

possible to solve explicitly the equations of motion (as happens for instance in the Kepler problem involving two massive bodies which interact gravitationally according to Newton law). These 'constants of motions' are endowed with a suitable algebraic structure under Poisson bracketing which is related in turn to complete integrability owing to Arnold–Liouville theorem [1].

The issue of computability of the relevant quantities of quantum CS theory, and in particular of the Jones polynomial according to the identification (13), is clearly independent of solvability/finiteness and to some extent goes beyond the scope of this paper. However, looking for 'effective' (possibly efficient) computational protocols might help in sheding light on the open question concerning the validation of the heuristic procedure associated with the path integral quantization scheme.

References

1. V. I. Arnold: *Mathematical Methods of Classical Mechanics*, 2nd edition (Springer, Berlin Heidelberg New York 1989)
2. M. Atiyah: *The geometry and physics of knots* (Cambridge Univ. Press, Cambridge 1990)
3. J. S. Birman and T. E. Brendle: Braids: a survey (*eprint* math.GT/0409205)
4. M. Daniel and C. M. Viallet: The geometrical setting of gauge theories of the Yang -Mills type. *Rev. Mod. Phys.* **52**, 175 (1980)
5. T. Eguchi, P. B. Gilkey and A. J. Hanson: Gravitation, gauge theories and differential geometry. *Phys. Rep.* **66**, 213 (1980)
6. S. Garnerone, A. Marzuoli and M. Rasetti: Quantum automata, braid group and link polynomials. *Quant. Inform. Comp.* **7**, 479 (2007)
7. S. Garnerone, A. Marzuoli and M. Rasetti: Quantum knitting. *Laser Physics* **16**, 1582 (2006)
8. D. J. Gross: Gauge theory - past, present, and future? *Chinese J. Phys.* **30**, 955 (1992)
9. E. Guadagnini: *The link invariants of the* Chern–Simons field theory (W. de Gruyter, Berlin New York 1993)
10. R. Jackiw: Introduction to the Yang-Mills quantum theory. *Rev. Mod. Phys.* **52**, 661 (1980)
11. F. Jaeger, D. L. Vertigan and D. J. A. Welsh: On the computational complexity of the Jones and Tutte polynomials. *Math. Proc. Cam. Phil. Soc.* **108**, 35 (1990)
12. A. Jaffe and F. Quinn: "Theoretical Mathematics": toward a cultural synthesis of mathematics and theoreical physics. *Bull. Amer. Math. Soc.* **29**, 1 (1993)
13. V. F. R. Jones: A polynomial invariant for knots via von Neumann algebras. *Bull. Amer. Math. Soc.* **12**, 103 (1985)
14. L. H. Kauffman: *Knots and physics* (World Scientific, Singapore 1991)
15. R. K. Kaul, T. R. Govindarajan and P. Ramadevi: *Schwarz type topological quantum field theories* Encycl. Math. Phys. vol. 4, ed by J-P. Francoise et al. (Academic Press Elsevier 2006) pp 494-502
16. W. B. R. Lickorish: *An Introduction to Knot Theory* (Springer, Berlin Heidelberg New York 1997)

17. G. Perelman: The entropy formula for the Ricci flow and its geometric applications (*eprints* arXiv: math/0211159); Ricci flow with surgery on three–manifolds (*eprints* arXiv: math/0303109); Finite extinction time for the solutions to the Ricci ow on certain three–manifolds (*eprints* arXiv: math/0307245)
18. W. P. Thurston: *Three–dimensional Geometry and Topology*, ed by S. Levy (Princeton Univ. Press, Princeton 1997)
19. E. Witten: Quantum field theory and the Jones polynomial. *Commun. Math. Phys.* **121**, 351 (1989)

Experimental Methods in Proofs

Gabriele Lolli

Dipartimento di Matematica, Università di Torino (Italy)
gabriele.lolli@unito.it

The presence of experimental methods in mathematics has been the *leit-motiv* of the so called, by Imre Lakatos in [12], renaissance of empiricism in the philosophy of mathematics.

Among the efforts of reviving the philosophy of mathematics, as urged by Reuben Hersh in [9], in the Seventies the empiricist trend was the most vociferous and fashionable. It could rely on, and was enhanced by the novelties introduced by the computer, namely much larger searches and the formation of conjectures, either by numerical computations or by computer graphics. But as a thesis on the nature of mathematics it had to find confirmation and roots in the history of mathematics.

Prominent among others were Hilary Putnam [14] and Imre Lakatos [11]; they didn't rescue from oblivion old philosophies claiming the empirical nature or origin of mathematical objects; they proposed the term "quasi-empirical" to characterize non-deductive methods of discovery and validation in mathematics. The term was taken from Euler, *via* Pólya.

It had actually been George Pólya, in [13, pp. 17-22], some twenty years before to stress the importance of the heuristic non-deductive moment in the search for a proof, and to call attention to Euler's unorthodox methods. The same Eulerian examples were quoted by Putnam and by Mark Steiner in [16].

Empirical methods are loosely meant by the empiricists, with differences among them, to be procedures of discovery and validation which are similar to those of the natural sciences.

> By "quasi-empirical" methods I mean methods that are analogous to the methods of the physical sciences, except that the singular statements which are "generalized by induction", used to test "theories", etc., are themselves the product of proof or calculations rather than being "observation reports" in the usual sense [14, pp. 49–65].

The reminder of the presence of such procedures has been beneficial to the philosophy of mathematics. The purpose of this essay however is to give elements, through the analysis of a few examples, to dismantle the often connected thesis that mathematics should thereby forego its pretension to a unique position among sciences. If the "statements used to test theories" are themselves mathematical, the similarity is limited to the forms of argumentation. But the fact that mathematicians use many strategies and various instruments to attack a problem does not mean that the solutions, when found, are not rooted in a deductive setting: they purport to be logical consequences of the data and of the formal specification of the notions involved. If not, they are sooner or later recognized defective, and the search goes on, aiming at a finer or deeper analysis.[1]

There are many degrees of involvement of empirical methods in mathematics.

The first, more widespread and less interesting, is that of inductive formation of conjectures through observations.

The second is the presence of empirical facts or elements in the very body of an argument proposed as a proof.

The strongest is represented by proofs found by machines, but it has in a way turned topsy-turvy, since the formal proofs thus produced are pure logic, the opposite of a liberated notion of proof.

Given its many facets, we will not dwell on this last topic, which began to be discussed with Tymoczko's remarks on the Four-Color Theorem[2] and is revived with every new computer performance.[3]

1 Euler's observations

Let us have a look at Euler's paper [4] quoted by Pólya, *Specimen de usu observationum in mathesi pura*, where he says that many notable properties are first observed and worked upon before being proved.

> Inter tot insignes numerorum proprietates, quae adhunc sunt inventae ac demonstratae, nullum est dubium, quin pleraeque primum ab inventoribus tantum sunt observatae et in multiplici numerorum tractatione animadversae, antequam de iis demonstrandis cogitaverint.[4]

A fuller presentation is in the summary:[5]

---

[1] The author is grateful to a careful and subtle referee for many corrections; not all of her suggestions for improvements could be met in such a short and unpretentious essay.

[2] See [19], [17], [18], [2].

[3] Such as the non-existence of a projective plane of order 10, the sphere packaging problem, the Robbins conjecture. See [6], [1].

[4] "Among so many notable properties of numbers, which up to now have been found and proved, there is no doubt that many have been observed and worked upon by their discoverers well before they thought of a proof thereof".

[5] "Summarium", in [4, pp. 459-60]. We quote from the *Opera omnia* edition. Italics added.

Haud parum paradoxum videbitur etiam in Matheseos parte, quae pura vocari solet, multum observationibus tribui, quae vulgo nonnisi in obiectis externis sensus nostros afficientibus locum habere videntur. Cum igitur numeri per se unice ad intellectum purum referri debeant, quid observationes et *quasi experimenta* in eorum natura exploranda valeant, vix perspicere licet. Interim tamen hic solidissimis rationibus ostensum est plerasque nomerorum proprietates, quas quidem adhux agnovimus, primum per solas observationes nobis innotuisse, idque plerumque multo antequam veritatem earum rigidis demostrationibus confermaverimus. Quin etiam adhuc multae numerorum proprietates nobis sunt cognitae, quas tamen nondum demonstrare valemus; ad earum igitur cognitionem solis observationibus sumus perducti. Ex quo perspicuum est in scientia numerorum, quae etiamnunc maxime est imperfecta, plurimum ad observationibus esse expectandum, quippe quibus ad novas proprietates numerorum continuo deducimur, in quarum demonstratione deinceps sit elaborandum. Talis cognitio solis observationibus innixa, quandiu quidem demonstratione destituitur, a veritate sollicite est discernenda atque ad inductionem referri solet. Non desunt autem exempla, quibus inductio sola in errores praecipitaverit. Quascumque ergo numerorum proprietates per observationes cognoverimus, quae idcirco sola inductione innituntur, probe quidem cavendum est, ne eas pro veris habeamus, sed ex hoc ipso occasionem nanciscimur eas accuratius explorandi earumque vel veritatem vel falsitatem ostendendi, quorum utrumque utilitate non caret.[6]

In this research Euler wants to characterize the numbers which can be written as $2aa + bb$, a and b relatively prime. Similar questions arise in con-

---

[6] "It is not a little paradox that in the part of mathematics which is usually called pure so much depends on observations, which people think to have to do only with external objects affecting our senses. Since numbers in themselves must refer uniquely to pure intellect, it i does not seem worth investigating the value of observations and quasi experiments in their study. However it can be shown with strong reasons that the greatest part of the properties of numbers we have come to know have been noted at first only through observations, long before their truth has been confirmed by strict proofs. And there are even many of them which we know but we are not yet able to prove: we have come to their knowledge only through observations. So it is clear that in the science of numbers, which is still greatly incomplete, a lot has to be expected from observations, to find new properties of numbers for which later a proof has to be worked out ... Such knowledge depending only on observations, in case a proof is lacking, must be carefully distinguished from truth, and based only on induction. There is no lack of examples in which induction alone has led to errors. Whenever we come to know a property of numbers through observations, based only on induction, beware not to take it as true, but take the opportunity to investigate it carefully and show its truth or falsity, in any case a useful deed".

nection with the study of Pythagorean triples. So Euler lists all such numbers up to 500, and begins to make observations on the table he is considering:

- that if a prime number is there, that is it if it can be written as sum of a square and twice a square, this presentation is unique;
- that if a prime number n can be so written, the same holds for $2n$;
- that if an odd number can be so written, the same is true for its double, and conversely, for an even number and its half;
- that if two numbers can be so written, the same is true for their product;
- that the prime divisors of such a number are of the same form;

and so on, to arrive through this series of remarks to the fact that

- the prime divisors of such a number, if it has any, are of the form $8n+1$ or $8n+3$,

and so on, finally arriving at a proposition affirmed by Fermat, without proof, that

- prime numbers of the form $8n+1$ or $8n+3$ can be written as $2aa+bb$ and they only.

Having arrived at such a conjecture, Euler according to the summary first checks it up to 1000, then proceeds to give a prove of it, through a series of theorems paralleling the observations. Some of the proofs are easy, in Euler's opinion, treading in the very observations steps, some require algebraic arguments, only one is profound and uses the method of infinite descent.

Euler did not underrate proofs. At the beginning of the paper he wrote:

> Quamvis autem huiusmodi proprietas per assiduam observationem fuerit animadversa, quae per se menti non parum esse iucunda, tamen, nisi demonstratio solida accrescerit, de eius veritate non satis certi esse possumus; exempla enim non desunt, quibus sola inductio in errorem praecipitaverit. Tum vero ipsa demonstratio non solum omnia dubia tollit, sed etiam naturae numerorum penetralia non mediocriter recludit nostramque numerorum cognitionem continuo magis promovet, a cuis certe doctrinae perfectione adhuc longissime sumus remoti. Verum si cui haec forte non magni momenti esse videantur, quod vix unquam ullum in Mathesi applicata usum habitura putentur, usus, quem inde in ratiocinando adipiscimur, non est contemnendus.[7]

---

[7] "Although such property has been controlled by careful observations, to the full satisfaction of our mind, this notwithstanding we cannot be certain of its truth, if some strong proof is not added; there is no lack of examples in which induction alone has led to errors. In fact the proof not only eliminates all doubts, but also illuminates the mysteries of their nature and greatly increases our knowledge of numbers, whose theory is still far from perfect. And if this does not appear of great importance in the uses which are considered relevant in applied mathematics, it must not be underrated the utility we get in reasoning" [4, pp. 461-62].

Proofs do not only eliminate doubts, they illuminate and improve our knowledge.

There are other features of Euler's methods which are attractive to the empiricists, for example the use of analogies (from finite to infinite) in the work on series. But Euler's attitude with respect to their proofs was the same. It is true that he said that his method for finding the sum of the series of the inverses of squares,

$$\sum_{k=1}^{\infty} \frac{1}{k^2} = \frac{\pi^2}{6},$$

which to some could appear not enough reliable, had had a strong confirmation in that he could prove with it a known result of Leibniz (the sum of the alternating series of inverses). So we should not have doubts about other results. But he nevertheless continued to search for a proof, until he found one, as Pólya, though not Putnam or Steiner, took care to remind, in [13, p. 21].

So it seems out of place the question that continues to be asked,[8] as in [5]: "What shall we say, then? Are we wrong to insist on rigorous proofs? Is there a special category of argument, something less than full proofs, something more than blowing smoke? Or do truly great mathematicians get special dispensation?".

There is nothing to add to Euler's own comments above on the utility and the limits of observations in pure mathematics.

2 Archimedes' mechanical method

The idea of using mechanical devices to perform operations or graphical representations of curves is very old. Eudoxus and Architas among others used machines to draw higher order curves to solve problems such as the trisection of an angle and the duplication of a cube.

> Eudoxus and Archytas had been the first originators of this far-famed and highly prized art of mechanics, which they employed as an elegant illustration of geometrical truth, and as a means of sustaining experimentally, to the satisfaction of the senses, conclusions too intricate for proofs by words and diagrams. In the solution of the problem, so often required in constructing geometrical figures, given two extremes, to find the two mean lines of a proportion, both these mathematicians had recourse to the aid of an instrument, adapting to their purpose certain curves and sections of lines.
> But what of Plato's indignation at it, and his invective against it as a mere corruption and annihilation of the one good in geometry, which was thus shamefully turning its back upon the unembodied objects

---

[8] Something more will be said in § 3.

of pure intelligence to recur to sensation, and to ask help (not to be obtained without base supervisions and depravations) from matter; so it was that mechanics came to be separated from geometry, and, repudiated and neglected by philosophers, took place as a military art.[9]

Fortunately not everybody yielded to Plato's curse. Prohibitions are never welcome in science.

In the dedication of the *Method* to Eratosthenes, Archimedes[10] explains that he is going to present the peculiarities of a method which confers a certain easiness to treat mathematical questions with mechanical considerations. He has obtained with this method a few results he has already communicated to Eratosthenes. He wants the method to be known because he is sure that it will be fruitful and produce other results.

Archimedes is convinced that this method will be useful also for the proofs. The results obtained with it do not come with a real proof, but it is easier to look for one when one has acquired a certain familiarity with the matter using the method. He promises that at the end he is going to give geometrical proofs

Then he begins with the first result, namely that the area of a parabolic segment is equal to 4/3 of the inscribed triangle (same base and same height).

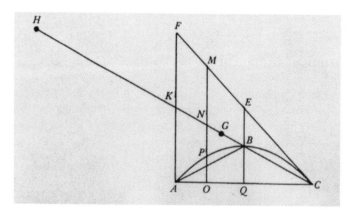

Fig. 1. Archimedes' mechanical proof

In the Fig. 1
- Q is the midpoint of AC,
- QBE is parallel to the axis of the parabola

---

[9] indexPlutarchPlutarch, *Life of Marcellus*, c 14.5, quoted from [10, p. 146].
[10] An english translation of *The Method of Archimedes* is published as an appendix to [7].

- AF is parallel to QBE
- CF is tangent in C
- HK = KC.

Think of K as the fulcrum of a scale with arms HK and (the other along) KC.

Let MO be any line parallel to EQ. By known geometric properties

$$EB = BQ, \; FK = KA, \; MN = NO$$

and moreover

$$CA : AO = MO : OP$$
$$CA : AO = CK : KN$$

hence

$$HK : KN = MO : OP.$$

Now, since N is the center of gravity of line MO, if we take a segment TS equal to OP and we put TS with its center of gravity in H, THS will balance MO, because HN is divided by K in parts which are inversely proportional to the weights TS, MO, that is because

$$HK : KN = MO : TS,$$

For the same reason, the other parallels to EQ balance the segments intercepted on them by the parabola, between A and C, (transported in H) so that the sum of the former balances, with respect to K, the sum of the latter.

Now the segments obtained as OP compose the parabolic segment ABC, while the vertical lines drawn in the triangle CFA make the triangle CFA. This, where it is, balances with respect to K, the parabolic segment transferred with its center of gravity in H. If G is the point of CK such that

$$CK = 3KG$$

G will be the center of gravity of CFA, hence, denoting by sgm.ABC the parabolic segment,

$$CFA : sgm.ABC \text{ in } H = HK : GK.$$

But

$$HK = 3KG$$

from which it follows

$$CFA = 3 \text{ sgm.ABC}.$$

As

$$CFA = 4 \text{ ABC}$$

it follows that

$$\text{sgm.}ABC = 4/3\ ABC.$$

According to Archimedes the argument above does not prove the result, and this will not do, since "those who boast of many discoveries, without ever giving a proof, sometimes can be caught out, having claimed to have impossible things".

But the argument bestows on the result an appearance of truth. When one rightly suspects the truth, one can more confidently look for a proof, and Archimedes had in fact found it, actually two. In the *Quadrature of the Parabola* he had first trasformed the argument of the *Method* based on the infinitesimals in a proof by exaustion. Here ironically he was again in trouble, because the exhaustion method was far from being accepted and he had to expostulate for it. It was based on Eudoxus' (now Archimedes') axiom, or the "lemma": given two unequal areas, it is possible by adding the difference between the two, to surpass any given bounded area. Archimedes is at pain to recall propositions in Euclid's *Elements* proved by this lemma, and asks for the same reliability to be accorded to his own uses.

But then in the final propositions of the *Quadrature*, 18-24, he gives a purely geometrical proof.[11]

3 Mathematics *in statu nascenti*

We claim that such arguments as constructed with Archimedes' mechanical method are more than heuristic suggestions as to the truth of the statements involved. They could be accepted as conclusive; this is not because empirical methods are allowed in mathematics, but because in fact such arguments, with a grain of liberality, can be considered proofs.

There are no physical operations involved, such as weighing tin plates, though such operations can be and are, or used to be performed in early maths education with children. The whole argument is a thought experiment based on notions which are still imperfectly mathematized, but on their way to become object of mathematical theories.

Archimedes himself contributed to the founding and development of Statics. In *On the Equilibrium of Planes* and *The Centers of Gravity of Planes* he laid the principles and the theory of the lever, stating a few postulates, such as:[12]

> - Equal weights at equal distances are in equilibrium, and equal weights at unequal distances are not in equilibrium but incline toward the weight which is at the greatest distance.
> - If, when weights at certain distances are in equilibrium, something is added to one of the weights, they are not in equilibrium ...

---

[11] See also Enrico Rufini in [15, pp. 216-19].
[12] The following quotations are taken from [3].

> - If magnitudes at certain distances are in equilibrium, other magnitudes equal to them will be also in equilibrium at the same distances...

From these the first laws of the lever are then proved, e.g: the first: Weights which balance at equal distances are equal.

Later Greek texts give a definition of the center of gravity:

> We say that the center of gravity of any body is a point within the body such that, if the body can be conceived to be suspended from the point, the weight carried thereby remains at rest and preserves the original position.

Archimedes did not refer to such definition, instead again he gave axioms for the center of gravity, such as

> - If from a magnitude some other magnitude is subtracted, and if the same point is the center of gravity of the original magnitude and of that subtracted, then that same point is the center of gravity of the remaining magnitude.
> - If from a magnitude some other magnitude is subtracted, and if the whole magnitude and the subtracted one do not have the same center of gravity, the center of gravity of the remaining magnitude is found by extending the line joining the two centers of gravity beyond the center of gravity of the original magnitude, and taking on it a segment which has to the segment joining the two centers of gravity the same proportion which holds between the weight of the subtracted magnitude and the weight of the remaining magnitude.
> - If the centers of gravity of any number of magnitudes lie on the same line, also the center of gravity of their sum will lie on the same line.
> - The center of gravity of a straight line is its middle point.
> - The center of gravity of a triangle is the point of intersection of the lines drawn from the vertices to the middle points of the sides,

and other similar, enough to calculate the center of gravity of plane polygonal figures.

Once the theory of momentum and center of mass is fully mathematized, there is no obstacle to vindicate Archimedes' proofs *ab omni naevo* and to recast them in perfectly acceptable mathematical proofs.

The same can be said for other similar examples. We recall only Lakatos' analysis of Euler's theorem $V - E + F = 2$ on polyedra. Lakatos' analysis is too well known, and perhaps obsolete, to dwell on it. The first proof presented in Lakatos' dialogue is inspired by that of Cauchy in 1813 and it is based on the following picture of a polyhedron to which a face has been subtracted and which is flattened on a plane

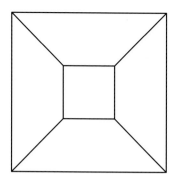

and on a triangulation

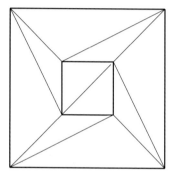

of it. If we cancel a triangle, such as

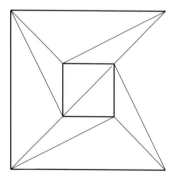

the faces and the edges are reduced by 1, some different subtraction being realized in the other cases, always preserving the relation $V - E + F = 1$.

We remember only the following crucial remark by Lakatos, relevant for our discussion:

> ALPHA: But then we are worse off than before! Instead of one conjecture we now have at least three! And this you call a 'proof'?
> ...
> DELTA: What does it do then? What do you think a mathematical proof proves?
> TEACHER: This is a subtle question which we shall try to answer later. Till then, I propose to retain the time-honoured technical term 'proof' for a *thought-experiment – or 'quasi-experiment' – which suggests a decomposition of the original conjecture into subconjectures or lemmas*, thus *embedding it* in a possibly quite distant body of knowledge. Our 'proof', for instance, has embedded the original conjecture – about crystals, or, say, solids – in the theory of rubber sheets [11, p. 9].

To call such a thought-experiment both a proof and a quasi-experiment though linking together these two terms deprives the latter of any meaning. A decomposition of a conjecture into subconjectures is a move quite legitimate and common also in a deductive setting; *per se* it has nothing to to with the possible distance of the conjured knowledge. Talking of crystals and rubber sheets does not place the discourse outside pure mathematics. The theory of rubber sheets of course is nothing else than what will be called topology.

4 Mechanical devices

The instruments mentioned in Plutarch's report were however real machines, not just methods inspired by mechanical considerations. Let us recall a few examples.

For the trisection of an angle Nicomedes made use of the conchoid (or cochloid according to Pappus) which was drawn by the contrivance shown in Fig. 2.

It was composed of two fixed perpendicular rulers, and one revolving around a peg in C; D is a fixed peg on PC which can move in the slot in AB.

This apparatus was used when one had to insert a segment of given length between two lines one of which was straight. The characteristic property of the conchoid in fact is that the *distance DP* is constant.

Just as the constructions by ruler and compass have a mathematical equivalent, so it is for this case. The equation of the conchoid is actually in polar coordinates

$$r = a + b\sec\theta.$$

As Pappus has shown, the solution could be obtained as the intersection of two conics, and from their equations the problem was reduced to the solution of a cubic equation.

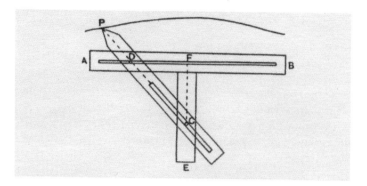

Fig. 2. Nicomedes' conchoid

For the trisection of an angle $A\hat{B}C$, with reference to Fig. 3, one had to find E on AF parallel to BC and draw BE in such a way that $DE = 2AB$.

With the artifact above, one would then use B for pole, AC for the ruler (the horizontal ruler) and $2AB$ for the distance.

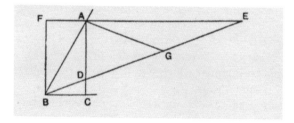

Fig. 3. The trisection of $A\hat{B}C$

Ironically, one of these devices, which we are going to consider next, was attributed to Plato.[13]

Hippocrates had shown that the doubling of the cube can be reduced to finding two mean proportionals in continued proportion: if $a : x = x : y = y : b$ then

---

[13] By Eutocius, but apparently only by him. See the discussion in [8, pp. 255-8].

$$a^3 : x^3 = a : b.$$

Menaechmus had discovered that since

$$x^2 = ay, y^2 = bx, xy = ab$$

and these curves exist, as section of right circular cones, one had to find the point P of intersection of the two conics shown in fig. 4, and its coordinates.

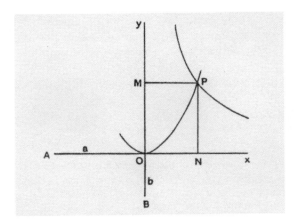

Fig. 4. Two mean proportionals between AO and OB

For the solution, one could use the device shown in Fig. 5.

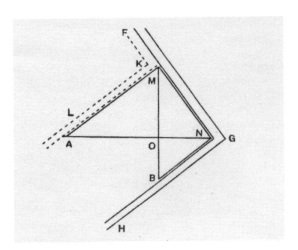

Fig. 5. Plato's alleged instrument

FGH is a rigid right angle, and FKL another one which slides along GF so that KL is always parallel to GH. It is positioned so that GH passes through B and it is rotated until the vertex G lies on the prolongation of AO. Then one slides FKL until KL passes through A. If K is not on BO the device has to be repositioned again with G on AO until the desired configuration shown in picture is reached.

Greek mathematical machines are a fascinating argument, which would be worth pursuing, but what we have said so far should be sufficient for a general provisional conclusion: that if we carefully look at the cases which are known, the mystery about the unholy wedding of experimental methods and mathematical proofs disappears. Or: much ado about nothing. These examples show that mathematics is not, contrary to the empiricists' contention, contaminated by empirical procedures, but that it grows by absorbing and symbolically transforming the physical experience in formal models.

References

1. B. Cipra: *What's Happening in the Mathematical Sciences*, vol. 1 (1993), vol. 2 (1994), vol. 3 (1995-6), vol. 4 (1998-9), (Providence, R.I., AMS 1993-1999)
2. M. Detlefsen and M. Luker: The Four-Color Theorem and Mathematical Proof. *The Journal of Philosophy* **77** (1980) pp 803-20
3. E. J. Dijksterhuis: *Archimedes* (Princeton University Press, Princeton 1987)
4. L. Euler: Specimen de usu observationum in mathesi pura. *Novi commentarii academiae scientiarum Petropolitanae*, 6 (1756-7), (1761) pp 185-230, with a summary *ibidem*, pp. 19-21, both in: *Opera omnia*, ser. I, vol. 2, *Commentationes arithmeticae* (Teubner, Leipzig 1915) pp. 459-92
5. F. Q. Gouvea: Euler's Convincing Non-Proofs. *Focus*, 27, n. 1 (2007) pp. 10-11
6. T. C. Hales: Cannonballs and Honeycombs. *Notices AMS* **47**, 4 (2000) pp 440-9
7. T. Heath: *The Works of Archimedes* (Dover, New York 1953)
8. T. Heath: *A History of Greek Mathematics* (1921), vol. 1 (Dover, New York 1981)
9. R. Hersh: Some Proposals for Reviving the Philosophy of Mathematics. *Advances in Mathematics* **31** (1979) pp. 31-50. Reprinted in [20] pp 9-28
10. M. Kac and S. M. Ulam: *Mathematics and Logic* (1968), (Penguin Books, London 1979)
11. I. Lakatos: *Proofs and Refutations* (Cambridge Univ. Press, Cambridge 1976) Previously published in four parts in *The British Journal for the Philosophy of Science* **14** (1963-64)
12. I. Lakatos: A Renaissance of Empiricism in the Recent Philosophy of Mathematics. In: *Philosophical Papers*, 2 voll (Cambridge Univ. Press, Cambridge 1978) Reprinted in [20] pp 29-48
13. G. Pólya: *Mathematics and Plausible Reasoning*, vol. 1: *Induction and Analogy in Mathematics* (Princeton Univ. Press, Princeton 1954)
14. H. Putnam: What is Mathematical Truth. In: *Philosophical Papers*, vol. 1, *Mathematics, Matter and Method* (Cambridge Univ. Press, Cambridge 1975) Reprinted in [20] pp 49-65

15. E. Rufini: *Il "Metodo" di Archimede* (Zanichelli, Bologna 1926) Reprinted: (Feltrinelli, Milano 1961)
16. M. Steiner: *Mathematical Knowledge* (Cornell Univ. Press, Ithaca 1975)
17. E. R. Swart: The Philosophical Implications of the Four-Color Problem. *Amer. Math. Monthly* **87** (1980) pp 697-707
18. P. Teller: Computer Proof. *The Journal of Philosophy* **77** (1980) pp 797-803
19. T. Tymoczko: The Four-Color Problem and Its Philosophical Significance. *The Journal of Philosophy*, 76 (1979) pp 57-83
20. T. Tymoczko (ed): *New Directions in the Philosophy of Mathematics* (Princeton Univ. Press, Princeton 1998)

Proofs Verifying Programs and Programs Producing Proofs: A Conceptual Analysis

Dag Prawitz

Filosofiska Institutionen, Stockholms Universitet (Sweden)
dag.prawitz@philosophy.su.se

I shall deal here with conceptual questions concerning two related phenomena: 1) the use of deductive machinery to verify the correctness of computer programs, and 2) the running of programs on computers to produce proofs.

1 Proofs verifying programs

How to verify the correctness of programs has become a serious problem with the growing use of electronic devices to control industrial processes and big systems of great public concern, like bank transactions, railway systems and atomic power plants. Faults in the control units may cause commercial losses but also hazards of safety. The faults may be traceable to the electronic hardware, but more often they depend on errors in the programming of the hardware.

When a program has been written, one usually tries to improve it by trial and error. The program is run in various ways, and bugs are removed as one finds them. But there is a growing concern that such inductive methods for testing programs are not sufficient, especially not when great damages may result from malfunctioning of the programs. An error in a program that controls the signals and switches in a railway system may cause a collision between two trains, and an error in a program that controls the pumps and vaults of an atomic power plant may cause an even greater catastrophe. That a program has worked as far as it has been tested cannot be enough. It has therefore been demanded that one proves deductively that the program is correct, that it works as intended.

This raises the question what, in principle, the difference is between, on the one hand, testing the correctness of a program inductively by running it for various choices of parameters and, on the other hand, proving the correctness deductively. The idea of proving the correctness of a program deductively may seem foreign in our relativistic time. Some people doubt that there is anything like conclusive proofs and that there can be any difference in principle between inductive and deductive methods.

A simple and traditional answer to the question about the difference is of course that a deductive proof guarantees the truth of what is proved, while an inductive proof does not do so. In particular, if a program is proved to be correct, then it is correct; it then works as intended and any malfunctioning must depend on the hardware, not on the program, while if the program is tested only inductively there is no such guarantee. It is this traditional view that is sometimes questioned. Let us therefore look at the question of the difference between deductive and inductive methods more carefully from some different perspectives.

1.1 Commercial point of view

Let us first take the perspective of an insurance company that offers insurances paying for damages caused by improper behaviour of electronic systems. Say that one of its policyholders asks for a reduction of the insurance premium on the ground that the correctness of the program of the insured system is now verified deductively. The company has to estimate how this novelty affects the risk that something goes wrong with the system. Even attempts to verify the correctness deductively may of course go wrong - they can never be foolproof. Although the company may find that the total risk has gone down, it may find instead that it stays the same, or even that the risk increases - in other words that the old way of carefully testing the program inductively was safer; some critics maintain that the latter is often the case. So from the point of view of the insurance company nothing has changed in principle. At least from a commercial perspective, proving deductively that a program is correct is just one way among others to verify correctness. This seems to speak in favour of the critics of the idea of deductive proofs as guarantees of truths and as being different in principle from inductive methods.

1.2 Experiences of verifiers

For people who are acquainted with verifications of programs, it is obvious that to verify a program by proving its correctness deductively is to proceed in a radically different way from testing it inductively, which does not mean that they can offer a theoretical description of what this difference consists in.[1] Their experience is that safety can be drastically improved by verifying the

---

[1] My own acquaintance with this field is limited to some co-operation with a former student of mine, Gunnar Stålmarck, who has started a commercially successful firm that specializes in proving the correctness of programs deductively. According to the experience of this firm, which has employed most logicians in Sweden and is now expanding abroad, programs that control very important functions of for instance traffic systems, airplanes or atomic power plants quite often turn out to contain errors, and their security is drastically improved by deductive

correctness of programs deductively, but of course, they must admit that it is a theoretical possibility that something goes wrong even in their procedures. Furthermore, the proofs by which the correctness of programs is established tend to be very long, and they are therefore usually constructed by computers. That takes us to problems that I shall discuss in the second part of the paper - what to say about proofs produced by machines. Without anticipating that discussion, it is obvious again that it is not in respect of safety that we can find a difference in principle between deductive and inductive approaches.

1.3 Philosophical perspectives

1.3.1. Why a proof guarantees the truth of what is proved

The fact that we may make mistakes when trying to prove something has been used philosophically as an argument against the idea that a proof guarantees the truth of what is proved. There is a well-known passage by David Hume that is sometimes construed as such an argument. Hume remarks that we have a tendency "to make mistakes when we apply rules of deductive sciences and therefore to fall in error". Hume finds that no mathematician, however skilful, places entire confidence in a truth immediately after having found a proof of it. "But his confidence increases every time he runs over his proofs", Hume says, and continues: "This gradual increase of assurance is nothing but the addition of new probabilities. Therefore all knowledge resolves itself into probability, and become at last of the same nature with that evidence, which we apply in common life."[2]

However, what Hume says here, when formulated a little more carefully, is quite compatible with the idea that a proof guarantees the truth of what is proved. Hume says himself that deductive rules are "certain and infallible". It is our attempts to apply them that are fallible.

If it turns out that we have made a mistake when trying to prove an assertion and that, in fact, what is asserted is false, then we say that we did not really have a proof. We only thought that we had a proof, but we did not really have one. This is how we use the notion of proof. We simply take it to be a conceptual truth that if we have deductively proved something, then what we have proved is true.

However, it is one thing that we take it to be conceptually true that a proof guarantees the truth of what is proved, and another thing to explain philosophically how there can be such things as proofs that guarantee truth.

---

verifications. That a program works as intended, for instance never leads to a situation in which two trains are allowed to run against each other on the same track, can often be shown to be equivalent to a certain formula in propositional logic being a tautology. To show that the formula is a tautology the firm uses a special, automated algorithm for which it has obtained a patent.

[2] Quotations from Hume's *A Treatise of Human Nature*, Book 1 (the beginning of part 4).

That we use a notion of proof in a particular way does not make it certain that there is really a coherent concept of proof that fits this use. What is needed is an analysis of the notion of proof, or rather the formation or creation of a concept of proof, which, if possible, agrees with such pre-theoretical ideas that we have about proofs.

The idea of formal proofs is of no great help here. It does contain the reasonable idea of seeing the overall structure of a proof as a chain of inferences. But whether the derivations of an interpreted formal system really represent proofs depends on the nature of the inference rules. The intention when setting up a formal system is of course that applications of the inference rules will yield representations of valid inferences. But since the idea of a formal system in itself does not imply anything to that effect, it does not relieve us of analysing what properties inferences are to have in order to make up proofs.

Someone may think that by just requiring that the inference rules of a formal system preserve truth and hence that the system is sound, one obtains the desired result that the formal derivations represent proofs and that the truth of what is proved is guaranteed. But to think so would be to overlook two simple facts. Firstly, an inference rule may preserve truth but the epistemic distance between premisses and conclusion may be so great that no one would think of using it in a proof. Secondly, the conclusiveness of a proof should come from the proof itself or, at least, it must do so at some level. If the conclusiveness of an interpreted formal derivation comes from a proof on the meta-level, showing that the used inference rules preserve truth, then the question remains where the conclusiveness of the meta-proof comes from; obviously once cannot always answer this question by referring to another proof.

To develop a precise concept of proof with desired properties one must go deeper, beyond formal derivations, which should be seen as codifications of proofs and not as answering the question what a proof is. Now as well as later, one may concentrate on the most crucial epistemological function of deductive inferences and proofs: the fact that they furnish us with conclusive grounds for asserting the conclusions. We should then ask what kind of things can amount to a conclusive ground for an assertion.

In the constructive tradition, there are ideas about what a ground for an assertion is and how it is connected with the meaning of the assertion.[3] These ideas can be used to specify recursively the grounds for assertions of different forms. A simple example is the idea that the meaning of a conjunction is such that a ground for asserting its truth is formed, e.g. by pair formation, from grounds for asserting the truth of its conjuncts. We may then specify that g is a ground for asserting the truth of a conjunction $A1 \wedge A2$ if and only if g is

---

[3] They are made explicit in [5] and are in effect suggested already by Gentzen [4], who remarks that an introduction rule for a logical constant in his system of natural deduction "gives, so to say, a definition of the constant in question", in other words, specifies the meaning of the sentences formed by using that constant.

of the form $\langle g1, g2\rangle$ where g_1 and g_2 are grounds for asserting the truth of A_1 and A_2 respectively. Ideas of this kind apply in particular to a constructive reading of the sentences, but can be extended to a classical reading to the extent that such a reading can be explained constructively.[4]

Once one has specified what constitutes grounds for assertions of different forms, there is the possibility to understand an inference not just as the assertion of a conclusion on the basis of premisses asserted earlier, but rather as the transformation of grounds. An inference is then individuated not only by its premisses and conclusion but also by the operation by which we make this transformation. We may illustrate the idea with two very simple examples of inferences, conjunction introduction and conjunction elimination, say of the form

$$\frac{A_1 \quad A_2}{A_1 \wedge A_2} \qquad \frac{A_1 \wedge A_2}{A_i}$$

for $i = 1, 2$. Having taken a ground for the assertion of the truth of a conjunction to be something of the form $\langle g_1, g_2 \rangle$, where gi is a ground for the assertion of the truth of A_i ($i = 1, 2$), we get immediately an operation that transforms grounds for the premisses of a conjunction introduction to a ground for its conclusion, namely pair formation. Furthermore, we get an operation i (i = 1 or 2) that transforms a ground for the premiss of one of the two forms of a conjunction elimination to a ground for its conclusion by the definition

$$\varphi(\langle g_1, g_2 \rangle) = g_i.$$

An inference that is individuated in this way by its premisses, conclusion, and operation applicable to grounds for the premisses may be defined as valid, if the operation when applied to the given grounds for the premisses actually yields a ground for the conclusion.[5] A proof may then naturally be understood as built up of successive applications of such operations.

This is a very quick sketch of a way of determining the concept of proof that does provide us with an answer to the question why a proof of an assertion guarantees the truth of the assertion: A proof of an assertion A produces a conclusive ground for A in virtue of what the assertion A means and how the inference operations that the proof is built up of are defined. Of course, much more has to be said to support this analysis, but my main point here is that

---

[4] As discussed for instance in Dag Westerståhl's contribution to this volume about the relation between classical and intuitionistic logic.

[5] For a short indication of how this can be done see my paper "Validity of Inferences" [14]. To see an inference as an operation on grounds is a further development of Gentzen's idea that the elimination rule for a logical constant is justified in terms of its meaning given by the introduction rule (see note 3). The reductions defined for applications of elimination rules, which give rise to normalizations of natural deductions, can be seen as a first articulation of that idea (see [12] and [13] - for a development of this idea see [11]). Analogues to these reductions now reappear as the very operations that individuate inferences.

an analysis of the concept of proof is needed if one is to defend the idea that a proof delivers a conclusive ground for its conclusion - it is my contention that one cannot arrive at such an analysis by starting instead from the concept of logical consequence in the model theoretical sense or, at least, it has not been shown that one can; but this is another question that I shall not go into here.[6]

1.3.2. The correspondence between proofs and computations

Before passing on to the second topic of the paper, I want to make a remark on the fact a proof becomes in principle like a computation when inferences are seen in the way suggested above as operations on grounds for assertions. This correspondence between proofs and computations is of especial interest in this context and can be developed so as to get a somewhat different perspective on the deductive approach to program verification.

Curry[7] was the first to note an isomorphism between proofs in the implicational fragment of minimal propositional logic and terms in certain computational systems (combinatory logic and lambda calculus). Howard [6] extended it to an isomorphism between deductions in a Gentzen system for intuitionistic predicate logic and an enriched lambda calculus; the result is often referred to as the Curry-Howard correspondence.

This formal correspondence gets a deeper significance when it is combined with foundational questions of the kind discussed above. Especially relevant for the connection with computations and programs is the observation first made by Kolmogorov [7] that propositions in intuitionistic systems can be interpreted as problems and intuitionistic proofs can be interpreted as methods or programs for solving problems. This interpretation of intuitionistic propositions was worked out systematically by Martin-Löf [8] and [9], showing that in the type-theory developed by him, expressions written $a \in A$ can be understood alternatively as saying that a is proof of the proposition A, a is an object of type A, and a is a program (algorithm) for solving the problem A.

In Martin-Löf's type theory, the task of verifying the correctness of a program, showing that it does what it is supposed to do, is thus of the same nature as that of checking that an alleged proof is correct, that it proves the proposition it claims to prove. Furthermore, as is to be expected since one of the readings of $a \in A$ is that a is a proof of A, it is decidable whether $a \in A$ holds in the system. The problem of verifying the correctness of a program is thereby reduced to a mechanically decidable problem.

A specific attempt to check whether $a \in A$ holds may of course go wrong. But if the expression $a \in A$ is provable in the system, then the program indicated by a does solve the task indicated by A. One may ask how we know that this is so, and this is again something that has to be answered in the form of a theory of meaning.[8]

---

[6] For a discussion of this issue see [14].
[7] See for instance [2] (from p. 321 onwards).
[8] Worked out for the case of Martin-Löf's type theory already in [10].

Characteristic of a type theory is of course that we do not consider objects in general but only objects of specific types. When the type theory is seen as a programming language, we similarly do not consider programs in general but only programs for solving specific problems. When building up a program in the type theory, the program therefore comes together with the problem that it is solving. If we consider programs generated in a programming language where the problem that the program solves is not indicated along with the generation of the program, and state separately that such a program p solves the task t, for instance in the form of a first order assertion $Solve(p, t)$, then we cannot expect to be able to decide mechanically whether $Solve(p, t)$ holds but may hope to be able to prove it deductively.

2 Proofs produced by computers

Computers have been used to produce proofs since the 1950's.[9] In the beginning only short proofs of already known theorems were produced in that way, and there was no philosophical discussion of the question whether real proofs were obtained in that way. This changed in the 1970's, when computers were used in attempts to verify the correctness of programs deductively, as discussed in the above, and especially when computers were reported to be able to prove so far unestablished mathematical theorems by delivering proofs that were too long to be taken in by a human, even if he or she devoted a lifetime to it. The most well known case is the proof of the four-colour theorem in 1977, which caused a lively debate hinted to in the invitation to the workshop on which the present volume is based. The question was and is whether the acceptance of such proofs constitutes a philosophically significant shift of mathematical practice, perhaps amounting to the beginning of a new paradigm in mathematics.

2.1 The debate

In the debate, which started shortly after the publication of the proof of the four-colour theorem, there were interesting arguments both for and against the view that a significant shift had occurred. One argument for that view, presented by Thomas Tymoczko [17], maintained that the acceptance of the proof of the four-colour theorem meant a radical change of the concept of proof, allowing it to rest on empirical evidence, namely the outfall of an experiment in the form of running a computer and reading off its output, which output was not a proof but only a message saying that there is a proof. Those who argued against that view were divided as to the reason for saying that a

---

[9] My own first-hand experience of automated deduction is limited to its childhood in the 50's and 60's, when we were a group [15] who implemented on a computer the first in principle complete proof procedure for 1st order predicate logic.

radical shift had not occurred. One view was that the concept of proof had not changed because empirical evidence was not a proper part of the proof of the four-colour theorem. Another view was that there was no important shift because even traditional proofs produced by humans rested sometimes on empirical evidence. I shall summarize the various positions without attempting to exhaust the whole debate.

2.1.1. Claims supporting the view that there is a philosophically significant change

Connected with his claim that the acceptance of the computer-produced proof of the four-colour theorem meant a change of the concept of proof, Tymoczko made three more specific claims that I list here:

1. From a traditional point of view, the accepted proof of the four-colour theorem is like a proof with a lacuna, a gap, because of one key lemma being left unproved. This gap is now regarded to have been filled by the report of a computer saying that it has produced a proof of this lemma.
2. Since no one can survey the computer-produced proof, the acceptance of it means that mathematics become like physics in resting on empirical evidence. The reliability of its results is then no longer certain but rests on a complex of empirical factors concerning the functioning of computers.
3. Some truths of mathematics are then no longer a priori, they are no longer resting on reason alone, but depend on sense experience concerning the results of experiments.

2.1.2. Arguments for the view that no significant change has occurred

I let counter-arguments given by Paul Teller [16] and Michael Detlefsen and Martin Luker [3] represent the two different kinds of positions as to why the acceptance of the computer-produced proof of the four-colour theorem does not constitute a significant shift.

(a) Teller counters all the three claims 1-3 stated above by saying that they depend on confusing proofs with checking or verifying the correctness of proofs. The *concept* of a proof has not changed; no empirical evidence has entered into the mathematical proofs. The loss of surveyability called attention to by Tymoczko means only a shift in methods of checking proofs, not a shift in conception of the things checked, Teller claims. But this shift in methods for *verifying* that a purported proof is a proof is not significant, because there is no difference in principle between letting a computer and a mathematician check a proof; both are fallible, and sometimes the computer is more reliable than the mathematician.

(b) Detlefsen and Luker take the quite different standpoint that the verification of the correctness of a proof, in particular when it comes to verifying the correctness of computations that the proof depends on, may constitute a part of the proof. They claim however that such verifications have sometimes rested on empirical evidence even in the case of traditional human proofs of mathematical theorems. Although it is true that empirical evidence is used in the proof of the four-colour theorem, the use of such evidence is thus nothing novel, they claim, and therefore, to rely on computers making the job of verification means no fundamental change of mathematical practice.

It seems to me that there is something reasonable in what is said by all the parties in this debate. No progress can be made concerning these issues, I think, without clarifying some of the key concepts that occur in the arguments. As an attempt in that direction I shall make a few conceptual points.

2.2 Proofs and verification of proofs

The distinction between proofs and verifications of proofs, which Teller rests his argument on, is certainly important. Proofs are what establish theorems, and traditionally, since the Greeks, the requirement in mathematics for asserting a theorem is the possession of a proof of it. To have verified that something is a proof is the prerequisite for a quite different assertion: that so and so is a proof.

To avoid a lurking regress, it is important to delimit what justifies an assertion; or to use my earlier terminology in the first part of the paper, to delimit what counts as a ground for an assertion. To formulate the point in general terms, one can say that if g is a ground for an assertion A, that is, if the possession of g is what is required in order to be justified in making the assertion A, then it cannot be required that one has also verified that g is a ground for A. Justifications must end somewhere. If the assertion of A required a verification v of the fact that g has the property of being a ground for A, then, after all, g would not be the proper ground for A, but it would instead be v or the pair (g, v) that constituted the real ground for A; if in addition one required some v' that verified that v is a verification of the fact that g is a "ground" of A, then it would be v or (g, v, v) that constituted the real ground for A, and so on.[10]

That one has verified that a proof is a proof or, more particularly, that an inference step in a proof is valid or that a computation made in a proof is correct, is therefore not a part of the proof. That is not to say, of course, that it is not wise to check one's proof; as Hume rightly remarks, the confidence in a proof increases when one runs over it. But the checking does not add anything to the proof itself.

---

[10] A similar argument for requiring that the property of being a proof in a logic calculus is decidable is given by Church [1].

However, having recognized the distinction between proofs and the verification or checking of proofs, we have not automatically answered the question whether the acceptance of the four-colour theorem involves a change of the concept of proof or, in other words, of what counts as evidence for a mathematical theorem. Neither Tymoczko nor Teller says much about the concept of proof, which is not surprising in view of the fact that there have been few attempts to analyse this concept and that there is thus little to build on. Of course, there is no problem concerning what we mean by a formal proof, but the issue does not concern that notion. Tymoczko agrees that there is good evidence for the *existence* of a formal proof of the key lemma, which he claims has nevertheless *not been proved* in a traditional sense, thus leaving a lacuna in the proof of the theorem.

The question we have to ask is not just what a proof is (about which I made a proposal in § 1.3), but rather what it is to have proved an assertion, which I have regarded as the same as having got in possession of a proof of the assertion. It is to be in *possession* of a proof of an assertion, not the mere existence of such a proof, which is the condition for being justified in making the assertion. Now it must be noted that to be in possession of an interpreted *formal proof* of a formula, representing an assertion A, is not in itself to have proved A. A formal proof is a syntactic object, like a numeral, and, as already noted in § 1.3, knowing such an object is not to know a proof of the assertion A. Hence, it is irrelevant for the issue that we are discussing whether we want to say that we are in possession of a formal proof of a formula representing the key lemma, because, even if we were in possession of such an object (which assumption I think is contrary to the actual situation), this in itself would not be enough to claim that we are in possession of a proof of that lemma.

Admittedly, having proved certain meta-theoretical facts about the formal system in which the key lemma is formulated, it would be enough to be in possession of a *deductive* proof of the *existence* of a formal proof of the lemma, because then, without being in possession of the formal proof itself, we could easily construct a real deductive proof of the lemma. To illustrate that idea, we may note that we may prove deductively the existence of a formal derivation of the equation $10^{10} \times 10^{10} = 10^{20}$ in a calculus containing just laws of identity and recursive definitions of exponentiation, multiplication, and addition. Without being in possession of this very long formal derivation, but having proved the soundness of the calculus in question, we can prove deductively (in a somewhat roundabout way) that $10^{10} \times 10^{10} = 10^{20}$. However, as everyone must agree, in contrast to the case just considered, we have not proved deductively that there is a formal proof of the key lemma; we have good evidence for the existence of a formal proof of the lemma, but that is something else.

The crucial question that remains to consider is whether one can say for other reasons that we are in possession of a deductive proof of the lemma. Without any determinate concept of proof it is difficult to achieve any precision here. Most people would agree however that if we are in possession of an interpreted formal proof, and, in addition, are seeing that, for each step in the formal proof, the assertion represented by the conclusion follows from the assertions represented by the premises, then we are in possession of a real proof. Dropping the reference to formal proofs, one may say simply that one is in possession of a deductive proof, if one has made a chain of inferences and has seen directly at each step that the conclusion does follow from the premises.

The subjective phrase "are seeing" or "have seen directly" cannot simply be dropped here, because, for instance, as already remarked in the first part of the paper, a one step proof consisting of some axioms as premises and a difficult theorem as conclusion is not considered to be a proof, although the conclusion does follow from the premises. The point of the concept of proof that I sketched in § 1.3 is that it offers a way to avoid such subjective phrases. Instead of saying metaphorically that when making an inference we "see" that the conclusion follows from the premises, I suggested that we should say that we apply an operation on the given grounds for the premises that transforms them to a ground for the conclusion. By carrying out such an operation we get in possession of a ground for the conclusion, and are then doubtlessly justified in asserting the conclusion (*nota bene*, without having in addition to verify that we are in possession of such a ground).

Regardless of whether we stay with an entirely intuitive notion of deductive proof, relying on an unanalyzed notion of inferring the conclusion from the premises or of seeing it to follow, or try to analyse an inference as the performance of an operation in the way I suggested, there is clearly nothing in support of saying that we have made inferences that amount to being in possession of a deductive proof of the key lemma in the proof of the four-colour theorem. One could say that it *in principle* it is possible to construct a deductive proof of that lemma, understanding the phrase "in principle" to mean as usual that human limitations with respect to time and space are disregarded. We typically say that it is possible in principle to give a proof of $10^{10} \times 10^{10} = 10^{20}$ using only laws of identity and recursive definitions of exponentiation, multiplication, and addition. In the same sense, it is reasonable to say that in principle it is possible to prove deductively the key lemma. But the ground for saying so is in this case empirical, not deductive. It is true that we may prove deductively that the program used by the computer to derive the key lemma was correct, but we cannot prove deductively that the computer executed the program correctly.

We must therefore conclude that the grounds that we have for asserting the four-colour theorem is not of the traditional kind in mathematics, but is partly of an inductive kind, resting as it does on the observation of the result of running a particular program on a computer, giving empirical evidence for the existence of a proof of the key lemma.

Detlefsen and Luker may be right that even before the use of computers there was often an inductive element in mathematical proofs in the form of a reference to the result of a human computation that was not incorporated as an integral part of the proof. They are also right in saying that in one respect there is no fundamental difference between referring to the result of human computation and referring to the result of electronic computation, in both cases we refer to an empirical event. Similarly, it is of course also true that most of us have, individually, merely inductive ground for difficult theorems: we simply trust other people who claim that some person is in possession of a deductive proof; it is only collectively that "we" are in possession of deductive proofs.

On the other hand, when a computations can be made in the usual human way, calculating the values of the occurring terms step-by-step, we do not need to refer to the fact that the computation has been carried out somewhere else, but can instead incorporate the computation in the main proof - there is no interesting difference between the two cases, it is simply a matter of organizing the presentation in different ways. But when a computation carried out electronically is too big for this to be possible, we are in the same situation as in the case of the four colour-theorem that we are forced to refer to the outcome of an empirical process.

It is to be noted that the question of how the correctness of a proof or a computation is verified has not entered at all in this discussion. What matters is how our theorems are established, what kind of grounds for them that we are in possession of, not how we verify that something is a ground. If a theorem has been established only by relying on computers as in the proof of the four-colour theorem or in a proof that involves big computations, then the proof is not entirely deductive, and there is the undeniably epistemological consequence that the theorem is known only a posteriori.

It is an entirely different matter whether our results become less safe when they are based on empirical grounds as compared to when we try to establish them deductively. That a deductive proof guarantees the truth of what is proved does not mean in any way that an assertion of a theorem is safer when based on the belief of having found a deductive proof of it than when based on observing the outcome of the running a particular program on a computer, interpreted as indicating the existence of a formal proof of a formula representing the theorem. We can be more or less convinced that we have really obtained a deductive proof of a particular theorem. Our confidence may gradually increase, as Hume says, when we run over the purported proof and find no error or when our result is corroborated by comparisons with other results. It is in this connection that verification of an alleged proof is relevant.

Our conviction that a computer has really found a proof is based on our belief that the program is correct and that the hardware is reliable. Since we often have good inductive grounds for the reliability of the hardware and may be able to prove deductively that our program is correct (or to check mechanically that it is correct in case the judgement to that effect is formulated as sketched in § 1.3.2.), we are often in the situation that a computer report of having found a specific proof may rightly be deemed as very trustworthy and as more trustworthy than a corresponding report from a human of having found a deductive proof. The deductive approach to the verification of programs is therefore not necessarily impaired because of relying on computers to construct formal proofs interpreted as showing that the programs behave as intended. On the contrary, for the reasons stated, there are good reasons to expect that the deductive approach increases the safety of the programs even more when it is combined with the use of computers for finding formal proofs - although it is true that the deductive approach will then rely partly on inductive evidence.

References

1. A. Church: *Introduction to Mathematical Logic* (Princeton University Press, Princeton 1956)
2. H. Curry, R. Feys: *Combinatory Logic*, vol. 1 (North Holland, Amsterdam 1958)
3. M. Detlefsen, M. Luker: The four-color theorem and mathematical proof. *The Journal of Philosophy* **77** (1980) pp 803-820
4. G. Gentzen: Untersuchungen ber das logische Schliessen. *Mathematische Zeitschrift* **39** (1935) pp 176-210, 405-431
5. A. Heyting: *Intuitionism, An Introduction* (North Holland, Amsterdam 1956)
6. W. A. Howard: The formulae-as-types notion of constructions. In: *To H. B. Curry: Essays on Combinatory Logic, Lambda Calculus and Formalism*, ed by J. P. Seldin and J. R. Hindley (Academic Press 1980) pp 479-490
7. A. N. Kolmogorov: Zur Deutung der intuitionistischen Logik. *Mathematische Zeitschrift* **35** (1932) pp 58-65
8. P. Martin-Löf: An intuitionistic theory of types: Predicative part. In: *Logic Colloquium '73*, ed by H. E. Rose and J. C. Shepherdson (North Holland, Amsterdam 1975) pp 73-118
9. P. Martin-Löf: Constructive mathematics and computer programming. In: *Logic, Methodology and Philosophy of Science* IV, ed by L. J. Cohen et al. (North-Holland, Amsterdam 1982) pp 153-175
10. P. Martin-Löf: *Intuitionistic Type Theory* (Bibliopolis, Napoli 1984)
11. P. Martin-Löf: On the meanings of the logical constants and the justifications of the logical laws. In: *Atti degli Incontri di Logica Matematica* **2** Dipartimento di Matematica, Universit di Siena (1985) pp 203-281. Reprinted in *Nordic Journal of Philosophical Logic* **1** (1996) pp 11-60
12. D. Prawitz: *Natural Deduction. A Proof-Theoretical Study* (Almqvist and Wiksell, Stockholm 1965). Reprinted (Dover Publications, New York 2006)
13. D. Prawitz: Ideas and results in proof theory. In: *Proceedings of the Second Scandinavian Logic Symposium*, ed by J. Fenstad (North-Holland, Amsterdam, 1971) pp 237-309
14. D. Prawitz: Validity of inference. In: *Proceedings from the 2nd Launer Symposium on the Occasion of the Presentation of the Launer Prize at Bern 2006* (2009) To appear
15. D. Prawitz, H. Prawitz and N. Voghera: A mechanical proof procedure and its realization in an electronic computer. *Journal of the Association for Computing Machinery* **7** (1960) pp 102-128

16. P. Teller: Computer proof. *The Journal of Philosophy* **77** (1980) pp 797-803
17. T. Tymoczko: The four-color problem and its philosophical significance. *The Journal of Philosophy* **76** (1979) pp 57-83

The Logic of the Weak Excluded Middle: A Case Study of Proof-Search

Giovanna Corsi

Dipartimento di Filosofia, Università di Bologna (Italy)
giovanna.corsi@unibo.it

1 Introduction

The logic J of the weak excluded middle, known also as Jankov's logic, is an extension of the intuitionistic logic obtained by adding the schema $\neg A \vee \neg\neg A$. This logic, we believe, offers a good case study for some metatheoretical properties: Is there a cut-free calculus for this logic? Is it analytic? Is there a proof-search procedure that answers the question whether a formula is a theorem or not, and if not, does it give us a strategy to build a countermodel? It is a well known result [4] that

Lemma 1. (Hosoi) *If a wff A contains the propositional letters $p_1, \ldots, p_n$, then A is a theorem of J iff $(\neg p_1 \vee \neg\neg p_1) \wedge \cdots \wedge (\neg p_n \vee \neg\neg p_n) \to A$ is a theorem of the intuitionistic logic.*

Hosoi's lemma already answers in the positive the problem of a proof-search procedure for J and we do not need any further investigation since many proof-search procedures are at our disposal for intuitionistic logic. A quite effective one is the one given by the rules of SIC, stack-based intuitionistic calculus, see [3]. The main features of SIC can be summarized in four points:

- Proof-search never enters into loops: due to the presence of the *a fortiori* rule.
- There is no need of back-tracking: due to the rules *push* and *pop*.
- If proof-search is successful, we can get a Gentzen style proof in the calculus IG.
- If proof-search is not successful, we can get a Kripke model based on a finite tree that falsifies the formula (sequent) we started with.

Therefore the question "is A a theorem of J?" becomes "is the proof-search for $(\neg p_1 \vee \neg\neg p_1) \wedge \cdots \wedge (\neg p_n \vee \neg\neg p_n) \to A$ in the calculus SIC successful?", where $p_1, \ldots, p_n$ are the propositional letters occurring in A. If yes, then we get a proof of $(\neg p_1 \lor \neg\neg p_1) \land \cdots \land (\neg p_n \lor \neg\neg p_n) \to A$ in IG and consequently a proof of A in IG plus the rule cut_{wem}:

$$\frac{\neg A \lor \neg\neg A,\ \Gamma \Rightarrow \Delta}{\Gamma \Rightarrow \Delta}\ cut_{wem}$$

For the sake of the reader here are the rules of IG, see [3].[1]

1.1 IG

$$\frac{}{p, \Gamma \Rightarrow \Delta, p}\ Identity$$

$$\frac{A, B, \Gamma \Rightarrow \Delta}{A \land B, \Gamma \Rightarrow \Delta}\ L_\land \qquad \frac{\Gamma \Rightarrow \Delta, A \quad \Gamma \Rightarrow \Delta, B}{\Gamma \Rightarrow \Delta, A \land B}\ R_\land$$

$$\frac{A, \Gamma \Rightarrow \Delta \quad B, \Gamma \Rightarrow \Delta}{A \lor B, \Gamma \Rightarrow \Delta}\ L_\lor \qquad \frac{\Gamma \Rightarrow \Delta, A, B}{\Gamma \Rightarrow \Delta, A \lor B}\ R_\lor$$

$$\frac{A \to B, \Gamma \Rightarrow \Delta, A \quad B, \Gamma \Rightarrow \Delta}{A \to B, \Gamma \Rightarrow \Delta}\ L_\to \qquad \frac{\Gamma, A \Rightarrow B}{\Gamma \Rightarrow A \to B, \Delta}\ R_\to$$

$$\frac{\Gamma \Rightarrow \Delta, A}{\neg A, \Gamma \Rightarrow \Delta}\ L_\neg \qquad \frac{\Gamma, A \Rightarrow}{\Gamma \Rightarrow \neg A, \Delta}\ R_\neg$$

$$\frac{\Gamma \Rightarrow \Delta, B}{\Gamma \Rightarrow \Delta, A \to B}\ a\ fortiori_\to \qquad \frac{\Gamma \Rightarrow \Delta}{\Gamma \Rightarrow \Delta, \neg A}\ a\ fortiori_\neg$$

Proof-search for a formula such as $(\neg p_1 \lor \neg\neg p_1) \land \cdots \land (\neg p_n \lor \neg\neg p_n) \to A$ in the calculus SIC is highly unsatisfactory for one main reason: if the proof-search fails, and so the formula A is not a theorem of J, we do not get any indication as to the construction of a model *based on a frame for J* that falsifies A. A Kripke frame $\langle W, R \rangle$ is a *frame for J* if R is a convergent (or directed) partial order, where R is *convergent* if $\forall x \forall y \forall z (zRx \land zRy \to \exists v(xRv \land yRv))$. The main goal of the present investigation is indeed that of introducing *ad hoc* proof-search procedures with the property that if proof-search fails we get countermodels based on convergent partial orders. The countermodels we will end up with are based on a special kind of convergent partial orders: trees with a final point. We will reach this goal by setting up a procedure that produces countermodels based on trees but in such a way that those models

---

[1] We present here a variation of IG in which negation is taken as primitive instead of being defined as $\neg A =_{df} A \to \bot$.

admit of the addition of a final point. This means that when the final point is added we still have a well defined model and this model is "conservative" with respect to the original one: what is true(false) at a point of the original model based on a tree remains true(false) in the extended model. Obviously, our proof-search strategy will also have the property that if it is successful, we then get a proof in a Gentzen-style calculus equivalent to $IG + \text{cut}_{wem}$. For reasons of convenience that will become clear in a moment we have chosen the Gentzen-style calculus JG, described in the next section.

2 The calculus JG

The main motivation for JG is that its rules are specular to the rules of SJC, the calculus in which the proof-search for J is performed, see § 3. Looking at the rules of JG, we see that they capture most of the properties of the negation of J and they keep "to the left" as much information as possible. This is crucial because the intended meaning of a formula A in the antecedent of a sequent - to the left - is that A is true, so for example from knowing that $\neg(B \vee C)$ is true, we get to know - via the rule $L_{\neg\vee}$ that $\neg B$ and $\neg C$ are both true and not simply that $(B \vee C)$ is false. In brief, the rules for negated formulas reflect the fact that negation commutes with the binary connectives. Are theorems of J:

- $\neg(A \wedge B) \leftrightarrow (\neg A \vee \neg B)$
- $\neg(A \vee B) \leftrightarrow (\neg A \wedge \neg B)$
- $\neg(A \rightarrow B) \leftrightarrow (\neg\neg A \wedge \neg B)$

   as well as the following implication:

- $(A \rightarrow B) \rightarrow (\neg A \vee \neg\neg B)$

A further peculiarity of J is the asymmetry between the schemata that hold for the negated formulas and those that hold for the implicative ones:

- $J \vdash (B \rightarrow \neg A) \vee (\neg A \rightarrow B)$
- $J \not\vdash (B \rightarrow (A \rightarrow C)) \vee ((A \rightarrow C) \rightarrow B)$

   again

- $J \vdash A \rightarrow (B \vee \neg C) \rightarrow (A \rightarrow B) \vee \neg C$
- $J \not\vdash A \rightarrow (B \vee (C \rightarrow D)) \rightarrow (A \rightarrow B) \vee (C \rightarrow D)$.

This asymmetry is reflected in the rules for right-negation and right-implication. A less standard rule of JG is left-implication. Its motivation will become clearer when we will examine it inside SJC. For the moment let us simply say that from the knowledge that $A \rightarrow B$ is true we want to come to know which (proper)subformulas (or negation) of (proper)subformulas of $A \to B$ are true and not simply that B is false and $A \to B$ remains true. As a matter of fact, from the truth of $A \to B$ we are entitled to assert that: either B is true or $\neg A$ is true or A and $\neg B$ are both false, and so that B is going to be true at a future stage, see the proof of $L^J_\to$ in the next paragraph.

2.1 Rules of JG

$$\frac{}{A, \Gamma \Rightarrow \Delta, A} \, Identity \qquad \frac{}{A, \neg A, \Gamma \Rightarrow \Delta} \, Non\text{-}contradiction$$

$$\frac{A, B, \Gamma \Rightarrow \Delta}{A \wedge B, \Gamma \Rightarrow \Delta} \, L_\wedge \qquad \frac{\Gamma \Rightarrow \Delta, A \quad \Gamma \Rightarrow \Delta, B}{\Gamma \Rightarrow \Delta, A \wedge B} \, R_\wedge$$

$$\frac{A, \Gamma \Rightarrow \Delta \quad B, \Gamma \Rightarrow \Delta}{A \vee B, \Gamma \Rightarrow \Delta} \, L_\vee \qquad \frac{\Gamma \Rightarrow \Delta, A, B}{\Gamma \Rightarrow \Delta, A \vee B} \, R_\vee$$

$$\frac{\Gamma, A \to B \Rightarrow A, \neg B, \Delta \quad \Gamma, B \Rightarrow \Delta \quad \Gamma, \neg A \Rightarrow \Delta}{\Gamma, A \to B \Rightarrow \Delta} \, L^J_\to$$

$$\frac{\neg A, \Gamma \Rightarrow \Delta \quad \neg B, \Gamma \Rightarrow \Delta}{\neg (A \wedge B), \Gamma \Rightarrow \Delta} \, L_{\neg \wedge} \qquad \frac{\neg A, \neg B, \Gamma \Rightarrow \Delta}{\neg (A \vee B), \Gamma \Rightarrow \Delta} \, L_{\neg \vee}$$

$$\frac{\Gamma \Rightarrow \neg A, \Delta}{\neg \neg A, \Gamma \Rightarrow \Delta} \, L_{\neg \neg} \qquad \frac{\neg B, \Gamma \Rightarrow \neg A, \Delta}{\neg (A \to B), \Gamma \Rightarrow \Delta} \, L_{\neg \to}$$

$$\frac{\Gamma \Rightarrow \Delta, B}{\Gamma \Rightarrow \Delta, A \to B} \, a\,fortiori_\to \qquad \frac{\Gamma \Rightarrow \Delta}{\Gamma \Rightarrow \Delta, \neg A} \, a\,fortiori_\neg$$

$$\frac{\Gamma, C_1, \ldots, C_k \Rightarrow}{\Gamma \Rightarrow \neg C_1, \ldots, \neg C_k, \Delta} \, R^J_\neg \qquad \frac{\Gamma, A \Rightarrow B, \neg C_1, \ldots, \neg C_k}{\Gamma \Rightarrow A \to B, \neg C_1, \ldots, \neg C_k, \Delta} \, R^J_\to$$

$$\frac{\neg A \vee \neg \neg A, \Gamma \Rightarrow \Delta}{\Gamma \Rightarrow \Delta} \, cut_{wem}$$

Here is a proof of the rule $L^J_\to$ in IG + cut_{wem}

$$\cfrac{\cfrac{\cfrac{\cfrac{\cfrac{A \Rightarrow A \quad B \Rightarrow B}{A \rightarrow B, A \Rightarrow B} L_\rightarrow}{A \rightarrow B, A, \neg B \Rightarrow} L_\neg}{A \rightarrow B, \neg B \Rightarrow \neg A} R_\neg}{A \rightarrow B, \neg B, \neg\neg A \Rightarrow} L_\neg \quad \cfrac{\Gamma, \neg A \Rightarrow \Delta}{\neg A \vee \neg\neg A, A \rightarrow B, \neg B, \Gamma \Rightarrow \Delta} L_\vee}{A \rightarrow B, \neg B, \Gamma \Rightarrow \Delta} cut_{wem}$$

$$\cfrac{\cfrac{\Gamma \Rightarrow \neg B, A, \Delta \quad \Gamma, B \Rightarrow \Delta}{\Gamma, A \rightarrow B \Rightarrow \neg B, \Delta} L_\rightarrow}{\Gamma, A \rightarrow B, \neg\neg B \Rightarrow \Delta} L_\rightarrow$$

$$\cfrac{\neg B \vee \neg\neg B, \Gamma, A \rightarrow B \Rightarrow \Delta}{A \rightarrow B, \Gamma \Rightarrow \Delta} cut_{wem}$$

2.2 Trees

Let N be the set of natural numbers, zero excluded, and N^* the set of finite lists of natural numbers (the empty list is denoted by ϵ).

- A *tree* τ is a not empty subset of N^* such that:
  1. if $\alpha \in \tau$ and $\alpha = \beta\gamma$, then $\beta \in \tau$, $\alpha, \beta \in N^*$;
  2. if $\alpha i \in \tau$ and $j < i$, then $\alpha j \in \tau$, $i, j \in N$.
- ϵ, the empty list, is said to be *the root*.
- For each node α, the nodes of the form αi are said to be *immediate successors* of α and we write $\alpha \lessdot \alpha i$.
- a *segment* $\alpha_1, \ldots \alpha_n$ is a sequence of nodes such that $\alpha_i \lessdot \alpha_{i+1}$, $1 \le i \le n$.
- β is a *successor* of α if $\alpha < \beta$, where $<$ is the transitive closure of $\lessdot$. $\alpha \le \beta$ iff ($\alpha = \beta$ or $\alpha < \beta$).
- α is said to be a *leaf* if it has no successors.
- An α-*branch* is a segment $\alpha_0 \ldots \alpha_n$ such that α_0 is the root and $\alpha_n = \alpha$.
- The *length* of an α-branch $\alpha_0 \ldots \alpha_n$ is n, i.e. the number of its nodes minus one.
- The *height* of a node α is the length of the α-branch.

Definition 1. *An JG-derivation is a triple $\mathcal{D} = \langle T, \phi, \rho \rangle$ where T is a finite tree, ϕ is a function that associates a sequent to every node of T and ρ is a function that associates a rule of JG to every node of T in such a way that:*

if $\alpha 1, \ldots, \alpha n$ are all and only the immediate successors of α, then

$$\cfrac{\phi(\alpha 1), \ldots, \phi(\alpha n)}{\phi(\alpha)} \rho(\alpha)$$

is a rule of JG.

- *If $\phi(\epsilon)$ is the sequent $\Gamma \Rightarrow \Delta$, $\mathcal{D}$ is said to be an JG-derivation of $\Gamma \Rightarrow \Delta$.*

- If $\rho(\alpha) = R$, then (that occurrence of) R is said to be of level α.

Definition 2. *A JG-derivation is a proof if $\phi(\alpha)$ is either identity or non-contradiction for any leaf α.*

3 The calculus SJC

As to the general motivation of the calculus SJC, stack-based Jankov calculus, we refer the reader to [3], where a twin calculus, SIC, is presented. Analogously to SIC, to each node of a SJC-derivation is associated a sequent-list [2]

$$\Gamma_1 \Rightarrow \Delta_1 \parallel \Gamma_2 \Rightarrow \Delta_2 \parallel \ldots \ldots \parallel \Gamma_n \Rightarrow \Delta_n$$

Before examining the rules of SJC, we want to point out a feature of SJC which is most important and very peculiar of the system we are about to describe: two distinct sequents can be *mutually closed*.

As to be expected, the *rationale* behind the rules of SJC is semantical: the rules of SJC are just to be read as steps leading to the construction of a countermodel for any non-theorem A of J. By the application of the rules we are bound to end up with a tree of sequents which can be readily transformed into a countermodel for A. The problem is that we need a model based on a convergent frame, not just a tree. In general, if we add a final point to a given model based on a tree, we get a contradictory situation. Take the simple case of $(q \to \neg p) \lor \neg p \to q)$. A countermodel for such a formula, if there were one, would be something like

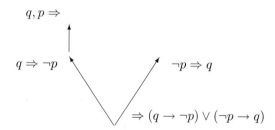

Now, no final point can possibly be added to this countermodel, in fact at the final point both p and $\neg p$ should be true, and this is impossible. The lesson to be learned from this is that when we deal with trees as given by the rules of SJC the notion of *mutually closed sequents* has to come into play. The sequents $q, p \Rightarrow$ and $\neg p \Rightarrow q$ taken separately are not contradictory, but

---

[2] A sequent-list can be read as a disjunction of implications $(\bigwedge \Gamma_1 \to \bigvee \Delta_1) \lor \ldots \ldots \lor (\bigwedge \Gamma_n \to \bigvee \Delta_n)$.

The Logic of the Weak Excluded Middle: A Case Study of Proof-Search   101

if we consider them as bearing information that has to merge together, they contradict each other. In SJC, the search for a countermodel (or for a proof) of $(q \to \neg p) \vee (\neg p \to q)$ takes the following form:

$$\cfrac{\cfrac{\cfrac{\cfrac{q, p \Rightarrow}{q \Rightarrow \neg p} R_\neg^J}{q \Rightarrow \neg p \,\|\, \neg p \Rightarrow q} pop}{\Rightarrow q \to \neg p,\ \neg p \to q} push^J}{\Rightarrow (q \to \neg p) \vee (\neg p \to q)} R_\vee$$

Let us examine this tree from below. We start by applying the rule $R_\vee$, then the rule $push^J$ breaks the sequent $\Rightarrow q \to \neg p,\ \neg p \to q$ into two sequents $q \Rightarrow \neg p$ and $\neg p \Rightarrow q$, and so it generates a sequent-list. From a semantical point of view, the rule $push^J$ corresponds to the splitting in a Kripke model. The rules of SJC apply only to the last sequent of a sequent-list. No rule can be applied to $\neg p \Rightarrow q$ since negated atoms are left unexamined. Therefore we have reached a sort of dead end and the only thing to do is to put into play the sequent before the last: this is done by the *pop* rule. The sequent $\neg p \Rightarrow q$ which doesn't occur in the premiss of the *pop*-rule is said to be a *pop-sequent*. Now the rule $R_\neg^J$ is enforced and we get $q, p \Rightarrow$. No sequent of the form $\Gamma, A, \neg A \Rightarrow \Delta$ or $\Gamma, A \Rightarrow A, \Delta$ occurs in the above tree, still there is a branch - the only branch of the tree - in which both p and $\neg p$ occur in the antecedents of a leaf-sequent and of a pop-sequent, respectively: $q, p \Rightarrow$ and $\neg p \Rightarrow q$ are then *mutually closed* and consequently the branch is closed.

From a proof-theoretical point of view it is interesting to see what two mutually closed sequents amount to: an application of the rule cut_{wem}. Let us see it in detail. Consider the sequents: $q, p \Rightarrow$ and $\neg p \Rightarrow q$. Add $\neg p$ where p is and $\neg\neg p$ where $\neg p$ is.

- In order to emphasize what the situation is, we underline the added formulas.
- Transcribe the underlined formulas till the first application of the $push^J$ rule below both the mutually closed sequents
- Apply the rule $R_\to$ in each branch separately
- Apply $L_\vee$ to the underlined formulas
- By the cut_{wem} rule eliminate the disjunction obtained by $L_\vee$.
- Then apply $R_\vee$.

$$\cfrac{\cfrac{\cfrac{\cfrac{\underline{\neg p}, q, p \Rightarrow}{\underline{\neg p}, q \Rightarrow \neg p} R_\neg}{\underline{\neg p} \Rightarrow q \to \neg p} R_\to \qquad \cfrac{\cfrac{\underline{\neg\neg p}, \neg p \Rightarrow q}{\underline{\neg\neg p} \Rightarrow \neg p \to q} R_\to}{}}{\underline{\neg p \vee \neg\neg p} \Rightarrow q \to \neg p,\ \neg p \to q} L_\vee}{\cfrac{\Rightarrow q \to \neg p,\ \neg p \to q}{\Rightarrow (q \to \neg p) \vee (\neg p \to q)} R_\vee} cut_{wem}$$

Let us now examine the role played by the rule $L_\to^J$ by the help of a further example. Take the sequent $\Rightarrow ((p \to q) \to r) \vee (\neg q \to s) \vee (p \to q)$ and analyse it by the rules of SJC, we obtain:

$$\cfrac{\neg p \Rightarrow r \quad \cfrac{\cfrac{p \to q, q \Rightarrow}{(p \to q)^\star \Rightarrow p, \neg q, r} R_\neg^J \quad q \Rightarrow r}{(p \to q) \Rightarrow r} L_\to^J}{\cfrac{\cfrac{\cfrac{\cfrac{(p \to q) \Rightarrow r \,\|\, \neg q \Rightarrow s}{(p \to q) \Rightarrow r \,\|\, \neg q \Rightarrow s \,\|\, p \Rightarrow q} pop}{\Rightarrow (p \to q) \to r, \neg q \to s, p \to q} push^J}{\Rightarrow ((p \to q) \to r) \vee (\neg q \to s) \vee (p \to q)} R_\vee}{}}$$

The leaf-sequent $p \to q, q \Rightarrow$ as well as $q \Rightarrow r$ is mutually closed with the pop-sequent $\neg q \Rightarrow s$, whereas the leaf-sequent $\neg p \Rightarrow r$ is mutually closed with the pop-sequent $p \Rightarrow q$. Therefore every branch is closed, the proof-search is successful and we can easily transform it into a JG-proof, see theorem 2. For the sake of clarity, here is the proof.

$$\small \cfrac{\cfrac{\cfrac{\cfrac{\cfrac{\cfrac{\neg\neg p, \neg p \Rightarrow r \quad \cfrac{\cfrac{\neg q, p \to q, q \Rightarrow}{\neg q, p \to q \Rightarrow p, \neg q, r} R_\neg^J \quad \neg q, q \Rightarrow r}{\neg q, p \to q \Rightarrow r} L_\to^J}{\neg\neg p, \neg q, p \to q \Rightarrow r} R_\to \quad \cfrac{\cfrac{\neg\neg q, \neg q \Rightarrow s}{\neg\neg q \Rightarrow \neg q \to s} R_\to}{\neg\neg q \Rightarrow \neg q \to s} L_\vee}{\cfrac{\neg q \vee \neg\neg q, \neg\neg p \Rightarrow ((p \to q) \to r), (\neg q \to s)}{\neg\neg p \Rightarrow ((p \to q) \to r), (\neg q \to s)} cut_{wem}} \quad \cfrac{\cfrac{\neg p, p \Rightarrow q}{\neg p \Rightarrow p \to q} R_\to}{\neg p \Rightarrow p \to q} L_\vee}{\cfrac{\neg p \vee \neg\neg p \Rightarrow ((p \to q) \to r), (\neg q \to s), (p \to q)}{\Rightarrow ((p \to q) \to r), (\neg q \to s), (p \to q)} cut_{wem}}}{\Rightarrow ((p \to q) \to r) \vee (\neg q \to s) \vee (p \to q)} R_\vee$$

If we had applied the standard rule $L_\to$ the situation would have been:

$$\cfrac{\cfrac{(p \to q)^\star \Rightarrow p, r \quad q \Rightarrow r}{(p \to q) \Rightarrow r} L_\to}{\cfrac{\cfrac{\cfrac{(p \to q) \Rightarrow r \,\|\, \neg q \Rightarrow s}{(p \to q) \Rightarrow r \,\|\, \neg q \Rightarrow s \,\|\, p \Rightarrow q} pop}{\Rightarrow (p \to q) \to r, \neg q \to s, p \to q} push^J}{\Rightarrow ((p \to q) \to r) \vee (\neg q \to s) \vee (p \to q)} R_\vee} pop$$

The leaf-sequent $(p \to q)^\star \Rightarrow p, r$ is neither closed nor mutually closed. The main weakness of the rule $L_\to$ in this context is that it prevents lemma 3 to hold, and so the incompatibility of sequents such as $\Gamma, A \to B \Rightarrow \Delta$ and $\Gamma', \neg(A \to B) \Rightarrow \Delta'$ is not transferred to the incompatibility of sequents containing (the negation of) subformulas of either A or B.

3.1 Rules of SJC

It is expedient to mark certain occurrences of formulas in order to avoid examining formulas unnecessarily.

Definition 3.

- Signed formulas, s-formulas, *are formulas marked by either* $\star$ *or* $\dagger$. *Only implicative or double negated formulas can be marked with* $\star$. *Atomic formulas and the negation of atomic formulas are never signed.*
- A sequent-list *is either the empty list*, $\langle \rangle$, *or a word of the form* $\zeta \| \Gamma \Rightarrow \Delta$ *where* Γ *is a multiset of formulas or s-formulas,* Δ *is a multiset of formulas and* ζ *is a sequent-list.*
- The length *of a sequent-list is defined by recursion:* $length(\langle \rangle) = 0$, $length(\zeta \| \Gamma \Rightarrow \Delta) = length(\zeta) + 1$.

Notational convention

- Greek letters $\zeta, \xi, \ldots$ denote sequent-lists. We write $\| \Gamma \Rightarrow \Delta$ instead of $\langle \rangle \| \Gamma \Rightarrow \Delta$.
- Capital greek letters signed by $\dagger$, $\Pi^\dagger \ldots$, denote multisets of formulas each of which is signed by $\dagger$.
- Capital greek letters signed by $\star$, $\Sigma^\star \ldots$, denote multisets of either implicative or double negated formulas each of which is signed by $\star$.
- $\mathbf{p}, \mathbf{q} \ldots$ denote multisets of atoms with no sign.

$$\frac{}{\zeta \| A, \Gamma \Rightarrow \Delta, A} \, Identity \qquad \frac{}{\zeta \| A, \neg A, \Gamma \Rightarrow \Delta} \, Non-contradiction$$

$$\frac{\zeta \| A, B, \Gamma \Rightarrow \Delta}{\zeta \| A \wedge B, \Gamma \Rightarrow \Delta} \, L_\wedge \qquad \frac{\zeta \| \Gamma \Rightarrow A, \Delta \quad \zeta \| \Gamma \Rightarrow B, \Delta}{\zeta \| \Gamma \Rightarrow A \wedge B, \Delta} \, R_\wedge$$

$$\frac{\zeta \| A, \Gamma \Rightarrow \Delta \quad \zeta \| B, \Gamma \Rightarrow \Delta}{\zeta \| A \vee B, \Gamma \Rightarrow \Delta} \, L_\vee \qquad \frac{\zeta \| \Gamma \Rightarrow A, B, \Delta}{\zeta \| \Gamma \Rightarrow A \vee B, \Delta} \, R_\vee$$

$$\frac{\zeta \| \Gamma, B \Rightarrow \Delta \quad \zeta \| \Gamma, \neg A \Rightarrow \Delta \quad \zeta \| \Gamma, (A \to B)^\star \Rightarrow A, \neg B, \Delta}{\zeta \| \Gamma, (A \to B) \Rightarrow \Delta} \, L_\to^J$$

$$\frac{\zeta\|\neg A, \Gamma \Rightarrow \Delta \quad \zeta\|\neg B, \Gamma \Rightarrow \Delta}{\zeta\|\neg(A \land B), \Theta \Rightarrow \Delta} L_{\neg \land} \qquad \frac{\zeta\|\neg A, \neg B, \Gamma \Rightarrow \Delta}{\zeta\|\neg(A \lor B), \Gamma \Rightarrow \Delta} L_{\neg \lor}$$

$$\frac{\zeta\|\neg\neg A^\star, \Gamma \Rightarrow \neg A, \Delta}{\zeta\|\neg\neg A, \Gamma \Rightarrow \Delta} L_{\neg\neg} \qquad \frac{\zeta\|\neg\neg A^\star, \neg B, \Gamma \Rightarrow \neg A, \Delta}{\zeta\|\neg(A \to B), \Gamma \Rightarrow \Delta} L_{\neg\to}$$

$$\frac{\zeta\|\Gamma, A^\dagger \Rightarrow B, \Delta}{\zeta\|\Gamma, A^\dagger \Rightarrow A \to B, \Delta} \text{ a fortiori}_\to \qquad \frac{\zeta\|\Gamma, C^\dagger \Rightarrow \Delta}{\zeta\|\Gamma, C^\dagger \Rightarrow \neg C, \Delta} \text{ a fortiori}_\neg$$

$$\frac{\zeta \| \Sigma, \Pi^\dagger, \mathbf{p}, C_1^\dagger, \ldots, C_k^\dagger, C_1, \ldots, C_k \Rightarrow}{\zeta\|\Sigma^\star, \Pi^\dagger, \mathbf{p} \Rightarrow \mathbf{q}, \neg C_1, \ldots, \neg C_k} R_\neg^J$$

$$\frac{\zeta\|\Sigma, \Pi^\dagger, \mathbf{p}, A_1^\dagger, A_1 \Rightarrow B_1, \neg C_1, \ldots, \neg C_k \| \ldots \ldots \| \Sigma, \Pi^\dagger, \mathbf{p}, A_n^\dagger, A_n \Rightarrow B_n, \neg C_1, \ldots, \neg C_k \Rightarrow}{\zeta\|\Sigma^\star, \Pi^\dagger, \mathbf{p} \Rightarrow \mathbf{q}, A_1 \to B_1, \ldots, A_n \to B_n, \neg C_1, \ldots, \neg C_k} push^J$$

$$\frac{\zeta}{\zeta \| \Sigma^\star, \Pi^\dagger, \mathbf{p} \Rightarrow \mathbf{q}} pop$$

Conditions on $push^J$:

- Σ contains exactly the formulas of $\Sigma^\star$ without the $\star$
- $A_1^\dagger \ldots A_n^\dagger$ do not occur in $\Pi^\dagger$

Definition 4. *An SJC-derivation is defined in the same way as a JG-derivation with the rules of SJC instead of the rules of JG.*

Definition 5. *Consider the conclusion of the pop-rule*

$$\frac{\zeta}{\zeta \| \Sigma^\star, \Pi^\dagger, \mathbf{p} \Rightarrow \mathbf{q}} pop$$

- *The last sequent $\Sigma^\star, \Pi^\dagger, \mathbf{p} \Rightarrow \mathbf{q}$ is said to be a pop-sequent.*
- *If $\zeta \| \Gamma \Rightarrow \Delta$ is the sequent-list associated to a leaf of a SJC-derivation, then the last sequent $\Gamma \Rightarrow \Delta$ is said to be a leaf-sequent.*
- *A leaf-sequent is said to be closed if it is either of the form $\Gamma', A \Rightarrow A, \Delta'$ or of the form $\Gamma', A, \neg A \Rightarrow \Delta$, for some A.*

- A leaf-sequent is said to be mutually closed *with a pop-sequent* if either it is of the form $\Gamma, r \Rightarrow \Delta$ and the pop-sequent of the form $\Sigma^{\star}, \Pi^{\dagger}, \mathbf{p}, \neg r \Rightarrow \mathbf{q}$ or it is of the form $\Gamma, \neg r \Rightarrow \Delta$ and the pop-sequent of the form $\Sigma^{\star}, \Pi^{\dagger}, \mathbf{p}, r \Rightarrow \mathbf{q}$, for some atom r.

Definition 6. *Consider a SJC-derivation $\mathcal{D}$.*

- *A branch is* closed *if its leaf-sequent is closed or mutually closed with a pop-sequent occurring in the branch.*
- *A branch is* open *if it is not closed.*
- *$\mathcal{D}$ is a* proof *if every branch is closed.*
- *$\mathcal{D}$ is* open *if it contains an open branch.*

What does an open branch of a SJC-derivation represents? A SJC-derivation splits only in the presence of rules with two or more premisses and not when the $push^J$-rule is applied, $push^J$ is a one-premiss rule. From a semantical point of view an open branch of a SJC-derivation represents one of the possible alternatives to falsify the sequent at the root. Now the falsehood of a formula is reduced by the rules of the calculus to the falsehood of a set of either implicative or negated or atomic subformulas. Each implicative formula has to be falsified, in general, at different points of a model. So the different implicative subformulas are put in a stack by the rule $push^J$ and examined sequentially as the branch develops. Therefore an open branch codifies, in general, a model based on a tree. Consider the following SJC-derivation:

$$
\cfrac{
  \cfrac{\vdots}{p \Rightarrow t \to s, q} \qquad
  \cfrac{
    \cfrac{
      \cfrac{
        \cfrac{p, t \Rightarrow s}{p, t \Rightarrow s \,\|\, p, y \Rightarrow v} pop
      }{p \Rightarrow t \to s, y \to v} push^J
    }{p \Rightarrow (t \to s), (q \land (y \to v))} R_\land
  }{p \Rightarrow (t \to s) \lor (q \land (y \to v))} R_\lor
}{p \Rightarrow (t \to s) \lor (q \land (y \to v)) \,\|\, x \Rightarrow z} pop
\qquad
\cfrac{
  \cfrac{\vdots}{p \Rightarrow (t \to s) \lor (q \land (y \to v)) \,\|\, x \Rightarrow w}
}{\;} R_\land
$$

$$
\cfrac{
  \cfrac{
    \cfrac{
      p \Rightarrow (t \to s) \lor (q \land (y \to v)) \,\|\, x \Rightarrow z \land w
    }{\Rightarrow p \to (t \to s) \lor (q \land (y \to v)), x \to (z \land w)} push^J
  }{\Rightarrow (p \to ((t \to s) \lor (q \land (y \to v)))) \lor (x \to (z \land w))} R_\lor
}{}
$$

The branch ending with the sequent $p, t \Rightarrow s$ provides a countermodel for the sequent at the root of the tree. The proof of this fact is given in lemma 4. The countermodel is as follows, where both the leaf-sequents and the pop-sequents are falsified at the leaves of the model.

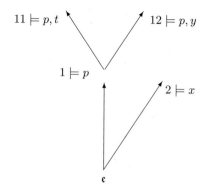

It is easy to see that
$11 \not\models t \to s$
$12 \not\models y \to v$
$1 \not\models p \to ((t \to s) \vee (y \to v))$
$1 \not\models p \to ((t \to s) \vee (q \wedge (y \to v)))$
$2 \not\models x \to z \wedge w$
$\mathfrak{e} \not\models p \to ((t \to s) \vee ((q \wedge (y \to v)))) \vee (x \to (z \wedge w))$

3.2 SJC versus JG

Some questions are in order : is the SJC-proof of $(q \to \neg p) \vee (\neg p \to q)$ cut-free? Does it contain cuts in disguise? Is there any reason to prefer a JG-proof to a SJC-proof? Proofs in JG present themselves in the received Gentzen-style way: a tree of sequents, with just one sequent associated to each node of the tree. The first price to pay for this layout is that we can not dispense with the cut-rule, at least we can not dispense with the special form of the cut-rule that we call cut_{wem}. A second price is the loss of the subformula property, as we have seen $\neg p \vee \neg\neg p$ occurs in the JG-proof $(q \to \neg p) \vee (\neg p \to q)$. On the contrary, in SJC we can still mantain that the subformula property is satisfied in so far as a generalized notion of subformula is put into play to the effect that $\neg A$ and $\neg B$ are also subformulas of $\neg (A \wedge B)$ and $\neg (A \vee B)$, and $\neg A, \neg\neg A, \neg B$ are counted among the subformulas of $\neg (A \to B)$. With SJC we associate a sequent-list to each node of a proof-tree, but we apply the rules only to the last sequent of a sequent-list, so this is a minor deviation from the Gentzen-style layout of proofs. Other calculi differ more profoundely: with the hypersequent calculi the inference rules can be applied to any sequents of a hypersequent and not only to the last one, and with the labelled calculi a proof contains notions - natural numbers, relations, etc - external to the language of the formulas we want to prove. The novelty with SJC is the notion of mutually closed sequents. In order to know if a sequent is closed it is not enough to look at the sequent itself, but one has to look also at some sequent occurring below it, in particular to the conclusion of an occurrence of the *pop*-rule. Our tenet is that the calculus SJC has good metatheoretical properties at a very low cost, if it is a cost, i.e. the use of sequent-lists instead of sequents and the notion of mutually closed sequents. Here are some good metatheoretical properties fulfilled by SJC:

1. Proof-search for a formula is performed by a single tree, as for classical logic.
2. Proof-search always terminates since never enters into loops, thanks to the *a fortiori* rules.
3. There is no need of back-tracking thanks to the rules $push^J$ and *pop*.
4. SJC-proofs satisfy the sub-formula property, where the notion of subformula is taken in a slightly generalized version.
5. SJC-proofs are cut-free.
6. SJC-proofs can be transformed, if we wish, into JG-proofs.
7. If proof-search in SJC fails we get a Kripke countermodel based on a convergent frame for the formula we started with.

4 Soundness and completeness

Given an open branch, a segment $\langle \zeta \,\|\, \Gamma_1 \Rightarrow \Delta_1, \ldots, \zeta \,\|\, \Gamma_k \Rightarrow \Delta_k \rangle$ is said to be a *pop-segment* if the first sequent-list $\zeta \,\|\, \Gamma_1 \Rightarrow \Delta_1$ is the premiss of an occurrence of the *pop*-rule or is the initial sequent of the tree, the last sequent-list, $\zeta \,\|\, \Gamma_k \Rightarrow \Delta_k$ is the conclusion of an occurrence of the *pop*-rule and the *pop*-rule never occurs inside the segment.

$$\frac{\zeta}{\zeta \,\|\, \Gamma_k \Rightarrow \Delta_k}\, pop$$

$$\vdots$$

$$\frac{\zeta \,\|\, \Gamma_1 \Rightarrow \Delta_1}{\zeta \,\|\, \Gamma_1 \Rightarrow \Delta_1 \,\|\, \Theta \Rightarrow \Theta'}\, pop$$

Lemma 2. *Let* $\sigma = \langle \zeta \,\|\, \Gamma_1 \Rightarrow \Delta_1, \ldots, \zeta \,\|\, \Gamma_k \Rightarrow \Delta_k \rangle$ *be a pop-segment of an open branch. Define* $\mathbb{L}_\sigma = \Gamma_1 \cup \cdots \cup \Gamma_k$. *Then*

- $\Delta_k = \mathbf{p}$
- *If* $\neg\neg B \in \mathbb{L}_\sigma$, *then* $B \in \mathbb{L}_\sigma$
- *If* $(B \to C) \in \mathbb{L}_\sigma$, *then either* $C \in \mathbb{L}_\sigma$ *or* $\neg B \in \mathbb{L}_\sigma$
- *If* $\neg(B \to C) \in \mathbb{L}_\sigma$, *then* $B \in \mathbb{L}_\sigma$ *and* $\neg C \in \mathbb{L}_\sigma$

Given a pop-segment $\sigma = \langle \zeta \,\|\, \Gamma_1 \Rightarrow \Delta_1, \ldots, \zeta \,\|\, \Gamma_k \Rightarrow \Delta_k \rangle$, let us denote by $\mathbb{L}_\sigma^{top}$ the formulas of Γ_k.

Lemma 3. *Let $\sigma_1 \ldots \sigma_n$ be the sequence of pop-segments that constitute an open branch. If there is a formula A such that $A \in \mathbb{L}_{\sigma_i}$ and $\neg A \in \mathbb{L}_{\sigma_j}$ then there is an atomic formula p such that $p \in \mathbb{L}_{\sigma_i}^{top}$ and $\neg p \in \mathbb{L}_{\sigma_j}^{top}$ or there is an atomic formula q such that $\neg q \in \mathbb{L}_{\sigma_i}^{top}$ and $q \in \mathbb{L}_{\sigma_j}^{top}$.*

Proof. By induction on A. If $p \in \mathbb{L}_{\sigma_i}$, then $p \in \mathbb{L}_{\sigma_i}^{top}$, because atomic formulas are never cancelled along a segment. Analogously for $A = \neg p$.

- $A = (B \wedge C)$. If $(B \wedge C) \in \mathbb{L}_{\sigma_i}$, then $\neg B \in \mathbb{L}_{\sigma_i}$ and $\neg C \in \mathbb{L}_{\sigma_i}$; if $\neg (B \wedge C) \in \mathbb{L}_{\sigma_i}$, then either $B \in \mathbb{L}_{\sigma_i}$ or $C \in \mathbb{L}_{\sigma_i}$, so by induction hypothesis the lemma holds.

- Analogously for $A = (B \vee C)$.

- $A = (B \to C)$. Let $(B \to C) \in \mathbb{L}_{\sigma_i}$ and $\neg(B \to C) \in \mathbb{L}_{\sigma_j}$. Then $\neg\neg B^\star \in \mathbb{L}_{\sigma_j}$ and $\neg C \in \mathbb{L}_{\sigma_j}$. Since σ_j is a pop-segment, by lemma 2.2, $B \in \mathbb{L}_{\sigma_j}$. Since $(B \to C) \in \mathbb{L}_{\sigma_i}$, by lemma 2.3 either $\neg B \in \mathbb{L}_{\sigma_i}$ or $C \in \mathbb{L}_{\sigma_i}$.

- $A = \neg B$. Let $\neg B \in \mathbb{L}_{\sigma_i}$ and $\neg\neg B \in \mathbb{L}_{\sigma_j}$. Then $B \in \mathbb{L}_{\sigma_j}$ by lemma 2.2.

4.1 Labelling sequent-lists

As with SIC, we label SJC-derivations. A labelling function $\phi^\star(\alpha)$ associates labels to every sequents occurring in a SJC-derivation $\mathcal{D} = \langle T, \phi, \rho \rangle$. Each label is a list of natural numbers, and we use the letters $\mathfrak{a}, \mathfrak{b}, \mathfrak{c}$, to denote them so as to avoid confusion with the labels of the nodes of T. The empty list of natural numbers is denoted by $\mathfrak{e}$.

In this section $\zeta_1 \ldots \zeta_n$ denote lists of labelled sequents.

Definition 7. *The function $\phi^\star$ is so defined:*

- *if $\phi(\epsilon) = \langle \| \Gamma \Rightarrow \Delta \rangle$ then $\phi^\star(\epsilon) = [\| \Gamma \Rightarrow_\mathfrak{e} \Delta]$,*

- *if $\phi^\star(\alpha) = \langle \zeta \| \Gamma \Rightarrow_\mathfrak{a} \Delta \rangle$ and $\rho(\alpha) \in \{L_\wedge, R_\wedge, L_\vee, R_\vee, L_\to, L_{\neg\wedge}, L_{\neg\vee}, L_{\neg\neg}, L_{\neg\to}, a\,fortiori_\to, a\,fortiori_\neg\}$, then $\phi^\star(\alpha i) = \langle \zeta \| \Gamma' \Rightarrow_\mathfrak{a} \Delta' \rangle$, where $\Gamma' \Rightarrow_\mathfrak{a} \Delta'$ is a premiss of $\rho(\alpha)$*

- *if $\phi^\star(\alpha) = \langle \zeta \| \Sigma^\star, \Pi^\dagger, \mathbf{p} \Rightarrow \mathbf{q}, \neg C_1, \ldots, \neg C_k \rangle$ and $\rho(\alpha) = R_\neg$, then $\phi^\star(\alpha 1) = \langle \zeta \| \Sigma, \Pi^\dagger, \mathbf{p}, C_1, \ldots, C_k \Rightarrow_{\mathfrak{a}1} \rangle$*

- *if $\phi^\star(\alpha) = \langle \zeta \| \Sigma^\star, \Pi^\dagger, \mathbf{p} \Rightarrow \mathbf{q}, A_1 \to B_1, \ldots, A_n \to B_n, \neg C_1, \ldots, \neg C_k \rangle$ and $\rho(\alpha) = push^J$, then $\phi^\star(\alpha 1) = \langle \zeta \| \Sigma, \Pi^\dagger, \mathbf{p}, A_1^\dagger, A_1 \Rightarrow_{\mathfrak{a}1} B_1, \neg C_1, \ldots, \neg C_k \rangle, \ldots, \langle \zeta \| \Sigma, \Pi^\dagger, \mathbf{p}, A_n^\dagger, A_n \Rightarrow_{\mathfrak{a}n} B_n, \neg C_1, \ldots, \neg C_k \rangle$*

- *if $\phi^\star(\alpha) = \langle \zeta \| \Gamma \Rightarrow_\mathfrak{a} \Delta \rangle$ and $\rho(\alpha) = pop$, then $\phi^\star(\alpha 1) = \langle \zeta \rangle$.*

Consider now the following open SJC-derivation and let us add labels to its sequents.

$$\dfrac{\dfrac{\dfrac{\dfrac{\dfrac{\langle\rangle}{q \Rightarrow p}\,pop}{q \Rightarrow p \,\|\, p \Rightarrow q}\,pop}{\Rightarrow q \rightarrow p,\ p \rightarrow q}\,push^J}{\Rightarrow (q \rightarrow p) \vee (p \rightarrow q)}\,R_\vee \qquad \dfrac{\dfrac{\dfrac{\dfrac{\langle\rangle}{q \Rightarrow_1 p}\,pop}{q \Rightarrow_1 p \,\|\, p \Rightarrow_2 q}\,pop}{\Rightarrow_\mathfrak{e} q \rightarrow p,\ p \rightarrow q}\,push^J}{\Rightarrow_\mathfrak{e} (q \rightarrow p) \vee (p \rightarrow q)}\,R_\vee$$

The labelled derivation can be transformed into the countermodel based on the tree to the left hand side and then into the countermodel based on the same tree but with a final point. The fact that the addition of the final point does not alter the truth values of the formulas at the various point of the initial tree is guaranteed by lemmas 3 and 5.

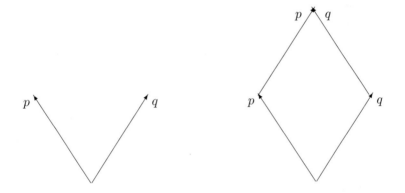

Consider the function λ that gives the label of *the last sequent of a sequent-list*: if $\phi^\star(\alpha) = \langle \zeta \| \Gamma \Rightarrow_\mathfrak{a} \Delta \rangle$, then $\lambda(\alpha) = \mathfrak{a}$.

Take an open branch $\zeta_1 \ldots \zeta_n$ of $\mathcal{D}$. For every $\mathfrak{a} = \lambda(\zeta_j)$, $1 \leq j \leq n$, we define

$$\mathcal{T}_\mathfrak{a} = \bigcup \{\Gamma : \zeta_j = \langle \xi \| \Gamma \Rightarrow_\mathfrak{a} \Delta \rangle \text{ for some } j, 1 \leq j \leq n\}$$
$$\mathcal{F}_\mathfrak{a} = \bigcup \{\Delta : \zeta_j = \langle \xi \| \Gamma \Rightarrow_\mathfrak{a} \Delta \rangle \text{ for some } j, 1 \leq j \leq n\}$$

$\mathcal{T}_\mathfrak{a}$ e $\mathcal{F}_\mathfrak{a}$ are *saturated*, i.e. they satisfy the following conditions:

1. if $(A \wedge B) \in \mathcal{T}_\mathfrak{a}$ then $A \in \mathcal{T}_\mathfrak{a}$ and $B \in \mathcal{T}_\mathfrak{a}$
2. if $(A \wedge B) \in \mathcal{F}_\mathfrak{a}$ then either $A \in \mathcal{F}_\mathfrak{a}$ or $B \in \mathcal{F}_\mathfrak{a}$
3. if $(A \vee B) \in \mathcal{T}_\mathfrak{a}$ then either $A \in \mathcal{T}_\mathfrak{a}$ or $B \in \mathcal{T}_\mathfrak{a}$
4. if $(A \vee B) \in \mathcal{F}_\mathfrak{a}$ then $A \in \mathcal{F}_\mathfrak{a}$ and $B \in \mathcal{F}_\mathfrak{a}$
5. if $(A \to B) \in \mathcal{T}_\mathfrak{a}$ then either $\neg A \in \mathcal{T}_\mathfrak{a}$ or $B \in \mathcal{T}_\mathfrak{a}$ or $(A \in \mathcal{F}_\mathfrak{a}$ and $\neg B \in \mathcal{F}_\mathfrak{a})$
6. if $(A \to B) \in \mathcal{F}_\mathfrak{a}$ and $A^\dagger \in \mathcal{T}_\mathfrak{a}$ then $B \in \mathcal{F}_\mathfrak{a}$
7.1 if $\neg(A \wedge B) \in \mathcal{T}_\mathfrak{a}$ then $\neg A \in \mathcal{T}_\mathfrak{a}$ or $\neg B \in \mathcal{T}_\mathfrak{a}$
7.2 if $\neg(A \vee B) \in \mathcal{T}_\mathfrak{a}$ then $\neg A \in \mathcal{T}_\mathfrak{a}$ and $\neg B \in \mathcal{T}_\mathfrak{a}$
7.3 if $\neg(A \to B) \in \mathcal{T}_\mathfrak{a}$ then $\neg B \in \mathcal{T}_\mathfrak{a}$ and $\neg A \in \mathcal{F}_\mathfrak{a}$
7.4 if $\neg\neg A \in \mathcal{T}_\mathfrak{a}$ then $\neg A \in \mathcal{F}_\mathfrak{a}$

Let us consider the model $\mathcal{M} = \langle W, \leq, I \rangle$, where $W = \{\lambda(\zeta_1), \ldots, \lambda(\zeta_n)\}$, $\leq$ is the usual order between lists of natural numbers (see § 2.2) and $I(p) = \{\mathfrak{b} \in W : p \in \mathcal{T}_\mathfrak{b}\}$. $\leq$ is a partial order and it holds that if $\mathfrak{a} \leq \mathfrak{b}$ and $\mathfrak{a} \in I(p)$ then $\mathfrak{b} \in I(p)$.

Lemma 4. *Let $\mathfrak{b} \in W$. For any formula E (signed or unsigned), (1) if $E \in \mathcal{T}_\mathfrak{b}$ then $\mathcal{M} \models_\mathfrak{b} E$; (2) if $E \in \mathcal{F}_\mathfrak{b}$ then $\mathcal{M} \not\models_\mathfrak{b} E$.*

Proof. By induction on the weighted length of E.

Definition 8. *The* weighted length *of a formula A, $wl(A)$, is so defined by induction on A.*

- $wl(p) = 0$
- $wl(A \wedge B) = wl(A \vee B) = wl(A \to B) = wl(A) + wl(B) + 2$
- $wl(\neg A) = wl(A) + 1$

If E is a propositional variable the lemma holds by the definition of $I(p)$. If E is $A \wedge B$ or $A \vee B$ the lemma holds because $\mathcal{T}_\mathfrak{b}$ is saturated and by the induction hypothesis.
Let $E = A \to B$.

- If $A \to B \in \mathcal{F}_\mathfrak{b}$ we distinguish two cases.

  - $A^\dagger \in \mathcal{T}_\mathfrak{b}$. Then $B \in \mathcal{F}_\mathfrak{b}$, so by induction hypothesis $\mathcal{M} \models_\mathfrak{b} A$ and $\mathcal{M} \not\models_\mathfrak{b} B$ therefore $\mathcal{M} \not\models_\mathfrak{b} A \to B$.
  - $A^\dagger \notin \mathcal{T}_\mathfrak{b}$. Then $A \to B$ is a principal formula of $push^J$. Therefore $A \in \mathcal{T}_{\mathfrak{b}j}$ and $B \in \mathcal{F}_{\mathfrak{b}j}$, for some $1 \leq j \leq m$. Since the branch is open, $\mathfrak{b}j = \lambda(\zeta_k)$, for some k, $1 \leq k \leq n$. By induction hypothesis $\mathcal{M} \models_{\mathfrak{b}j} A$ and $\mathcal{M} \not\models_{\mathfrak{b}j} B$, therefore $\mathcal{M} \not\models_\mathfrak{b} A \to B$.

- If $(A \to B) \in \mathcal{T}_\mathfrak{b}$, then for any $\mathfrak{b}' \in W$, $\mathfrak{b} \leq \mathfrak{b}'$, either $\neg A \in \mathcal{T}_{\mathfrak{b}'}$ or $B \in \mathcal{T}_{\mathfrak{b}'}$ or $((A \to B) \in \mathcal{T}_\mathfrak{b}, A \in \mathcal{F}_{\mathfrak{b}'}$ and $\neg B \in \mathcal{F}_{\mathfrak{b}'})$, therefore by induction hypothesis either $\mathcal{M} \models_{\mathfrak{b}'} \neg A$ or $\mathcal{M} \models_{\mathfrak{b}'} B$ or $\mathcal{M} \not\models_{\mathfrak{b}'} A$, hence $\mathcal{M} \models_\mathfrak{b} A \to B$.

Let $E = \neg A$.

- If $\neg A \in \mathcal{F}_{\mathfrak{b}}$ we distinguish two cases.
  - $A^\dagger \in \mathcal{T}_{\mathfrak{b}}$. Then by induction hypothesis $\mathcal{M} \models_{\mathfrak{b}} A$, therefore $\mathcal{M} \not\models_{\mathfrak{b}} \neg A$.
  - $A^\dagger \notin \mathcal{T}_{\mathfrak{b}}$. Then $\neg A$ is a principal formula of $push^J$. Therefore $A \in \mathcal{T}_{\mathfrak{b}j}$ for some $1 \leq j \leq m$. Since the branch is open, $\mathfrak{b}j = \lambda(\zeta_k)$, for some k, $1 \leq k \leq n$. By induction hypothesis $\mathcal{M} \models_{\mathfrak{b}j} A$, therefore $\mathcal{M} \not\models_{\mathfrak{b}} \neg A$.

- If $\neg A \in \mathcal{T}_{\mathfrak{b}}$ we distinguish five cases:
  - If $\neg p \in \mathcal{T}_{\mathfrak{b}}$, then for any $\mathfrak{b}' \in W$ (in particular those $\mathfrak{b}' \geq \mathfrak{b}$), $p \notin \mathcal{T}_{\mathfrak{b}'}$ because the branch is open, therefore for any $\mathfrak{b}'$, $\mathcal{M} \not\models_{\mathfrak{b}'} p$, so $\mathcal{M} \models_{\mathfrak{b}} \neg p$.
  - If $\neg(A \wedge B) \in \mathcal{T}_{\mathfrak{b}}$, then either $\neg A \in \mathcal{T}_{\mathfrak{b}}$ or $\neg B \in \mathcal{T}_{\mathfrak{b}}$, so by induction hypothesis either $\mathcal{M} \models_{\mathfrak{b}} \neg A$ or $\mathcal{M} \models_{\mathfrak{b}} \neg B$, then the lemma follows.
  - If $\neg(A \vee B) \in \mathcal{T}_{\mathfrak{b}}$, then $\neg A \in \mathcal{T}_{\mathfrak{b}}$ and $\neg B \in \mathcal{T}_{\mathfrak{b}}$, so by induction hypothesis either $\mathcal{M} \models_{\mathfrak{b}} \neg A$ and $\mathcal{M} \models_{\mathfrak{b}} \neg B$, then the lemma follows.
  - If $\neg\neg A \in \mathcal{T}_{\mathfrak{b}}$, then for any $\mathfrak{b}'$, $\mathfrak{b}' \geq \mathfrak{b}$, $\neg A \in \mathcal{F}_{\mathfrak{b}}$, so for any $\mathfrak{b}'$, $\mathfrak{b}' \geq \mathfrak{b}$, there is a $\mathfrak{c}$, $\mathfrak{c} \geq \mathfrak{b}'$, such that $A \in \mathcal{T}_{\mathfrak{c}}$. By induction hypothesis for any such $\mathfrak{b}'$ there is a $\mathfrak{c}$, $\mathfrak{c} \geq \mathfrak{b}'$, such that $\mathcal{M} \models_{\mathfrak{c}} A$, so $\mathcal{M} \models_{\mathfrak{b}} \neg\neg A$.
  - If $\neg(A \rightarrow B) \in \mathcal{T}_{\mathfrak{b}}$, then $\neg\neg A^\star \in \mathcal{T}_{\mathfrak{b}}$, $\neg B \in \mathcal{T}_{\mathfrak{b}}$ and $\neg A \in \mathcal{F}_{\mathfrak{b}}$, so for every $\mathfrak{b}'$, $\mathfrak{b}' \geq \mathfrak{b}$, there is a $\mathfrak{c}$, $\mathfrak{c} \geq \mathfrak{b}'$, such that $A \in \mathcal{T}_{\mathfrak{c}}$ and $B \in \mathcal{F}_{\mathfrak{c}}$, therefore by induction hypothesis $\mathcal{M} \models_{\mathfrak{c}} A$ and $\mathcal{M} \not\models_{\mathfrak{c}} B$, hence $\mathcal{M} \models_{\mathfrak{c}} \neg(A \rightarrow B)$.

The theorem follows.

Definition 9. *Given an open branch* $\zeta_1), \ldots, \zeta_n$, *let* $\Omega = \bigcup \{\Gamma : \zeta_j = \langle \xi \| \Gamma \Rightarrow \Delta \rangle$ *for some* $j, 1 \leq j \leq n\}$

Let us consider the model $\mathcal{M}'$ obtained by adding to $\mathcal{M}$ a final point ω, so $\mathcal{M}' = \langle W \cup \{\omega\}, \leq, I' \rangle$, where

- $\mathfrak{a} \leq \omega$ for all $\mathfrak{a} \in W$ and $\omega \leq \omega$.
- $\mathfrak{a} \in I'(p)$ iff $\mathfrak{a} \in I(p)$, for all $\mathfrak{a} \in W$.
- if for some $\mathfrak{a} \in W$, $p \in \mathcal{T}_{\mathfrak{a}}$, then $\omega \in I'(p)$.

I' is well defined because $\zeta_1 \ldots \zeta_n$ is an open branch.

Lemma 5. *For any formula* $A \in \Omega$ *and* $\mathfrak{c} \in W$,

1. *if* $\mathcal{M} \models_{\mathfrak{c}} A$ *then* $\mathcal{M}' \models_{\mathfrak{c}} A$ *and* $\mathcal{M}' \models_{\omega} A$
2. *if* $\mathcal{M} \not\models_{\mathfrak{c}} A$ *then* $\mathcal{M}' \not\models_{\mathfrak{c}} A$.

Proof. By induction on the weighted length of A.

- If $\mathcal{M} \models_{\mathfrak{c}} p$, then $\mathcal{M}' \models_{\mathfrak{c}} p$ and $\mathcal{M}' \models_{\omega} p$ by definition of I'.

- If $\mathcal{M} \not\models_{\mathfrak{c}} p$, then $\mathcal{M}' \not\models_{\mathfrak{c}} p$ by definition of I'.

- If $\mathcal{M} \models_{\mathfrak{c}} \neg p$, then $p \notin \Omega$ since the branch is open, therefore for all $\mathfrak{a} \in W$, $\mathcal{M}' \not\models_{\mathfrak{a}} p$, hence $\mathcal{M}' \not\models_{\omega} p$, consequently $\mathcal{M}' \models_{\mathfrak{c}} \neg p$ and $\mathcal{M}' \models_{\omega} \neg p$.

- If $\mathcal{M} \not\models_{\mathfrak{c}} \neg D$, where D is any wff, then there is a $\mathfrak{d} \in W$, $\mathfrak{c} \leq \mathfrak{d}$ such that $\mathcal{M} \models_{\mathfrak{d}} D$. Whence by induction hypothesis $\mathcal{M}' \models_{\mathfrak{d}} D$ and $\mathcal{M}' \models_{\omega} D$, therefore $\mathcal{M}' \not\models_{\mathfrak{c}} \neg D$.

- If $\mathcal{M} \models_{\mathfrak{c}} \neg(B \wedge C)$, then $\mathcal{M} \models_{\mathfrak{c}} \neg B$ or $\mathcal{M} \models_{\mathfrak{c}} \neg C$, so by induction hypothesis either ($\mathcal{M}' \models_{\mathfrak{c}} \neg B$ and $\mathcal{M}' \models_{\omega} \neg B$) or ($\mathcal{M}' \models_{\omega} \neg C$ and $\mathcal{M}' \models_{\omega} \neg C$), whence $\mathcal{M}' \models_{\mathfrak{c}} \neg(B \wedge C)$ and $\mathcal{M}' \models_{\omega} \neg(B \wedge C)$.

- If $\mathcal{M} \models_{\mathfrak{c}} \neg(B \vee C)$, then $\mathcal{M} \models_{\mathfrak{c}} \neg B$ and $\mathcal{M} \models_{\mathfrak{c}} \neg C$, so by induction hypothesis $\mathcal{M}' \models_{\mathfrak{c}} \neg B$, $\mathcal{M}' \models_{\mathfrak{c}} \neg C$, $\mathcal{M}' \models_{\omega} \neg B$ and $\mathcal{M}' \models_{\omega} \neg C$. Whence $\mathcal{M}' \models_{\mathfrak{c}} \neg(B \vee C)$ and $\mathcal{M}' \models_{\omega} \neg(B \vee C)$.

- If $\mathcal{M} \models_{\mathfrak{c}} \neg\neg B$, then $\mathcal{M} \models_{\mathfrak{d}} B$ for some $\mathfrak{d} \geq \mathfrak{c}$. By induction hypothesis $\mathcal{M}' \models_{\mathfrak{d}} B$ and $\mathcal{M}' \models_{\omega} B$, therefore $\mathcal{M}' \models_{\mathfrak{c}} \neg\neg B$ and $\mathcal{M}' \models_{\omega} \neg\neg B$.

- If $\mathcal{M} \models_{\mathfrak{c}} \neg(B \to C)$, then $\mathcal{M} \models_{\mathfrak{c}} \neg\neg B$ and $\mathcal{M} \models_{\mathfrak{c}} \neg C$, whence for some $\mathfrak{d} \geq \mathfrak{c}$, $\mathcal{M} \models_{\mathfrak{d}} B$. By induction hypothesis $\mathcal{M}' \models_{\mathfrak{d}} B$, $\mathcal{M}' \models_{\mathfrak{c}} \neg C$, $\mathcal{M}' \models_{\omega} B$ and $\mathcal{M}' \models_{\omega} \neg C$, therefore $\mathcal{M}' \models_{\mathfrak{c}} \neg(B \to C)$ and $\mathcal{M}' \models_{\omega} \neg(B \to C)$.

- If $\mathcal{M} \models_{\mathfrak{c}} (B \to C)$, then $\mathcal{M} \models_{\mathfrak{c}} \neg B$ or there is a $\mathfrak{d}$, $\mathfrak{d} \geq \mathfrak{c}$, such that $\mathcal{M} \models_{\mathfrak{d}} C$. Then by induction hypothesis either ($\mathcal{M}' \models_{\mathfrak{c}} \neg B$ and $\mathcal{M}' \models_{\omega} \neg B$) or ($\mathcal{M}' \models_{\mathfrak{d}} C$ and $\mathcal{M}' \models_{\omega} C$), therefore $\mathcal{M}' \models_{\mathfrak{c}} B \to C$ and $\mathcal{M}' \models_{\omega} B \to C$.

- If $\mathcal{M} \not\models_{\mathfrak{c}} (B \to C)$, then there is a $\mathfrak{d} \in W$, $\mathfrak{c} \leq \mathfrak{d}$, and $\mathcal{M} \models_{\mathfrak{d}} B$ and $\mathcal{M} \not\models_{\mathfrak{d}} C$. Whence by induction hypothesis $\mathcal{M}' \models_{\mathfrak{d}} B$ and $\mathcal{M}' \not\models_{\mathfrak{d}} C$, therefore $\mathcal{M}' \not\models_{\mathfrak{d}} B \to C$.

- If $\mathcal{M} \models_{\mathfrak{c}} (B \wedge C)$, then $\mathcal{M} \models_{\mathfrak{c}} B$ and $\mathcal{M} \models_{\mathfrak{c}} C$, so by induction hypothesis $\mathcal{M}' \models_{\mathfrak{c}} B$, $\mathcal{M}' \models_{\mathfrak{c}} C$, $\mathcal{M}' \models_{\omega} B$ and $\mathcal{M}' \models_{\omega} C$, whence $\mathcal{M}' \models_{\mathfrak{c}} (B \wedge C)$ and $\mathcal{M}' \models_{\omega} (B \wedge C)$.

- If $\mathcal{M} \not\models_{\mathfrak{c}} (B \wedge C)$, then $\mathcal{M} \not\models_{\mathfrak{c}} B$ or $\mathcal{M} \not\models_{\mathfrak{c}} C$, so by induction hypothesis $\mathcal{M}' \not\models_{\mathfrak{c}} B$ or $\mathcal{M}' \not\models_{\mathfrak{c}} C$, whence $\mathcal{M}' \not\models_{\mathfrak{c}} (B \wedge C)$.

- If $\mathcal{M} \models_{\mathfrak{c}} (B \vee C)$, then $\mathcal{M} \models_{\mathfrak{c}} B$ or $\mathcal{M} \models_{\mathfrak{c}} C$, so by induction hypothesis either ($\mathcal{M}' \models_{\mathfrak{c}} B$ and $\mathcal{M}' \models_{\omega} B$) or ($\mathcal{M}' \models_{\mathfrak{c}} C$ and $\mathcal{M}' \models_{\omega} C$), whence $\mathcal{M}' \models_{\mathfrak{c}} (B \vee C)$ and $\mathcal{M}' \models_{\omega} (B \vee C)$.

- If $\mathcal{M} \not\models_{\mathfrak{c}} (B \vee C)$, then $\mathcal{M} \not\models_{\mathfrak{c}} B$ and $\mathcal{M} \not\models_{\mathfrak{c}} C$, so by induction hypothesis $\mathcal{M}' \not\models_{\mathfrak{c}} B$ and $\mathcal{M}' \not\models_{\mathfrak{c}} C$, whence $\mathcal{M}' \not\models_{\mathfrak{c}} (B \vee C)$.

Theorem 1. *(Completeness of SJC) Given an open SJC-derivation $\mathcal{D}$ of a sequent-list $\|\Gamma \Rightarrow \Delta$ not containing signed formulas, there is a model $\mathcal{M}'$ based on a tree with a final point such that $\mathcal{M}' \not\models \bigwedge(\Gamma) \to \bigvee(\Delta)$.*

Theorem 2. *(Soundness of SJC) Given a SJC-proof $\mathcal{D}$ of a sequent-list $\|\Gamma \Rightarrow \Delta$ not containing signed formulas, there is a JG-proof of $\Gamma \Rightarrow \Delta$.*

Proof. By induction on the number n of the mutually closed leaf-sequents. $n = 0$. Proceed as for lemma 4.7 of [3]. Let $n > 0$.

First transformation. We transform $\mathcal{D}$ into a SJC-derivation $\mathcal{D}'$ with auxiliary formulas added to the antecedents of some of its sequents. Auxiliary formulas do not play any role whatsoever in the SJC-derivation $\mathcal{D}'$, and are double underlined, so $\mathcal{D}'$ is nothing but $\mathcal{D}$ with some double underlined formulas added to its sequents.

Consider a leaf-sequent $\Theta, p \Rightarrow \Phi$ and let $\Gamma, \neg p \Rightarrow \Delta$ be the pop-sequent mutually closed with $\Theta, p \Rightarrow \Phi$. Denote by $pop_{(\neg\neg p)}$ the occurrence of the *pop*-rule whose conclusion is $\zeta \| \Gamma, \neg p \Rightarrow \Delta$. (Analogously, if the leaf sequent is $\Theta, \neg p \Rightarrow \Phi$ and the pop-sequent is $\Gamma, p \Rightarrow \Delta$, denote the *pop*-rule by $pop_{(\neg p)}$.) Consider then the occurrence of the rule $push^J$ of maximum height that is below both $\Theta, p \Rightarrow \Phi$ and $\Gamma, \neg p \Rightarrow \Delta$. Denote such an occurrence by $push^J_{(\neg p, \neg\neg p)}$. Add $\underline{\underline{\neg p}}$ to the antecedent of $\Theta, p \Rightarrow \Phi$ and to the antecedents of all the sequents between $\Theta, p \Rightarrow \Phi$ and the premiss of $push^J_{(\neg p, \neg\neg p)}$, add $\underline{\underline{\neg\neg p}}$ to the antecedent of $\Gamma, \neg p \Rightarrow \Delta$ and to the antecedents of all the sequents between $\Gamma, \neg p \Rightarrow \Delta$ and the premiss of $push^J_{(\neg p, \neg\neg p)}$. $\mathcal{D}'$ is still a SJC-derivation since the double underlined formulas are not part of the derivation.

Let us show the transformation by way of an example. The SJC-proof below contains two mutually closed leaf-sequents and a closed leaf-sequent:

$$\dfrac{\dfrac{\dfrac{\dfrac{\dfrac{\dfrac{\dfrac{\dfrac{p, r \Rightarrow s \quad p, q \Rightarrow s}{p, r \vee q \Rightarrow s} L_\vee}{p, r \vee q \Rightarrow s \| p, \neg r \Rightarrow t} pop}{p \Rightarrow r \vee q \to s, \neg r \to t} push^J}{p \Rightarrow (r \vee q \to s) \vee (\neg r \to t)} R_\vee \quad \dfrac{\dfrac{p, \neg s \Rightarrow p}{p \Rightarrow \neg s \to p} push^J}{}}{p \Rightarrow [(r \vee q \to s) \vee (\neg r \to t)] \wedge (\neg s \to p)} R_\wedge}{p \Rightarrow [(r \vee q \to s) \vee (\neg r \to t)] \wedge (\neg s \to p) \| \neg q, z \Rightarrow} pop}{p \Rightarrow [(r \vee q \to s) \vee (\neg r \to t)] \wedge (\neg s \to p) \| \neg q \Rightarrow \neg z} R_\neg}{\Rightarrow p \to ((r \vee q \to s) \vee (\neg r \to t)) \wedge (\neg s \to p), \neg q \to \neg z} push^J$$

After the first transformation, it becomes

$$\dfrac{\dfrac{\underline{\neg r}, p, r \Rightarrow s \qquad \underline{\neg q}, p, q \Rightarrow s}{\underline{\neg r}, \underline{\neg q}, p, (r \vee q) \Rightarrow s} L_\vee}{\dfrac{\underline{\neg r}, \underline{\neg q}, p, (r \vee q) \Rightarrow s \parallel \underline{\neg\neg r}, p, \neg r \Rightarrow t}{\dfrac{\neg q, p \Rightarrow (r \vee q \to s), (\neg r \to t)}{\neg q, p \Rightarrow (r \vee q \to s) \vee (\neg r \to t)} R_\vee} push^J_{(\neg r, \neg\neg r)}} pop_{(\neg\neg r)}$$

$$\dfrac{p, \neg s \Rightarrow p}{p \Rightarrow (\neg s \to p)} push^J$$

$$\dfrac{\neg q, p \Rightarrow [(r \vee q \to s) \vee (\neg r \to t)] \wedge (\neg s \to p)}{\dfrac{\neg q, p \Rightarrow [(r \vee q \to s) \vee (\neg r \to t)] \wedge (\neg s \to p) \parallel \underline{\neg\neg q}, \neg q, z \Rightarrow}{\dfrac{\neg q, p \Rightarrow [(r \vee q \to s) \vee (\neg r \to t)] \wedge (\neg s \to p) \parallel \underline{\neg\neg q}, \neg q \Rightarrow \neg z}{\Rightarrow p \to [(r \vee q \to s) \vee (\neg r \to t)] \wedge (\neg s \to p), (\neg q \to \neg z)} push^J_{(\neg q, \neg\neg q)}} R_\neg} pop_{(\neg\neg q)}$$

Second transformation. Now the double underlined formulas become simply underlined and are treated on a par with the non underlined ones; we keep them underlined only for the sake of clarity. Therefore all the leaf-sequents are axioms and also the pop-sequents containing underlined formulas are axioms. Notice that the sequent at the root contains no underlined formulas. We transform the SJC-proof $\mathcal{D}'$ into a JG-proof $\mathcal{D}^\star$.

By induction on the height of the SJC-proof $\mathcal{D}'$ we show that
(1) every sequent-list contains at least a sequent provable in JG and
(2) all the sequents containing underlined formulas are provable in JG.
Let $\Gamma \Rightarrow \Delta$ be the sequent at the root of $\mathcal{D}'$.

1. The height of $\mathcal{D}'$ is 1. Then $\Gamma \Rightarrow \Delta$ is either identity or non-contradiction, so it is an axiom of JG. No underlined formulas can possibly occur in $\Gamma \Rightarrow \Delta$.
2. The height of $\mathcal{D}'$ is greater than 1. Consider the last rule.
   a) The last rule is pop. By induction hypothesis the premiss of the pop-rule satisfies conditions (1) and (2), so it does the conclusion. Notice that the last sequent of the conclusion contains no underlined formulas.
   b) The last rule is $pop_{(\neg p)}$ or $pop_{(\neg\neg p)}$. By induction hypothesis the premiss of the pop-rule satisfies conditions (1) and (2). Moreover the last sequent of the conclusion of $pop_{(\neg p)}$ ($pop_{(\neg p, \neg\neg p)}$) contains an underlined formula and it is an axiom of JG. So conditions (1) and (2) are again satisfied.
   c) The last rule is one of $L_\wedge, R_\vee, L_{\neg\vee}, L_{\neg\neg}, L_{\neg\to}, afortiori_\to, afortiori_\neg, R^J_\to$. Let $\Gamma' \Rightarrow \Delta'$ be a JG-provable sequent of the premiss ξ'.
   - if $\Gamma' \Rightarrow \Delta'$ is the last sequent of ξ', then also the last sequent of the conclusion ξ is JG-provable by the application of the JG-rule with the same name.
   - if $\Gamma' \Rightarrow \Delta'$ is not the last sequent of ξ', then ξ contains the same JG-provable sequents as ξ', since SJC-rules apply only to the last sequent.

d) The rule is one of $R_\wedge$, $L_\vee$, $L_\rightarrow^J$, $L_{\neg\wedge}$, $L.\neg\vee$. Analogously to the preceding case.

e) The last rule is $push^J$:

$$\frac{\varsigma\|\Sigma, \Pi^\dagger, \mathbf{p}, A_1^\dagger, A_1 \Rightarrow B_1, \neg C_1, \ldots, \neg C_k \| \ldots \ldots \| \Sigma, \Pi^\dagger, \mathbf{p}, A_n^\dagger, A_n \Rightarrow B_n, \neg C_1, \ldots, \neg C_k \Rightarrow}{\varsigma\|\Sigma^\star, \Pi^\dagger, \mathbf{p} \Rightarrow \mathbf{q}, A_1 \rightarrow B_1, \ldots, A_n \rightarrow B_n, \neg C_1, \ldots, \neg C_k} \; push^J$$

By induction hypothesis the premiss ξ' contains at least a JG-provable sequent, say $\Gamma' \Rightarrow \Delta'$.

- if $\Gamma' \Rightarrow \Delta'$ is a principal sequent of $push^J$ (i.e. it is one of the last n sequents of ξ), then we get a JG-proof of the conclusion of $push^J$ by an application of $R_\rightarrow$ to $\Gamma' \Rightarrow \Delta'$. The conclusion trivially satisfies conditions (1) and (2).
- if $\Gamma' \Rightarrow \Delta'$ is not a principal sequent of $push^J$, then $\Gamma' \Rightarrow \Delta'$ occurs also in the conclusion of $push^J$ and so conditions (1) and (2) are satisfied.

f) The last rule is $push^J_{(\neg p, \neg\neg p)}$. So the situation in $\mathcal{D}'$ is as follows

$$\frac{\xi\|\ldots\|\Gamma, \neg p, A_i \Rightarrow B_i\|\ldots\|\Gamma, \underline{\neg\neg p}, A_j \Rightarrow B_j\|\ldots}{\xi\|\ldots\|\Gamma \Rightarrow A_1 \rightarrow B_1 \ldots A_n \rightarrow B_n} \; push^J_{(\neg p, \neg\neg p)}$$

Since sequents containing underlined formula are provable in JG, this can be transformed into

$$\cfrac{\cfrac{\vdots}{\cfrac{\Gamma, \neg p, A_i \Rightarrow B_i}{\Gamma, \neg p \Rightarrow A_1 \rightarrow B_1 \ldots A_n \rightarrow B_n} R_\rightarrow} \qquad \cfrac{\cfrac{\vdots}{\cfrac{\Gamma, \neg\neg p \Rightarrow A_j \Rightarrow B_j}{\Gamma, \neg\neg p \Rightarrow A_1 \rightarrow B_1 \ldots A_n \rightarrow B_n} R_\rightarrow}}{\cfrac{\neg p \vee \neg\neg p, \Gamma \Rightarrow A_1 \rightarrow B_1 \ldots A_n \rightarrow B_n}{\Gamma \Rightarrow A_1 \rightarrow B_1 \ldots A_n \rightarrow B_n}} L_\vee}{} cut_{wem}$$

So the theorem is proved. Here is the JG-proof od the example we started with.

$$\dfrac{\dfrac{\dfrac{\dfrac{\dfrac{\dfrac{\dfrac{\neg r, p, r \Rightarrow s \qquad \neg q, p, q \Rightarrow s}{\neg r, \neg q, p, (r \vee q) \Rightarrow s} L_\vee}{\neg r, \neg q, p \Rightarrow (r \vee q \to s)} R_\to \qquad \dfrac{\neg\neg r, p, \neg r \Rightarrow t}{\neg\neg r, p \Rightarrow (\neg r \to t)} R_\to}{\neg r \vee \neg\neg r, \neg q, p \Rightarrow (r \vee q \to s), (\neg r \to t)} L_\vee}{\dfrac{\neg q, p \Rightarrow (r \vee q \to s), (\neg r \to t)}{\neg q, p \Rightarrow (r \vee q \to s) \vee (\neg r \to t)} R_\vee \qquad \dfrac{p, \neg s \Rightarrow p}{p \Rightarrow (\neg s \to p)} R_\to}{\dfrac{\neg q, p \Rightarrow [(r \vee q \to s) \vee (\neg r \to t)] \wedge (\neg s \to p)}{\neg q \Rightarrow p \to [(r \vee q \to s) \vee (\neg r \to t)] \wedge (\neg s \to p)} R_\to} R_\wedge \qquad \dfrac{\dfrac{\neg\neg q, \neg q, z \Rightarrow}{\neg\neg q, \neg q \Rightarrow \neg z} R_\neg}{\neg\neg q \Rightarrow (\neg q \to \neg z)} R_\to}{\dfrac{\neg q \vee \neg\neg q \Rightarrow p \to [(r \vee q \to s) \vee (\neg r \to t)] \wedge (\neg s \to p), (\neg q \to \neg z)}{\Rightarrow p \to [(r \vee q \to s) \vee (\neg r \to t)] \wedge (\neg s \to p), (\neg q \to \neg z)} cut_{wem}} L_\vee}{\Rightarrow p \to [(r \vee q \to s) \vee (\neg r \to t)] \wedge (\neg s \to p), (\neg q \to \neg z)} cut_{wem}$$

Acknowledgements

My deep gratitude to Agata Ciabattoni for having discussed with me over the internet various topics related to this paper.

References

1. G. Corsi: Semantic Trees for Dummett's Logic LC. *Studia Logica* XLV (1986) pp 99-206
2. G. Corsi: A cut-free calculus for Dummett's LC quantified. *Zeitschr. f. math. Logik und Grundlagen d. Math.* **35** (1989) pp 289-301
3. G. Corsi, G. Tassi: Intuitionistic logic freed of all metarules. *Journal of Symbolic Logic* **72** (2007) pp 1204-1218
4. T. Hosoi: Pseudo two-valued evaluation method for intermediate logics. *Studia Logica* XLV (1986) pp 3-8
5. T. Hosoi: Gentzen-type Formulation of the Propositional Logic LQ. *Studia Logica* XLVII (1988) pp 4-48
6. V. A. Jankov: The calculus of the weak "law of excluded middle". *Math. USSR Izvestija* **2** (1968) pp 997-1004

Automated Search for Gödel's Proofs

Wilfried Sieg and Clinton Field

Department of Philosophy, Carnegie Mellon University, Pittsburgh (US)
ws15@andrew.cmu.edu

We present *strategies* and *heuristics* underlying a search procedure that finds proofs for Gödel's incompleteness theorems at an *abstract axiomatic level*. As *axioms* we take for granted the representability and derivability conditions for the central syntactic notions as well as the diagonal lemma for constructing self-referential sentences. The *strategies* are logical ones and have been developed to search for natural deduction proofs in classical first-order logic. The *heuristics* are mostly of a very general mathematical character and are concerned with the goal-directed use of definitions and lemmata. When they are specific to the meta-mathematical context, these heuristics allow us, for example, to move between the object-and meta-theory. Instead of viewing this work as high-level proof search, it can be regarded as a first step in a proof-planning framework: the next refining steps would consist in verifying the axiomatically given conditions. Comparisons with the literature are detailed in Section 4. (The general mathematical heuristics are indeed general: in Appendix B we show that they, together with two simple algebraic facts and the logical strategies, suffice to find a proof of "$\sqrt{2}$ is not rational".)

1 Background

In a genuinely experimental spirit, we extended the *intercalation method for proof search* from pure first-order logic to parts of mathematics by interweaving general logical strategies with specific mathematical heuristics. The guiding question for our investigation was: What is needed, in addition to purely logical considerations, for finding proofs of significant theorems in a fully automated way? We answer the question for Gödel's incompleteness theorems [23]. When proved at an *abstract axiomatic level* they lend themselves naturally to such an investigation; they have intricate, yet not overwhelmingly difficult proofs, and they are obviously significant. During the academic years

1975–77, the first author had taken steps towards establishing them interactively. That work was done for a computer-based course on *Elementary Proof Theory*; a detailed report was given in [18] and a brief summary in [22].

Elementary Proof Theory presented the incompleteness theorems for ZF*, that is Zermelo–Fraenkel set theory without the axiom of infinity; see, for example, [7]. Its major innovation consisted in carrying out the meta-mathematical work in a formal theory of binary trees and elementary inductive definitions, called TEM.[1] Without the detour of their arithmetization, the inductively given syntactic notions were shown to be representable in ZF*; the diagonal lemma was established and the proof of the Hilbert–Bernays derivability conditions, central for the second theorem, was sketched. Within that high-level framework the standard material on the incompleteness theorems is compact and the proofs are direct. It was natural to ask, whether the proofs can be found via an appropriate extension of the intercalation method.

The arguments for the incompleteness theorems are carried out in the first-order theory TEM: instead of viewing syntactic objects as (having been coded as) natural numbers, we consider them as finitely branching trees; instead of defining syntactic notions recursively, we specify them by elementary inductive definitions, briefly, by eid's. In the language of TEM we have the constant S for the empty tree and the function symbol [,] for the binary operation of building a tree from two given ones. We use X, Y, Z—possibly with indices—as variables ranging over binary trees. The axioms for S and [,] are formulated in analogy to those of Dedekind–Peano arithmetic for zero and successor. The further axioms of TEM include the induction principle for binary trees, and closure and minimality conditions for the eid's. Instead of discussing these axioms in generality—the details do not matter for the current project—we specify some definitions that are actually needed to characterize the formal theory for which the incompleteness theorems are to be proved.

The theory to be considered is ZF*, Zermelo and Fraenkel's theory of sets without the axiom of infinity. The details of its axiomatic formulation do not matter either for the current project. Let us assume that it is formulated in a first-order language with x, y, z—possibly with indices—as variables ranging over sets. To indicate the general character of eid's we specify the generating clauses of the familiar notion of a formula (taking for granted the concepts of atomic formula and of variable); @ stands for any binary sentential connective, Q for the existential or universal quantifier:

If X is an atomic formula, X is a FORMULA;
If X is a FORMULA, $[\neg, X]$ is a FORMULA;
If X is a FORMULA and Y is a FORMULA, $[@, [X, Y]]$ is a FORMULA;
If X is a variable and Y is a FORMULA, $[[Q, X], Y]$ is a FORMULA.

---

[1] TEM abbreviates **T**heory for **E**lementary **M**eta-Mathematics. Feferman systematically investigates in his papers [10] and [11] the use of "finitary inductive" definitions in meta-mathematics.

We write also "FORM(X)" for "X is a FORMULA". TEM contains for such eid's a *closure* and a *minimality* principle. The first principle asserts that FORM is closed under the above clauses and is expressed by

FOR ALL X (if $\mathfrak{A}($FORM$, X)$ then FORM(X)).[2]

The minimality principle claims that FORM is the smallest such class. This is approximated in first-order logic by the usual principle of induction for formulas:

If FOR ALL X (if $\mathfrak{A}(P, X)$ then $P(X)$)
 then FOR ALL X (if FORM(X) then $P(X)$).

Formulas are binary trees built up from the empty tree using pairing. In a similar way one can generate inductively the relation X *is a proof of Y from assumptions* $Z_1, \ldots, Z_n$ or from a(n inductively generated) class of axioms; if X is a proof of Y using axioms of ZF*, this relation is denoted by PROOF(X, Y). To indicate that there is a ZF*-proof for Y, we write ZF$^* \vdash (Y)$, ZF$^* \vdash Y$ or THEO(Y).

Using the constant $\emptyset$ and the set-theoretic pairing operation $\langle,\rangle$ one can build up terms in the language of ZF* whose parse trees are isomorphic to the binary trees; they are used as names for the meta-mathematical trees in the same way as numerals in Dedekind–Peano arithmetic are used as names for natural numbers. With every meta-mathematical tree we can directly associate its set-theoretic name or *code*: CODE$(S) = \emptyset$ and CODE$([X,Y]) = \langle$CODE$(X),$ CODE$(Y)\rangle$. We also write $\lfloor X \rfloor$ for CODE(X) or indicate it by $\boldsymbol{X}$. This is the apparatus needed to formulate the *representability conditions* for the syntactic notions. We give them paradigmatically for FORM and PROOF:

If FORM(X) then ZF$^* \vdash$ form$(\boldsymbol{X})$, and
If NOT FORM(X) then ZF$^* \vdash \neg$form$(\boldsymbol{X})$.

"form" is a formula in the language of set theory for which these conditions are provable in TEM. Similarly, there is a formula "proof" in the language of ZF* that represents the proof relation PROOF:

If PROOF(X, Y) then ZF$^* \vdash$ proof$(\boldsymbol{X}, \boldsymbol{Y})$, and
If NOT PROOF(X, Y) then ZF$^* \vdash \neg$proof$(\boldsymbol{X}, \boldsymbol{Y})$.

Using the first representability condition for PROOF one can establish:

If THEO(Y) then ZF$^* \vdash$ theo$(\boldsymbol{Y})$,

---

[2] $\mathfrak{A}(P, X)$ is obtained from the generating clauses; it is the disjunction of the following TEM-formulas: (i) X is atomic; (ii) $(X)_0$ is $\neg$ and $P((X)_1)$; (iii) $(X)_0$ is @ and $P(((X)_1)_0)$ and $P(((X)_1)_1)$; (iv) $((X)_0)_0$ is Q and $((X)_0)_1$ is a variable and $P((X)_1)$. P can be viewed as either a meta-variable over TEM-formulas or as a free second-order variable; under the second reading we have an appropriate substitution rule in the logical calculus for TEM.

where theo(y) abbreviates $(\exists x)$ proof(x, y) Finally, we will use the *Self-reference Lemma* (or *Diagonal Lemma*) in the form: if F is a formula in the language of set theory (with one free variable), then there is a sentence D_F in that very language such that ZF* proves $(D_F \leftrightarrow F(\boldsymbol{D_F}))$. Applied to the formula ¬theo(y), the self-reference lemma yields the Gödel sentence G that expresses its own unprovability, i.e., ZF* proves $(G \leftrightarrow \neg\text{theo}(\boldsymbol{G}))$.

With this systematic background it is not difficult to prove that G is not provable in ZF* assuming, of course, that ZF* is consistent. So let us assume—in order to obtain a contradiction—that ZF* proves G; then, by the diagonal lemma concerning G, ZF* proves ¬theo($\boldsymbol{G}$). On the other hand, by the (semi-) representability of THEO, we can infer from the fact that ZF* proves G, that ZF* establishes theo($\boldsymbol{G}$). Thus, ZF* proves both ¬theo($\boldsymbol{G}$) and theo($\boldsymbol{G}$), and we have obtained a contradiction! The independence of G requires a proof that ¬G is not provable either; for that a stronger assumption concerning ZF*, stronger than mere consistency, has to be made. Gödel used for that purpose the notion of ω-consistency; the corresponding concept for the context of our meta-mathematical set-up is τ-consistency, thinking of τ as the class of (sets denoted by codes for) binary trees. ZF* is τ-consistent is defined by the condition: there is no formula $F(y)$ such that ZF* proves $(\exists y)(\tau(y) \& F(y))$ and also ¬F($\boldsymbol{Y}$) for all Y; or equivalently, for all formulas $F(y)$, if ZF* proves ¬F($\boldsymbol{Y}$) for all Y, then ZF* does not prove $(\exists y)(\tau(y) \& F(y))$.

Assuming that ZF* is τ-consistent, we show now that ZF* does not prove the negation of the Gödel sentence G. By what we established already (and the fact that τ-consistency implies ordinary consistency) we know that

FOR ALL X: NOT PROOF(X, G);

the representability of PROOF implies

FOR ALL X: ZF$^* \vdash \neg$proof($\boldsymbol{X}, \boldsymbol{G}$).

But then the τ-consistency of ZF* ensures

NOT ZF$^* \vdash (\exists y)$ proof($y, \boldsymbol{G}$).

As the formula $(\exists y)$ proof($y, \boldsymbol{G}$) is abbreviated by theo($\boldsymbol{G}$), we can use the self-reference lemma for G to infer that this formula is in ZF* provably equivalent to ¬G. Thus, NOT ZF$^* \vdash (\neg G)$, and the independence of G from ZF* has been established.

Given the axiomatic context provided by the representability of PROOF and THEO and the self-reference lemma applied to ¬theo(y), the proofs are direct, yet intricate. To take a first step towards describing the search algorithm that finds proofs of these and related theorems, we present briefly the basic ideas underlying the intercalation method for classical logic; for the theoretical underpinnings we refer to Sieg [19], Sieg and Byrnes [20] and Byrnes [6]. We should emphasize at this point that, in our view, logical formality per se does not facilitate the finding of proofs. However, logic within a natural deduction framework does help to bridge the gap between assumptions and conclusions by suggesting very rough structures for arguments, i.e., *logical structures* that depend solely on the syntactic form of assumptions and goals. This role of logic, though modest, is the crucial starting-point for moving up to subject-specific considerations that support a theorem. In the case study at hand we will show, how far these logical considerations go, and how they can be extended quite naturally by the *leading mathematical ideas* underlying Gödel's proofs.

2 Intercalation: broad strategies & special heuristics

The intercalation method is a proof search procedure that is goal-directed and guided by the possibly expanding syntactic context of the problem at hand. In first-order logic it is a complete procedure and a basis for broad logical strategies. The fundamental idea is straightforward. In order to bridge the gap between premises $A_1, \ldots, A_n$ and a goal B, one applies *systematically* the rules of the natural deduction calculus, i.e., the elimination rules are applied only from "above", whereas the introduction rules are inverted and applied from "below". Such systematic applications of the rules generate a search space that either contains a proof of B from the assumptions $A_1, \ldots, A_n$ or provides a semantic counterexample to the claim that B is a logical consequence of $A_1, \ldots, A_n$—tertium non datur; in addition, proofs contained in the search space are necessarily normal. The argument for this sharpened completeness theorem provides a method for searching directly for normal proofs; indeed, it yields also a semantic argument for normal form theorems in natural deduction. Such arguments concerning classical first-order logic were first given in [19], later also for intuitionistic logic and some modal logics in collaboration with Cittadini in [21].

Normal proofs satisfy a similar *subformula property* as cut-free derivations in the sequent calculus. That, of course, allows a restriction of the systematic search and is basic for broad strategies underlying our proof search: (i) extracting B via elimination rules—if B is a strictly positive subformula of an assumption, (ii) sub-goaling via the appropriate inverted introduction rule—if B is a logically complex formula, (iii) refuting B via the elimination rule for negation—if an appropriate pair of contradictory formulas is available.[3] In the latter case there must be a negation that is a strictly positive subformula of an assumption. It is evident that direct proof search is strongly and naturally constrained by the syntactic context of the problem, as only particular subformulas can be intercalated between assumptions and goals.

With these logical strategies in the background let us return to the proof of the first part of the first incompleteness theorem and examine, how the

---

[3] This condition was modified for the republication. The old formulation was "(iii) refuting B via the rules for negation—if B is a negation or an atomic formula and if an appropriate pair of contradictory formulas is available". Negated formulas are actually treated under (ii); the restriction to atomic formulas is too restrictive.

intercalation method might find it with "a little help" (when pure logic is unable to proceed any further). So we begin with the goal NOT (ZF* ⊢ (G)) and the premise ZF*CONS. We also have a definition and a lemma available, namely, the definition

ZF*CONS IFF NOT [ZF* ⊢ (G) AND ZF* ⊢ (¬G)]

and the consequence of the diagonal lemma for ¬theo(x), i.e.,

ZF* ⊢ (G ↔ ¬theo($\boldsymbol{G}$)).[4]

The goal cannot be extracted from the premises. Thus, the algorithm proceeds indirectly with the assumption ZF* ⊢ (G) and needs a pair of contradictory formulas as new goals. However, no negation occurs as a strictly positive subformula of the premise. As there is a negation in the definition of the premise, we use it and the premise to infer

NOT [ZF* ⊢ (G) AND ZF* ⊢ (¬G)].

This negation is one element of a contradictory pair, and the algorithm attempts to prove [ZF* ⊢ (G) AND ZF* ⊢ (¬G)]. This formula cannot be extracted: even though it is a subformula of a premise, it is not a strictly positive one. So the algorithm inverts the formula and attempts to prove the new goals ZF* ⊢ (G) and ZF* ⊢ (¬G). The former goal is already an assumption of the indirect proof, so we examine the latter goal.

It is here that we make the first significant change to the proof search procedure. ZF* ⊢ (¬G) cannot be extracted, but as an existential formula it can be inverted. Instead of searching for a term in the language of TEM describing a ZF*-proof of ¬G, the search proceeds "inside" ZF*. The claim ZF* ⊢ (¬G) can be justified, after all, by the presentation of a proof of ¬G within ZF*. The procedure tries now to find a ZF*-proof for the goal ¬G. As the formula ¬G cannot be extracted, indirect proof is applied to ¬G: assume G and find a contradictory pair. There is no negation immediately available in the premises, except through the diagonal lemma for G. Note that this lemma is formulated within TEM as a provability claim for ZF* and should be available for any ZF*-proof. In general, when attempting an extraction or looking for contradictory pairs within a ZF*-proof, strictly positive subformulas of ZF*-formulas A must be considered, where ZF* ⊢ (A) occurs as a strictly positive subformula of a premise or available assumption in TEM. So, the diagonal lemma makes available the formula ¬theo($\boldsymbol{G}$), which is used to construct the contradictory pair. This leaves theo($\boldsymbol{G}$) as a new goal, which cannot be extracted. The regular proof search procedure would attempt an inversion. But here an additional step can be considered, since theo is a semi-representable relation: we can justify theo($\boldsymbol{G}$) by establishing ZF* ⊢ (G) in TEM. ZF* ⊢ (G) is an assumption in TEM, so the proof is complete.

---

[4] We could have chosen one of the more general formulations of consistency, for example, NOT (EXISTS X)(ZF* ⊢ (X) AND ZF* ⊢ (¬X)). The quantificational search in the SH-expansion (see [20]) would find the appropriate instance quickly.

The expanded version of the proof search algorithm, which results from the careful examination of the above proof, interweaves mathematical and purely logical considerations in an intercalating and goal-directed manner. It has the following main steps:

Extraction

If the goal is in TEM, then extraction functions as described above for first-order logic. If the goal is in ZF*, then the set of formulas available for extraction is expanded by those formulas A, for which the claim ZF* ⊢ (A) is extractable in TEM and the goal is extractable from A. That is the inference Prov E, which is used to turn A into a part of the ZF*-proof.

Inversion

For the standard connectives inversion is applied as discussed earlier. There are two additional cases where "inversion" is applied. The first case occurs, when the goal in TEM is a statement of the form ZF* ⊢ (A). Here the algorithm tries to find a proof of A in ZF*; that is the inversion of the inference Prov I.[5] In the second case, when the goal is a formula like $[\neg]$ rel(X) in ZF*, and when the relation REL is represented by rel, the procedure tries to prove [NOT] REL(X) in TEM, after having explored indirect strategies in ZF*. For semi-representable relations such as ZF* ⊢ (X), this step is obviously not applied to the negation ¬ rel(X) in ZF*.

Extended extraction and inversion ("Meaning of premises and goals")

Definitional and other mathematical equivalences are used to obtain either a new available formula from which the current goal is extractable or to get an equivalent statement as a new goal. This we would like to do relative to a developing background theory; currently, we just add the definitions and lemmata explicitly to the list of premises.

Indirect strategies are pursued in the same way as in pure first-order logic, with one exception: the set of contradictory pairs for indirect proofs in ZF* is expanded by pairs whose negations are strictly positive subformulas of A in case ZF* ⊢ (A) (and this TEM-statement is itself extractable from an available TEM-claim).

This completes the informal description of the algorithm that searches for statements surrounding the first incompleteness theorem. The extensions of extraction and inversion mentioned have a very general mathematical character, whereas the extensions via ProvE and ProvI express most directly meta-mathematical content. The former rule reflects, in part, that theorems can be appealed to in proofs, and the latter rule expresses that the search mechanism provides syntactically correct object theoretic proofs.

---

[5] If the goal is of the form ZF* ⊢ ($[\neg]$ rel(X)), the algorithm tries first to prove [NOT] REL(X) directly.

The extended search procedure evolved out of a probing analysis of the standard proofs for the first incompleteness theorem and incorporates what we take to be the *leading mathematical ideas* for this part of meta-mathematics. It finds proofs not only for the first and second incompleteness theorems (after incorporating the derivability conditions), but also for a broader range of theorems and lemmata in this general area; cf. Appendix A for a proof of Löb's Theorem and Appendix D for two further examples. Even without the specifically meta-mathematical steps the algorithm is of real mathematical interest, as it discovers the structure of the proof for the irrationality of the square root of 2; see Appendix B.

3 Machine proofs & new heuristics

We present now the proofs of the first and second incompleteness theorem and start out by explaining the format of proofs. Proofs are presented in a modified Fitch-style format, which can be given using only plain text; cf. [12].[6] A line of dashes sets off the assumptions themselves. To distinguish the parts of the proof which occur in TEM and those which are embedded ZF^*-proofs, we mark every line in the object language with a star. Note that ZF^*-proofs retain the scope indications from the meta-language, and appeals to representability will use all available TEM-assumptions.

The rules include the standard natural deduction rules. For example, conjunction introduction has the name "And I", and the left and right-hand versions of conjunction elimination are named "And E L" and "And E R" respectively. To these basic rules we add special rule names for every heuristically applied theorem or lemma. "Rep" names the rule for representable or semi-representable relations, where the premise is a representable relation in TEM and the conclusion the corresponding relation in ZF^*. "Prov E" and "Prov I" indicate provability elimination and introduction.

We present first the machine proof of non-provability of the Gödel sentence G, assuming that ZF^* is consistent. In addition, the machine uses an instance of the diagonal lemma $ZF^* \vdash (G \leftrightarrow \neg(\text{theo}(G)))$ and the definition of consistency, ZF^* CONS IFF NOT $(ZF^* \vdash (G)$ AND $ZF^* \vdash (\neg(G)))$.

---

[6] Dawn McLaughlin modified the presentation of proofs in such a way that the next sentence in the original publication could be dropped. That sentence was: "We show the scope of assumptions by inserting bars between the number and formula on each line, with nested assumptions being noted by alternating bars and exclamation points."

Proof. [7]

1. $\mathsf{ZF}^* \vdash (G \leftrightarrow \neg(\mathsf{theo}(\boldsymbol{G})))$ Premise
2. $\mathsf{ZF}^*\mathsf{CONS}$ Premise
3. $\mathsf{ZF}^*\mathsf{CONS}$ IFF NOT $(\mathsf{ZF}^* \vdash (G)$ AND $\mathsf{ZF}^* \vdash (\neg(G)))$ Premise
4. | $\mathsf{ZF}^* \vdash (G)$ Assumption
* 5. | | G Assumption
* 6. | | $\mathsf{theo}(\boldsymbol{G})$ Rep 4
* 7. | | $(G \leftrightarrow \neg(\mathsf{theo}(\boldsymbol{G})))$ Prov E 1
* 8. | | $\neg(\mathsf{theo}(\boldsymbol{G}))$ Iff E R 7, 5
* 9. | | $\neg(G)$ Not I 5, 6, 8
10. | $\mathsf{ZF}^* \vdash (\neg(G))$ Prov I 9
11. | $\mathsf{ZF}^* \vdash (G)$ AND $\mathsf{ZF}^* \vdash (\neg(G))$ And I 4, 10
12. | NOT $(\mathsf{ZF}^* \vdash (G)$ AND $\mathsf{ZF}^* \vdash (\neg(G)))$ Iff E R 3, 2
13. NOT $(\mathsf{ZF}^* \vdash (G))$ Not I 4, 11, 12   □

To prove the independence of G we have also to establish the non-provability of $\neg G$. As remarked earlier, that requires the stronger hypothesis of τ-consistency. Here are the premises for the non-provability of $\neg G$:

- the diagonal lemma $\mathsf{ZF}^* \vdash (G \leftrightarrow \neg(\mathsf{theo}(\boldsymbol{G})))$,
- $\mathsf{ZF}^*_\tau \mathsf{CONS}$,
- $\mathsf{ZF}^*_\tau \mathsf{CONS}$ IMPLIES [(FOR ALL X) $(\mathsf{ZF}^* \vdash (\neg(\mathsf{proof}(\boldsymbol{X}, \boldsymbol{G}))))$ IMPLIES NOT $(\mathsf{ZF}^* \vdash (\mathsf{theo}(\boldsymbol{G}))))$],
- $\mathsf{ZF}^*_\tau \mathsf{CONS}$ IMPLIES $\mathsf{ZF}^* \mathsf{CONS}$,
- and a reformulation of what was established above, namely
  $\mathsf{ZF}^* \mathsf{CONS}$ IMPLIES (FOR ALL X)(NOT (PROOF$(X, G)))$.

---

[7] When following this argument and all the other machine proofs, the reader should keep in mind the intercalation strategies for bridging the gap between assumptions and goals. After all, they motivate the steps in the arguments.

Proof.

1.	$ZF^* \vdash (G \leftrightarrow \neg(theo(\boldsymbol{G})))$	Premise
2.	$ZF^*_\tau CONS$	Premise
3.	$ZF^*_\tau CONS$ IMPLIES	Premise
	$[(\text{FOR ALL } X)(ZF^* \vdash (\neg(proof(\boldsymbol{X},\boldsymbol{G})))$	
	IMPLIES NOT $(ZF^* \vdash (theo(\boldsymbol{G}))))]$	
4.	ZF^*CONS IMPLIES ZF^*CONS	Premise
5.	ZF^*CONS IMPLIES	Premise
	$(\text{FOR ALL } X)(\text{NOT }(PROOF(X,G))))$	
6.	$\;\;\;ZF^* \vdash (\neg(G))$	Assumption
* 7.	$\;\;\;\;\;\;\neg(theo(\boldsymbol{G}))$	Assumption
* 8.	$\;\;\;\;\;\;(G \leftrightarrow \neg(theo(\boldsymbol{G})))$	Prov E 1
* 9.	$\;\;\;\;\;\;G$	Iff E L 8, 7
* 10.	$\;\;\;\;\;\;\neg(G)$	Prov E 6
* 11.	$\;\;\;theo(\boldsymbol{G})$	Not E 7, 9, 10
12.	$\;\;\;ZF^* \vdash (theo(\boldsymbol{G}))$	Prov I 11
13.	$\;\;\;(\text{FOR ALL } X)(ZF^* \vdash (\neg(proof(\boldsymbol{X},\boldsymbol{G})))$	Imp E 3, 2
	$\;\;\;$ IMPLIES NOT $(ZF^* \vdash (theo(\boldsymbol{G})))$	
14.	$\;\;\;ZF^*CONS$	Imp E 4, 2
15.	$\;\;\;(\text{FOR ALL } X)(\text{NOT }(PROOF(X,G)))$	Imp E 5, 14
16.	$\;\;\;\text{NOT }(PROOF(X,G))$	All E 15
* 17.	$\;\;\;\neg(proof(\boldsymbol{X},\boldsymbol{G}))$	Rep 16
18.	$\;\;\;ZF^* \vdash (\neg(proof(\boldsymbol{X},\boldsymbol{G})))$	Prov I 17
19.	$\;\;\;(\text{FOR ALL } X)ZF^* \vdash (\neg(proof(\boldsymbol{X},\boldsymbol{G})))$	All I 18
20.	$\;\;\;\text{NOT }(ZF^* \vdash (theo(\boldsymbol{G})))$	Imp E 13, 19
21.	$\text{NOT }(ZF^* \vdash (\neg(G)))$	Not I 6, 12, 20 □

For the proof of the *second incompleteness theorem*, i.e., the non-provability of the formal consistency statement zf*cons under the assumption of the consistency of ZF^*, the formalism has to satisfy the *Hilbert–Bernays derivability conditions* D_1 and D_2. D_1 is the formalized semi-representability condition for the theorem predicate [theo(X) → theo(**theo(X)**)], whereas D_2 is the provable closure under modus ponens [theo($\boldsymbol{X} \to \boldsymbol{Y}$) → (theo($\boldsymbol{X}$) → theo($\boldsymbol{Y}$))]. The algorithm makes use of these conditions as rules with one additional heuristic to exploit D_2: if theo($\boldsymbol{F}$) is the goal and F, as a consequent of a conditional (or biconditional), is a strictly positive subformula of an available purely implicational formula, apply D_2 repeatedly and try to extract theo($\boldsymbol{F}$).

Proof.

	1.	ZF* ⊢ (theo(G) ↔ ¬G))	Premise[8]
	2.	ZF* ⊢ (zf*cons ↔ ¬(theo(G) & theo(¬G)))	Premise
	3.	NOT (ZF* ⊢ (G))	Premise
	4.	ZF* ⊢ (zf*cons)	Assumption
*	5.	¬(G)	Assumption
*	6.	(theo(G) ↔ ¬G)	Prov E 1
*	7.	theo(G)	Iff E L 6, 5
*	8.	theo(**theo**(G))	Der 1 7
*	9.	theo(**theo**(G)) → theo(¬G)	Der 2 6
*	10.	theo(¬G)	Imp E 9, 8
*	11.	theo(G) & theo(¬G)	And I 7, 10
*	12.	(zf*cons ↔ ¬(theo(G) & theo(¬G)))	Prov E 2
*	13.	zf*cons	Prov E 4
*	14.	¬(theo(G) & theo(¬G))	Iff E L 12, 13
*	15.	G	Not E 5, 11, 14
	16.	ZF* ⊢ (G)	Prov I 15
	17.	NOT (ZF* ⊢ (zf*cons))	Not I 4, 17, 3 □

This argument made use of the special character of the Gödel sentence G—in order to obtain the two conjuncts of line 11. Instead, one can exploit the elegant way of proceeding made possible by Löb's theorem in [14]:

For all sentences F: ZF* ⊢ (theo(F) →F) IFF ZF* ⊢ (F).

Löb's theorem expresses that a sentence F is provable in ZF* if and only if its *reflection formula* (theo(F) → F) can be established in ZF*. Consider a *refutable* sentence H (i.e., a sentence whose negation is provable in ZF*) and assume that ZF* is consistent; then H is not provable in ZF*. Löb's theorem implies that the corresponding reflection formula (theo(H) → H) is not provable either. Thus, the *second incompleteness theorem* amounts to establishing NOT (ZF* ⊢ (zf*cons)) from the premises NOT (ZF* ⊢ (theo(H) → H)), ZF* ⊢ (zf*cons ↔ ¬(theo(H) & theo(¬H))), and ZF* ⊢ (¬H). That is done in the next proof.

---

[8] Notice that the diagonal lemma is used here in a propositionally equivalent form; the current algorithm does not find the proof, when it is given in its standard form.

Proof.

	1.	NOT $(ZF^* \vdash (\text{theo}(\boldsymbol{H}) \to H))$	Premise
	2.	$ZF^* \vdash (zf^*\text{cons} \leftrightarrow \neg(\text{theo}(\boldsymbol{H}) \,\&\, \text{theo}(\neg\boldsymbol{H})))$	Premise
	3.	$ZF^* \vdash (\neg H)$	Premise
	4.	$ZF^* \vdash (zf^*\text{cons})$	Assumption
* 5.		$\text{theo}(\boldsymbol{H})$	Assumption
* 6.		$\neg(H)$	Assumption
* 7.		$\text{theo}(\neg\boldsymbol{H})$	Rep 3
* 8.		$\text{theo}(\boldsymbol{H}) \,\&\, \text{theo}(\neg\boldsymbol{H})$	And I 5, 7
* 9.		$(zf^*\text{cons} \leftrightarrow \neg(\text{theo}(\boldsymbol{H}) \,\&\, \text{theo}(\neg\boldsymbol{H})))$	Prov E 2
* 10.		$zf^*\text{cons}$	Prov E 4
* 11.		$\neg(\text{theo}(\boldsymbol{H}) \,\&\, \text{theo}(\neg\boldsymbol{H}))$	Iff E R 9, 10
* 12.		H	Not E 6, 8, 11
* 13.		$\text{theo}(\boldsymbol{H}) \to H$	Imp I 5, 12
	14.	$ZF^* \vdash (\text{theo}(\boldsymbol{H}) \to H)$	Prov I 13
	15.	NOT $(ZF^* \vdash (zf^*\text{cons}))$	Not I 4, 14, 1 □

This proof of the second incompleteness theorem uses Löb's Theorem only in the discussion leading up to the precise derivational problem. In Appendix A the preliminary considerations are incorporated into the proof; there we also show an elegant machine proof of Löb's Theorem.

4 Comparisons

A number of researchers have pursued goals similar to ours, but with interestingly different programmatic perspectives and strikingly different computational approaches. We focus on work by Ammon [1], Quaife [15], Bundy et al. [5] and Shankar [17]. We first discuss Ammon's and Quaife's work, as theirs is programmatically closest to ours: Ammon aims explicitly for a *fully automatic* proof of the first incompleteness theorem, and Quaife establishes the incompleteness theorems and Löb's theorem in a setting that is similarly "abstract" as ours.

In his 1993 Research Note *An automatic proof of Gödel's incompleteness theorem*, Ammon describes the SHUNYATA program and the proof it found for the first incompleteness theorem. SHUNYATA's proof is structurally identical with the proof in Kleene's book *Introduction to Metamathematics* (pp. 204–8); the latter proof is discussed in great detail in Sections 4 and 5 of Ammon's note. Two main claims are made: (i) Gödel's undecidable sentence is "constructed" by the program "on the basis of elementary rules for the formation of formulas", and this is taken as evidence for the subsidiary claim (on p. 305) that the program "implicitly rediscovered Cantor's diagonal method"; (ii) the proof of its undecidability is found by a heuristically guided *complete proof procedure* involving Gentzen's natural deduction rules for full first-order logic. The first claim (made on p. 291 and reemphasized on p. 295) is misleading: the Gödel sentence is of course constructible by the elementary rules for the (suitably extended) language of number theory, but that the formula so constructed expresses its unprovability has to be ensured by other means (and is "axiomatically" required to do so by Ammon's definition 3 and lemma 1).[9] As to the second claim (made on p. 294), the paper contains neither a logical calculus nor a systematic proof procedure using the rules of the calculus. What one finds are local heuristics for analyzing quantified statements and conditionals together with directions to prove the negation of a statement, i.e., to use the not introduction rule. These latter directions are quite open-ended, as there is no mechanism for selecting appropriate contradictory pairs. (Cf. Ammon's discussion of the "contradiction heuristic" on p. 296.)

In 1988 Quaife had already published a paper on *Automated proofs of Löb's Theorem and Gödel's two incompleteness theorems*. The paper presents proofs of the theorems mentioned in its title[10] "at a suitable level of abstraction"— as the author emphasizes on p. 219—"from the underlying details of Gödel numbering and of recursive functions". The suitable level of abstraction is provided by the provability logic $K4$. That well-known logic contains as special axioms the derivability conditions and as its special rule (beyond modus ponens) the rule of "necessitation"; the additional rule corresponds to the semi-representability of the theorem predicate. In order to make use of the resolution theorem proving system ITP, the first-order meta-theory of $K4$ is represented in ITP by five "clauses", which are listed in Appendix C. Four of the clauses correspond to the axioms and rules just mentioned, whereas the very first clause guarantees that all tautologies are obtained. The tautologies are established by "applying properly specified demodulators" and transforming given sentential formulas into conjunctive normal form; the underlying procedure is complex and involves particular weighting schemes. Quaife illustrates the procedure by presenting on pp. 226–7 a derivation of a "reasonably complex tautology"; the derivation uses a sequence of 73 demodulation steps. Quaife concludes the discussion of this derivation by saying: "ITP can also be asked to print out the line-by-line application of each demodulator, but that detailed proof is too long for this article". We present this tautology and its direct (and easily found) natural deduction proof in Appendix C.

---

[9] Our assessment of this claim is in full agreement with that found in the *Letter to the Editor* by Brüning et al. [3].

[10] Quaife establishes only the unprovability of G, not of its negation under the assumption of ω-consistency. On p. 229 he asserts, "With the right axioms, its proof [i.e., the other half of the first incompleteness theorem, S&F] could be reproduced about as easily as the principal half above".

In contrast to Ammon's paper, we find here a conceptually and technically straightforward meta-mathematical and logical set-up: representability and derivability conditions are axiomatically assumed, and the logical inference machinery is precisely and carefully described. However, it is very difficult to understand, how the syntactic context of axioms, theorems and assumptions directs the search in a way that is motivated by the leading ideas of the mathematical subject.[11] The proofs use in every case "axioms and previously proven theorems" in addition to the standard hypotheses for the theorem under consideration. It is clear that the "previously proven theorems" are strategically selected, and it is fair to ask, whether the full proof—from axioms through intermediate results to the meta-mathematical theorems—should be viewed as "automated" or rather as "interactive" with automated large logical steps. So the direct computational question is, would proofs of the main theorems be found, if only the axioms were available?

The answer is most likely "No". OTTER, the resolution theorem prover that developed out of ITP, was not able to prove, under appropriately similar conditions, the full first incompleteness theorem in 1996; that is reported in Bundy, Giunchiglia, Villafiorita and Walsh's paper *An incompleteness theorem via abstraction*.[12] It was precisely this computational problem that motivated their paper, namely to show how "abstraction" can be useful to attack it. They present a proof of Gödel's theorem, where the real focus is not on the particular meta-mathematical proof, but rather on the process of abstraction and refinement that aids proof planning. This process is not a fully automated one, since both the choice of the abstraction and the subsequent refinement of the abstract proof into the original language require external guidance. While we share the ultimate goal of limiting the search space for mathematical proofs by "abstraction", their semi-automated abstraction process is a very different, though complementary approach.

The three approaches we have been discussing are as "abstract" as ours in the sense that the diagonal lemma, the representability condition and, in Quaife's and our case, the derivability conditions are taken for granted. Shankar's book *Metamathematics, Machines, and Gödel's Proof* focuses on an interactive proof of (the Rosser version of) the first incompleteness theorem.[13] The explicit goal was to find out, whether the full proof could *in practice be checked* using a computer program, i.e., the Boyer–Moore theorem prover. In the preface to his book Shankar points out that "A secondary goal was to determine the effort involved in such a verification, and to identify the strengths and weaknesses of automated reasoning technology". The

---

[11] A similar reservation is articulated by Fearnley-Sander in his review [9] of Quaife's book [16].

[12] On p. 10 they write: "This proof [of the full first incompleteness theorem; S&F] turns out to be a considerable challenge to an unguided theorem prover. We have given these axioms to OTTER (v. 3.0) ... but it blew up".

[13] In addition, Shankar provides a "mechanical proof" of the Church–Rosser Theorem in Chapter 6.

crucial meta-mathematical task and most significant difficulty consisted in verifying the representability conditions—for a particular theory (the system Z_2 for number theory in Cohen's book) and a particular way of making computability precise (via McCarthy's Lisp). That required, of course, a suitable formalization of all meta-mathematical considerations within, what Shankar calls on p. 141, "a constructive axiomatization of pure Lisp". In Sections 5.4 and 5.5 Shankar gives a very informative analysis of, and an excellent perspective on, the work presented.

Moving back from interactive theorem proving to automated proof search, it is clear that the success of our search procedure results from carefully interweaving mathematical and logical considerations, which lead from explicitly formulated principles to a given conclusion. Proofs provide *explanations* of what they prove by putting their conclusions in a context that shows them to be correct. This need not be a global context providing a foundation for all of mathematics, but it can be a rather more restricted one as here for the presentation of the incompleteness theorems. Such a local deductive organization is *the* classical methodology of mathematics with two well-known aspects: the formulation of principles and the reasoning from such principles; we have illustrated only the latter aspect by using suitable strategic considerations and appropriate heuristic "leading mathematical ideas".

The task of considering a part of mathematics, finding appropriate basic notions, and explicitly formulating principles—so that the given part can be systematically developed—is of a quite different character. For Dedekind the need to introduce new and more appropriate notions arises from the fact that human intellectual powers are imperfect. The limitation of these powers leads us, Dedekind argues in [8], to frame the object of a science in different forms or different systems. To introduce a notion, "as a motive for shaping the systems", means in a certain sense to formulate a hypothesis concerning the inner nature of a science, and it is only the further development that determines the real value of such a notion by its greater or smaller *efficacy* [*Wirksamkeit*] in recognizing general truths. In the part of meta-mathematics we have been considering, Hilbert and Bernays did just that: their formulation of representability and derivability conditions ultimately led to more "abstract" ones and, in particular, to the principles for the provability logic $K4$ and related systems; see [2].[14]

5 Concluding remarks

No matter how one might mechanize an attempt of gaining such a principled deeper understanding of a part of mathematics, the considerations for a sys-

---

[14] In a different, though closely related case, Hilbert and Bernays succeeded in providing "recursiveness conditions" for the informal concept of calculability in a deductive formalism; that was done in a supplement of the second volume of their *Grundlagen der Mathematik*.

tematic and efficient automated development would still be central. In our given meta-mathematical context, there is an absolutely natural step to be taken next. As we emphasized earlier, there is no conflict or even sharp contrast between proof search and proof planning: proof search is hierarchically and heuristically organized through the use of "axioms" and their subsequent verification (or refutation). The guiding idea for verification in the intercalation approach is to generate sequences of formulas, reduce differences, and arrive ultimately at syntactic identities. Such difference reduction also underlies the techniques for inductive theorem proving that have been developed by Bundy et al. in their recent book [4]. We conjecture that those techniques can be seamlessly joined with the intercalation method to take the *next step* and prove the representability conditions. The strictly formal proof in TEM might then be transformed into a ZF* proof of the first derivability condition, automatically. From a different, more proof-theoretic perspective one might wish to compare the intercalation method for natural deduction calculi with appropriately formulated methods for sequent calculi with and without cuts. That might lead to interesting heuristics for choosing suitable cut formulas (to make proof search more efficient).[15]

Acknowledgements

Our work was supported by all the members of the current AProS team, in particular by Joseph Ramsey, Orlin Vakarelov, and Ian Kash; we are very grateful.[16]

Appendix A: Löb's theorem

The context of the theorem is given in Section 3. Here we present an argument obtained by our automated proof search and re-prove the second incompleteness theorem; in the latter proof, the appeal to Löb's theorem is explicitly

---

[15] This issue was suggested as a good research direction by an anonymous referee.

[16] This paper was first published in the *Annals of Pure and Applied Logic* **133**, 2005, pp. 319–338. It was dedicated to Helmut Schwichtenberg who shares our fascination with automated proof search. He also introduced, a long time ago in lectures at the University of Münster, the first author to the intricacies of Gödel's arguments.

The paper is republished with the permission of Elsevier; it was reset by Dawn McLaughlin, who improved the graphical presentation of text and proofs.

The AProS system—implementing the strategically informed proof search—can be downloaded from www.phil.cmu.edu/projects/apros/ The proofs of the theorems in this paper can be obtained, sometimes with slight modifications: the implementation of the search algorithm has been dramatically improved by Tyler Gibson; he also constructed the beautiful interface.

built into the argument. In order to prove Löb's theorem in TEM, one faces two claims, namely,

(i) $\mathsf{ZF}^* \vdash (\mathsf{theo}(\boldsymbol{F}) \to F)$ IMPLIES $\mathsf{ZF}^* \vdash (F)$

and

(ii) $\mathsf{ZF}^* \vdash (F)$ IMPLIES $\mathsf{ZF}^* \vdash (\mathsf{theo}(\boldsymbol{F}) \to F)$.

The last claim is immediate, whereas the first is difficult: its proof uses the instance of the diagonal lemma for the formula $(\mathsf{theo}(x) \to F)$. Here is the precise derivational problem at the heart of Löb's theorem: $\mathsf{ZF}^* \vdash (F)$ can be proved from the premises

$\mathsf{ZF}^* \vdash (\mathsf{theo}(\boldsymbol{F}) \to F)$

and

$\mathsf{ZF}^* \vdash (L \leftrightarrow (\mathsf{theo}(\boldsymbol{L}) \to F))$.

We actually have two proofs of Löb's theorem, which differ in the presentation of the derivability conditions. In the first proof the conditions are formulated as premises and are instantiated for this problem. They enter the search through the standard extraction procedure. In the second proof heuristics guide their application. The heuristics were described above and have a fairly general character; they are designed to apply each condition when it may be useful. The resulting proofs are very similar, differing mainly in the greater number of extraction rule applications necessary in the first proof to make use of the axiomatically given derivability conditions. We present only the first proof.

Proof.

1.	$\mathsf{ZF}^* \vdash (L \leftrightarrow (\text{theo}(\boldsymbol{L}) \rightarrow F))$	Premise
2.	$\mathsf{ZF}^* \vdash (\text{theo}(\boldsymbol{L}) \rightarrow (\text{theo}(\textbf{theo}(\boldsymbol{L})) \rightarrow \text{theo}(\boldsymbol{F})))$	Premise
3.	$\mathsf{ZF}^* \vdash (\text{theo}(\boldsymbol{L}) \rightarrow \text{theo}(\textbf{theo}(\boldsymbol{L})))$	Premise
4.	$\quad\mathsf{ZF}^* \vdash ((\text{theo}(\boldsymbol{F}) \rightarrow F))$	Assumption
* 5.	$\quad\quad\text{theo}(\boldsymbol{L})$	Assumption
* 6.	$\quad\quad\text{theo}(\boldsymbol{L}) \rightarrow (\text{theo}(\textbf{theo}(\boldsymbol{L})) \rightarrow \text{theo}(\boldsymbol{F}))$	Prov E 2
* 7.	$\quad\quad(\text{theo}(\textbf{theo}(\boldsymbol{L})) \rightarrow \text{theo}(\boldsymbol{F}))$	Imp E 6, 5
* 8.	$\quad\quad(\text{theo}(\boldsymbol{L}) \rightarrow \text{theo}(\textbf{theo}(\boldsymbol{L})))$	Prov E 3
* 9.	$\quad\quad\text{theo}(\textbf{theo}(\boldsymbol{L}))$	Imp E 8, 5
* 10.	$\quad\quad\text{theo}(\boldsymbol{F})$	Imp E 7, 9
* 11.	$\quad\quad(\text{theo}(\boldsymbol{F}) \rightarrow F)$	Prov E 4
* 12.	$\quad\quad F$	Imp E 11, 10
* 13.	$\quad(\text{theo}(\boldsymbol{L}) \rightarrow F)$	Imp I 5, 12
* 14.	$\quad(L \leftrightarrow (\text{theo}(\boldsymbol{L}) \rightarrow F))$	Prov E 1
* 15.	$\quad L$	Iff E L 14, 13
16.	$\quad\mathsf{ZF}^* \vdash (L)$	Prov I 15
* 17.	$\quad\text{theo}(\boldsymbol{L})$	Rep 16
* 18.	$\quad F$	Imp E 13, 17
19.	$\quad\mathsf{ZF}^* \vdash (F)$	Prov I 18
20.	$(\mathsf{ZF}^* \vdash ((\text{theo}(\boldsymbol{F}) \rightarrow F))$ IMPLIES $\mathsf{ZF}^* \vdash (F))$	Imp I 4, 19
21.	$\quad\mathsf{ZF}^* \vdash (F)$	Assumption
* 22.	$\quad\quad\text{theo}(\boldsymbol{F})$	Assumption
* 23.	$\quad\quad F$	Prov E 21
* 24.	$\quad(\text{theo}(\boldsymbol{F}) \rightarrow F)$	Imp I 22, 23
25.	$\quad\mathsf{ZF}^* \vdash ((\text{theo}(\boldsymbol{F}) \rightarrow F))$	Prov I 24
26.	$(\mathsf{ZF}^* \vdash (F)$ IMPLIES $\mathsf{ZF}^* \vdash ((\text{theo}(\boldsymbol{F}) \rightarrow F)))$	Imp I 21, 25
27.	$(\mathsf{ZF}^* \vdash ((\text{theo}(\boldsymbol{F}) \rightarrow F))$ IFF $\mathsf{ZF}^* \vdash (F))$	Iff I 20, 26 □

Now we present the proof of the second incompleteness theorem with the explicit use of Löb's Theorem.

Proof.

1.	ZF*CONS	Premise
2.	ZF* ⊢ (¬(H))	Premise
3.	(ZF*CONS IFF NOT ((ZF* ⊢ (H) AND ZF* ⊢ (¬(H)))))	Premise
4.	ZF* ⊢ (zf*cons ↔ ¬((theo(H) & theo(¬(H)))))	Premise
5.	(ZF* ⊢ (H) IFF ZF* ⊢ ((theo(H) → H)))	Premise
6.	ZF* ⊢ (zf*cons)	Assumption
7.	NOT ((ZF* ⊢ (H) AND ZF* ⊢ (¬(H))))	Iff E R 3, 1
*8.	theo(H)	Assumption
*9.	¬(H)	Assumption
*10.	(zf*cons ↔ ¬((theo(H) & theo(¬(H)))))	Prov E 4
*11.	zf*cons	Prov E 6
*12.	¬((theo(H) & theo(¬(H))))	Iff E R 10, 11
*13.	theo(¬(H))	Rep 2
*14.	(theo(H) & theo(¬(H)))	And I 8, 13
*15.	H	Not E 9,14, 12
*16.	(theo(H) → H)	Imp I 8, 15
17.	ZF* ⊢ ((theo(H) → H))	Prov I 16
18.	ZF* ⊢ (H)	Iff E L 5, 17
19.	(ZF* ⊢ (H) AND ZF* ⊢ (¬(H)))	And I 18, 2
20.	NOT (ZF* ⊢ (zf*cons))	Not I 6, 19, 7 □

Appendix B

The square root of 2 is not rational. The logical search algorithm uncovers directly the following proof of the claim from the premises:

(1) $\sqrt{2}$ is rational $\leftrightarrow (\exists x)(\exists y)(\sqrt{2} * x = y \,\&\, \neg(\exists z)(z|x \,\&\, z|y))$
(2) $(\forall x)(\forall y)(2 * x^2 = y^2 \to 2|x \,\&\, 2|y)$
(3) $(\forall x)(\forall y)(\sqrt{2} * x = y \to 2 * x^2 = y^2)$

The universe of discourse consists of the set of all reals or just the algebraic ones, but the range of the quantifiers consists just of the sort of positive integers. Here is the translation of the automatically generated proof; "translation", as the parser understands only a more restricted language.

Proof.

1.	$\sqrt{2}$ is rational $\leftrightarrow (\exists x)(\exists y)(\sqrt{2}*x = y \,\&\, \neg(\exists z)(z|x \,\&\, z|y))$	Premise
2.	$(\forall x)(\forall y)(2*x^2 = y^2 \rightarrow 2|x \,\&\, 2|y)$	Premise
3.	$(\forall x)(\forall y)(\sqrt{2}*x = y \rightarrow 2*x^2 = y^2)$	Premise
4.	$\quad \sqrt{2}$ is rational	Assumption
5.	$\quad (\exists x)(\exists y)(\sqrt{2}*x = y \,\&\, \neg(\exists z)(z|x \,\&\, z|y))$	Iff E R 1, 4
6.	$\quad\quad (\exists y)(\sqrt{2}*u = y \,\&\, \neg(\exists z)(z|u \,\&\, z|y))$	Assumption
7.	$\quad\quad\quad (\sqrt{2}*u = v \,\&\, \neg(\exists z)(z|u \,\&\, z|v))$	Assumption
8.	$\quad\quad\quad (\forall y)(2*u^2 = y^2 \rightarrow 2|u \,\&\, 2|y)$	All E 2
9.	$\quad\quad\quad (2*u^2 = v^2 \rightarrow 2|u \,\&\, 2|v)$	All E 8
10.	$\quad\quad\quad (\forall y)(\sqrt{2}*u = y \rightarrow 2*u^2 = y^2)$	All E 3
11.	$\quad\quad\quad (\sqrt{2}*u = v \rightarrow 2*u^2 = v^2)$	All E 10
12.	$\quad\quad\quad \sqrt{2}*u = v$	And E L 7
13.	$\quad\quad\quad 2*u^2 = v^2$	Imp E 11, 12
14.	$\quad\quad\quad 2|u \,\&\, 2|v$	Imp E 9, 13
15.	$\quad\quad\quad (\exists z)(z|u \,\&\, z|v)$	Ex I 14
16.	$\quad\quad\quad \neg(\exists z)(z|u \,\&\, z|v)$	And E R 7
17.	$\quad\quad\quad \bot$	$\bot$ I 15, 16
18.	$\quad\quad \bot$	Ex E 6, 7, 17
19.	$\quad \bot$	Ex E 5, 6, 18
20.	$\neg(\sqrt{2}$ is rational$)$	Not I 4, 19 □

$\bot$ is taken as a placeholder for an appropriate contradiction, say, $(P \,\&\, \neg P)$.

Appendix C

In [15, pp. 226–227], this "reasonably complex tautology" is presented:

$$[(P \rightarrow (Q \rightarrow R)) \rightarrow ((Q \rightarrow (R \rightarrow S)) \rightarrow (Q \rightarrow (P \rightarrow S)))].$$

Its proof, however, is considered to be too long for incorporation into the article. In our natural deduction framework the proof is absolutely canonical and direct; here it is—in twelve lines:

Proof.

1.	$(P \to (Q \to R))$	Assumption
2.	$(Q \to (R \to S))$	Assumption
3.	Q	Assumption
4.	P	Assumption
5.	$(R \to S)$	Imp E 2, 3
6.	$(Q \to R)$	Imp E 1, 4
7.	R	Imp E 6, 3
8.	S	Imp E 5, 7
9.	$(P \to S)$	Imp I 4, 8
10.	$(Q \to (P \to S))$	Imp I 3, 9
11.	$((Q \to (R \to S)) \to (Q \to (P \to S)))$	Imp I 2, 10
12.	$((P \to (Q \to R)) \to ((Q \to (R \to S)) \to (Q \to (P \to S))))$	Imp I 1, 11 □

As mentioned in Section 4, Quaife's framework is a formulation of the first-order meta-theory of $K4$ within ITP. The predicate $\mathsf{Thm}K4(x)$ expresses that the formula x is a theorem of $K4$. Here are the clauses generating theorems (from p. 223):

(ITP.A1) If $\mathsf{taut}(x)$ then $\mathsf{Thm}K4(x)$;
(ITP.A2) $\mathsf{Thm}K4((b(x \to y) \to (b(x) \to b(y))))$;
(ITP.A3) $\mathsf{Thm}K4(b(x) \to b(b(x)))$;
(ITP.R1) If $\mathsf{Thm}K4((x \to y))$ & $\mathsf{Thm}K4(x)$ then $\mathsf{Thm}K4(y)$;
(ITP.R2) If $\mathsf{Thm}K4(x)$ then $\mathsf{Thm}K4(b(x))$.

A1 guarantees that all tautologies are theorems; A2 and A3 correspond to the derivability conditions; R1 is modus ponens, and R2 expresses the semi-representability of the theorem predicate.

Appendix D

Here we present two further computer-generated proofs surrounding the incompleteness theorems. The first claim is a version of the first half of the first incompleteness theorem, asserting the unprovability of the reflection formula for the Gödel sentence.

(i) $\mathsf{ZF}^*\mathsf{CONS}$ IMPLIES NOT $(\mathsf{ZF}^* \vdash (\mathsf{theo}(\boldsymbol{G}) \to \boldsymbol{G}))$.

Proof.

	1.	(ZF*CONS IFF NOT ((ZF* ⊢ (G) AND ZF* ⊢ (¬(G)))))	Premise
	2.	ZF* ⊢ ((G ↔ ¬(theo(**G**))))	Premise
	3.	ZF*CONS	Assumption
	4.	ZF* ⊢ ((theo(**G**) → G))	Assumption
	5.	NOT ((ZF* ⊢ (G) AND ZF* ⊢ (¬(G))))	Iff E R 1, 3
* 6.		(G ↔ ¬(theo(**G**)))	Prov E 2
* 7.		theo(**G**)	Assumption
* 8.		(theo(**G**) → G)	Prov E 4
* 9.		G	Imp E 8, 7
* 10.		¬(theo(**G**))	Iff E R 6, 9
* 11.		¬(theo(**G**))	Not I 7, 7, 10
* 12.		G	Iff E L 6, 11
	13.	ZF* ⊢ (G)	Prov I 12
* 14.		G	Assumption
* 15.		theo(**G**)	Rep 13
* 16.		¬theo(**G**)	Iff E R 6, 14
* 17.		¬(G)	Not I 14, 15, 16
	18.	ZF* ⊢ (¬(G))	Prov I 17
	19.	(ZF* ⊢ (G) AND ZF* ⊢ (¬(G)))	And I 13, 18
	20.	NOT (ZF* ⊢ ((theo(**G**) → G)))	Not I 4, 19, 5
	21.	(ZF*CONS IMPLIES NOT (ZF* ⊢ ((theo(**G**) → G))))	Imp I 3, 20 □

The argument is perfectly canonical—up to the extraction step in line 12; at this point G could have been extracted from the formula (theo(**G**) → G) in line 4. The resulting proof is "symmetric" to the given one.

The second claim asserts that for any refutable sentence R, the formula expressing its unprovability, i.e., ¬(theo(**R**)), is in ZF* equivalent to its reflection formula (theo(**R**) → R)).

(ii) ZF* ⊢ (¬(R)) IMPLIES ZF* ⊢ (((¬(theo(**R**)) ↔ (theo(**R**) → R))).

Proof.

	1.	$\mathsf{ZF}^* \vdash (\neg(R))$	Premise
*	2.	$\neg(\mathsf{theo}(\boldsymbol{R}))$	Assumption
*	3.	$\mathsf{theo}(\boldsymbol{R})$	Assumption
*	4.	$\neg(R)$	Assumption
*	5.	R	Not E 2, 3
*	6.	$(\mathsf{theo}(\boldsymbol{R}) \to R)$	Imp I 5
*	7.	$(\neg(\mathsf{theo}(\boldsymbol{R})) \to (\mathsf{theo}(\boldsymbol{R}) \to R))$	Imp I 6
*	8.	$(\mathsf{theo}(\boldsymbol{R}) \to R)$	Assumption
*	9.	$\mathsf{theo}(\boldsymbol{R})$	Assumption
*	10.	$\neg(R)$	Prov E 1
*	11.	R	Imp E 8, 9
*	12.	$\neg(\mathsf{theo}(\boldsymbol{R}))$	Not I 10, 11
*	13.	$((\mathsf{theo}(\boldsymbol{R}) \to R) \to \neg(\mathsf{theo}(\boldsymbol{R})))$	Imp I 12
*	14.	$(\neg(\mathsf{theo}(\boldsymbol{R})) \leftrightarrow (\mathsf{theo}(\boldsymbol{R}) \to R))$	Iff I 7, 13
	15.	$\mathsf{ZF}^* \vdash ((\neg(\mathsf{theo}(\boldsymbol{R})) \leftrightarrow (\mathsf{theo}(\boldsymbol{R}) \to R)))$	Prov I 14 □

References

1. K. Ammon: An automatic proof of Gödel's incompleteness theorem. *Artificial Intelligence* **61** (1993) pp 291–306
2. G. Boolos: *The logic of provability* (Cambridge University Press, Cambridge 1993)
3. S. Brüning, M. Thielscher and W. Bibel: Letter to the editor. *Artificial Intelligence* **61** (1993) pp 353–4
4. A. Bundy, F. Giunchiglia, A. Villafiorita and T. Walsh: An incompleteness theorem via abstraction. Technical Report #9302-15 (Istituto per la ricerca scientifica e tecnologica, Trento 1996)
5. A. Bundy, D. Basin, D. Hutter and A. Ireland: *Rippling: Meta-level guidance for mathematical reasoning* (Book manuscript 2003)
6. J. Byrnes: *Proof search and normal forms in natural deduction.* Ph.D. Thesis (Department of Philosophy, Carnegie Mellon University 1999)
7. P. J. Cohen: *Set theory and the continuum hypothesis* (Benjamin, Reading Mass. 1966)
8. R. Dedekind: *Über die Einführung neuer Funktionen in der Mathematik* (Habilitationsrede 1854, pp 428–38) In: *Gesammelte mathematische Werke*, ed by Fricke, Noether and Ore, vol. 3 (Vieweg 1933)
9. D. Fearnley-Sander: Review of *Quaife* [16]. http://psyche.cs.monash.edu.au/

10. S. Feferman: Inductively presented systems and the formalization of metamathematics. In: *Logic Colloquium '80*, ed by D. van Dalen, D. Lascar and T. J. Smiley (North-Holland, Amsterdam 1982) pp 95–128
11. S. Feferman: Finitary inductively presented logics. In: *Logic Colloquium '88*, ed by R. Ferro et al. (North-Holland, Amsterdam 1988) pp 191-220
12. F. Fitch: *Symbolic Logic* (The Ronald Press Company, New York 1952)
13. K. Gödel: Über formal unentscheidbare Sätze der Principia mathematica und verwandter Systeme I. *Monatshefte für Mathematik und Physik* **38** (1931) pp 173–198
14. M. Löb: Solution of a problem of Leon Henkin. *J. Symbolic Logic* **20** (1955) pp 115–118
15. A. Quaife: Automated proofs of Löb's theorem and Gödel's two incompleteness theorems. *Journal of Automated Reasoning* **4** (1988) pp 219–231
16. A. Quaife: *Automated Development of Fundamental Mathematical Theories* (Kluwer, Dordrecht 1992)
17. N. Shankar: *Metamathematics, Machines, and Gödel's Proof.* Cambridge Tracts in Theoretical Computer Science 38 (Cambridge University Press, Cambridge 1994)
18. W. Sieg: *Elementary proof theory.* Technical Report 297 (Institute for Mathematical Studies in the Social Sciences, Stanford 1978) 104 pp
19. W. Sieg: *Mechanisms and Search* (Aspects of Proof Theory). AILA Preprint (1992)
20. W. Sieg and J. Byrnes: Normal natural deduction proofs (in classical logic). *Studia Logica* **60** (1998) pp 67–106
21. W. Sieg and S. Cittadini: Normal natural deduction proofs (in non-classical logics). Technical Report No. CMU-PHIL-130 (2002) 29 pp. The paper has since been published in: *Mechanizing Mathematical Reasoning*, ed by Hutter and Stephan (Springer, 2005) pp 169–191
22. W. Sieg, I. Lindstrom and S. Lindstrom: Gödel's incompleteness theorems - a computer-based course in elementary proof theory. In: *University-Level Computer-Assisted Instruction at Stanford 1968-80*, ed by P. Suppes (Stanford 1981) pp 183–193

Proofs as Efficient Programs

Ugo Dal Lago and Simone Martini

Dipartimento di Scienze dell'Informazione, Università di Bologna (Italy)
dallago@cs.unibo.it
martini@cs.unibo.it

> *There may, indeed, be other uses of the system than its use as a logic.*
> A. Church [8]

Logic and theory of computation have been intertwined since their first days. The formalized notion(s) of effective computation are at first technical tools for the investigation of first order systems, and only ten years later – in the hands of John von Neumann – become the blueprints of engineered physical devices. Generally, however, one tends to forget that in those same years, in the newly-born proof-theory of Gerhard Gentzen [20] there is an implicit, powerful notion of computation – an effective, combinatorial procedure for the simplification of a proof. However, the complexity of the rules for the elimination of cuts (especially the commutative ones, in the modern jargon) hid the simplicity and generality of the basic computational notion those rules were based upon. We had to wait thirty more years before realizing in full glory that Gentzen's simplification mechanism and one of the formal systems for computability (Church's λ-calculus) were indeed one and the same notion.

As far as we know, Haskell Curry is the first to explicitly realize [11] that the *types* of some of his basic combinators correspond to axioms of intuitionistic implicational calculus, and that, more generally, the types assignable to expressions made up of combinators are exactly the provable formulae of intuitionistic implicational logic. It is William Howard in 1969 to extend this *formulas as types* correspondence to the more general *proofs as programs* isomorphism ([27], published in 1980 but widely circulated before). Under this interpretation, the two dynamics – proof normalization on one hand, and β-reduction on the other – are identified, so that techniques and results from one area are immediately available to the other.

In this paper, we will discuss the use of the Curry-Howard correspondence in *computational complexity theory*, the area of theoretical computer science concerned with the definition and study of complexity classes and their relations. The standard approach to this discipline is to fix first a machine model (e.g., Turing machines) equipped with an explicit cost (e.g., number of transitions) and define then the (inherent) complexity of a function as the minimal cost needed to compute it with those machines. More formally, let $h : \mathcal{N} \to \mathcal{N}$ be a non-decreasing function and assume Turing machines (TM) with unitary cost per transition. We say first that a TM M works in time h iff for any input x, $M(x)$ terminates in at most $h(|x|)$ steps, where $|\ |$ is a suitable notion of size for the input. A function f has complexity h iff there exists a TM computing f working in time h. At this point one may define the *complexity class of h* as the set of all functions with complexity h. Main examples of such classes are the polynomial functions (FPTIME), the exponential functions (FEXPTIME), or the Kalmar-elementary ones (FELEMTIME). In building up this theory, one of the first tasks is to show that the definition of a complexity class is somehow independent from the machine model adopted at first. Here comes the notion of *reasonable* machine models [39]:

> *Reasonable machines can simulate each other within a polynomially-bounded overhead in time and a constant-factor overhead in space.*

Complexity classes like FPTIME and the others we mentioned before are clearly invariant with respect reasonable machine models, and are thus amenable to general theoretical treatment.

If these classes are so robust, however, there should be characterizations of them independent from explicit machine models. This is the subject of *implicit computational complexity* (ICC) whose main aim is the description of complexity classes based on language restrictions, and not on external measure conditions or on an explicit machine model. It borrows techniques and results from mathematical logic (model theory, recursion theory, and proof-theory) and in doing so it has allowed the incorporation of aspects of computational complexity into areas such as formal methods in software development and programming language design. The most developed area of implicit computational complexity is probably the model-theoretic one – finite model theory being a very successful way to describe complexity problems. In the design of programming language tools (e.g., type-systems), however, syntactical techniques prove more useful, and in the last twenty years much work has been devoted to restricted ways of formulating recursion theory, and to proof-theoretical techniques to enforce resource bounds on programs.

1 Proofs as programs

The Curry-Howard correspondence between natural deduction proofs and lambda-terms is summarized in Fig. 1. Formally, we start by defining λ-*terms* as:

$$M ::= x \mid \lambda x.M \mid MM ,$$

where x ranges over a denumerable set of *variables*. As usual, λ binds variables in its scope. *Formulas* (or *types*) are defined starting from a base type (o) as

$$A ::= o \mid A \to A \,.$$

The system in Fig. 1 defines judgments of the form $\Gamma \vdash M : A$, where M is a term, A is a formula, and Γ is a set of pairs $x : A$, where x is a variable and all the variables in Γ are distinct.

$$\Gamma, x : A \vdash x : A$$

$$\frac{\Gamma \vdash M : A \to B \quad \Gamma \vdash N : A}{\Gamma \vdash MN : B} \; (\to,E) \qquad \frac{\Gamma, x : A \vdash M : B}{\Gamma \vdash \lambda x.M : A \to B} \; (\to,I)$$

Fig. 1. Natural deduction proofs and typed lambda-terms

If we forget terms, we are left with the usual natural deduction system for propositional implicational intuitionistic logic. Terms may be seen just as a convenient, linearized way to write the proofs, instead of the usual two-dimensional tree-like notation. On the other hand, a computer scientist will recognise the rules assigning types to functional programs. This correspondence between proofs and programs is completed by observing that the notion of proof normalization (that is, the elimination of the detours composed of an introduction immediately followed by an elimination) is the same as the reduction of the corresponding term (see Fig. 1, where we have adopted for more clarity the standard two-dimensional notation for proofs); $M[x \leftarrow N]$ denotes the substitution of N for the free occurrences of x in M.

$$\frac{\genfrac{}{}{0pt}{}{[x:A]}{\genfrac{}{}{0pt}{}{\pi}{M:B}}}{\frac{\lambda x.M : A \to B \quad \genfrac{}{}{0pt}{}{\delta}{N:A}}{(MN):B}} \quad \Longrightarrow \quad \genfrac{}{}{0pt}{}{\genfrac{}{}{0pt}{}{\delta}{N:A}}{\genfrac{}{}{0pt}{}{\pi}{M[x \leftarrow N]:B}}$$

Fig. 2. Normalization is beta-reduction

We may now apply logical methods to the study of computation as term rewriting, and vice versa. From the point of view of computational complexity, however, λ-calculus is a rather awkward computational model, since it misses a basic notion of elementary step with bounded cost. Indeed, it is evident that a step in a Turing machine can be "actually performed" in constant time, thanks to the finiteness of the states and the symbols. Even for machine models with non-constant time step – like Unbounded Register Machines (URM, see, e.g., [12]), where in a single step one may store or increment an arbitrary large natural number – one can usually give a *natural* and plausible cost model (for URMs one can assume, for instance, that one step accounts for the logarithm of the length of the manipulated number). For λ-calculus all these considerations seem to vanish, since beta-reduction is inherently too complex to be considered as an atomic operation, at least if we stick to explicit representations of lambda terms. Indeed, in a beta step

$$(\lambda x.M)N \to M[x \leftarrow N],$$

there can be as many as $|M|$ occurrences of x inside M. As a consequence, $M[x \leftarrow N]$ can be as big as $|M| \times |N|$. And this applies to any step during a reduction, where the terms N and M have no longer a direct evident connection with the original term.

As a result, in the literature the actual cost of normalizing a lambda-term has been studied without any reference to the number of beta steps to normal form. One of the main results of this field is Statman's theorem [36], showing that there is no Kalmar-elementary function of the length of a typed λ-term bounding the work needed to compute its normal form. This result, in a sense, matches in λ-calculus the series of results in proof theory giving bounds on the size of a *normal* proof, as a function of the size of the proof before normalization (see, e.g., [21]). These results could be summarized by saying that, in any reasonable logic, if the (natural deduction) proof π normalizes to ρ, then the size of ρ can be (hyper-)exponential in the size of π. But then we are faced with another problem in the direct application of the Curry-Howard correspondence to computational complexity. How could we use a tool that, in any interesting case, has a hyper-exponential complexity? How is it possible to use it for studying interesting complexity classes, and especially the most important PTIME?

The solution comes from a different perspective, the natural one from the programming view-point. We should forget the natural symmetry, present in λ-calculus, between programs (terms) and data (just other terms), and recall that, in computing, the roles of programs and data *are not* symmetric. We are instead interested in the behavior of a single, fixed program as a function of the size of its data. That is, under the proofs-as-programs isomorphism, the behavior of a fixed proof (say $\Gamma \vdash M : A \to B$) when composed with all sensible proofs (that is when "cut" against – applied to – all proofs of the shape $\Gamma \vdash N : A$). In explicit terms, we should study the time to normalize MN, with M in normal form, as a function of $|N|$ only.

We may borrow some terminology from software engineering and distinguish two main ways of applying logical notions and techniques to computational complexity.

1. Implicit computational complexity *in-the-large*, using logic to study complexity classes. More specifically, given a complexity class $\mathcal{C}$, find a logical system G such that:

$$A \vdash A \ (Ax) \qquad \frac{\Gamma \vdash A \quad A, \Delta \vdash B}{\Gamma, \Delta \vdash B} \ (Cut)$$

$$\frac{\Gamma \vdash C}{\Gamma, !A \vdash C} \ (Weak.) \qquad \frac{\Gamma, !A, !A \vdash B}{\Gamma, !A \vdash B} \ (Contr.)$$

$$\frac{\Gamma \vdash A \quad B, \Delta \vdash C}{\Gamma, A \multimap B, \Delta \vdash C} \ (\multimap, l) \qquad \frac{\Gamma, A \vdash B}{\Gamma \vdash A \multimap B} \ (\multimap, r)$$

$$\frac{\Gamma, A, B \vdash C}{\Gamma, A \otimes B \vdash C} \ (\otimes_i, l) \qquad \frac{\Gamma \vdash A \quad \Delta \vdash B}{\Gamma, \Delta \vdash A \otimes B} \ (\otimes, r)$$

$$\frac{A_1, \ldots, A_n \vdash B}{!A_1, \ldots, !A_n \vdash !B} \ (!)$$

$$\frac{\Gamma, A \vdash B}{\Gamma, !A \vdash B} \ (\epsilon) \qquad \frac{\Gamma, !!A \vdash B}{\Gamma, !A \vdash B} \ (\delta)$$

$$\frac{\Gamma, T[S/t] \vdash C}{\Gamma, \forall t.T \vdash C} \ (\forall, l) \qquad \frac{\Gamma \vdash C}{\Gamma \vdash \forall t.C} \ t \notin FV(\Gamma) \ (\forall, r)$$

Fig. 3. Intuitionistic Multiplicative Exponential Linear Logic, MELL

Soundness: for any interesting type A and $M : A \to A$ in the system G, there is a function $f_A \in \mathcal{C}$ such that for any $N : A$, the cost of normalizing MN is bounded by $f_A(|N|)$.

Completeness: for any function F computable in complexity $\mathcal{C}$, there is a proof M_F (in system G) "representing" F.

2. Implicit computational complexity *in-the-small*, using logic to study single machine-free models of computation. More specifically, in the context we just described, giving *natural* cost models for λ-calculus reduction, showing that it is indeed a reasonable model (that is, polynomially related to Turing machines).

2 ICC in the large

Finding a system characterizing interesting complexity classes is far from obvious, all standard logics having a too high complexity of reduction. The breakthrough came with Girard's *light linear logic* [25] (LLL). Linear logic [22] put under scrutiny the structural rules (especially contraction, responsible for arbitrary duplication of subproofs during cut-elimination) and their role during normalization. Intuitionistic logic, however, may be embedded in linear logic, and therefore the latter cannot be used as a limited complexity formal system. One has to weaken linear logic, limiting the use of contraction.

Second order multiplicative linear logic (MELL) is summarized in Fig. 2, as a sequent calculus. Rule *(Contr.)* is the culprit for exponential blow-up, but since it may only be applied to !-marked formulas, the rules for (!) are our main source of concern. If we look at the logic from an axiomatic point of view, one of the fundamental properties of the logic is the law $!A \equiv !A \otimes !A$ ($A \equiv B$ is short for $A \multimap B$ and $B \multimap A$), which is obtained from rules (*Contr.*), (*Weak.*) and (!). Other important laws are of course derivable (! acts much the same as $\Box$ in modal logic S4), namely $!A \multimap A$ (from (ϵ), "dereliction") and $!A \multimap !!A$ (from (δ), "digging"). The interplay between these rules is the main source for complexity of normalization (and for the expressivity of the logic, of course). Girard's insight was to weaken the operator !, still maintaining enough expressivity to code interesting functions.

If we drop (ϵ) and (δ), we get *Elementary linear logic* (ELL), which is sound and complete (in the meaning explained above) for Kalmar-elementary functions, a complexity class still too big for computer science. To get just the polynomial functions we must drastically restrict (!) to have at most one formula to the left of $\vdash$. But now the system would be too weak (far from complete from polytime, that is) and to fix this Girard adds another modality (§), to compensate for the loss of a full (!)-rule. The system is best formulated in an affine form (i.e., full weakening is allowed), following [2]. The relevant rules for *Light affine logic* (LAL) are summarized in Fig. 2 (they replace rules (*Weak.*), (*Contr.*), (!), (ϵ), and (δ) of Fig. 2; in (§), $n, m \geq 0$; in (!u), A can be absent).

$$\frac{\Gamma \vdash C}{\Gamma, A \vdash C} \ (AWeak.) \qquad \frac{\Gamma, !A, !A \vdash B}{\Gamma, !A \vdash B} \ (Contr.)$$

$$\frac{A_1, \ldots, A_n, C_1, \ldots, C_m \vdash B}{!A_1, \ldots, !A_n, §C_1, \ldots, §C_m \vdash §B} \ (§) \qquad \frac{A \vdash B}{!A \vdash !B} \ (!u)$$

Fig. 4. Light affine logic, LAL

With this key idea, one obtains indeed a sound and complete system for polytime. The following theorem expresses the soundness result, the most delicate and interesting one. We observe, first, that there is natural way to code in the logic, as a type, interesting data types; the theorem may be stated with respect to $\ulcorner BS \urcorner$, the type coding binary strings[1].

Theorem 1. *Let* $\vdash_{LAL} M : \ulcorner BS \urcorner \to \ulcorner BS \urcorner$. *For any* $\vdash_{LAL} W : \ulcorner BS \urcorner$, *we can compute the normal form of* MW *in time* $O((d+1)|MW|^{2d})$, *where* d *depends only on* M.

Observe that, when M is fixed and in normal form, the theorem says that M is a program with complexity $O(|W|^{2d})$, that is, polynomial in its input.

---

[1] The statement uses Landau's big-oh notation: For f and g functions, f is in $O(g)$ iff $\exists c, d, n_0$ such that $\forall n > n_0 \ f(n) \leq cg(n) + d$.

Care should be taken, however, on the correct reading of the normalization result. In order the theorem to hold as stated, we cannot simply reduce MW as a lambda-term (because even terms typable in LAL could have esponentially long normal forms, though "computable" in polynomial time [7]). *Proof-nets* (the intended notation for proofs in linear logic) should be used instead.

Once discovered, the technology has been applied, *mutatis mutandis*, to several other formal systems. In addition to LLL and LAL, we have EAL (for elementary time), SLL (for polynomial time [29]), STA_B (for logarithmic space [19]), etc. Moreover, we now have also systems where, contrary to LAL, the soundness for polynomial time holds for lambda-calculus reduction, like DLAL [7] and other similar systems. As a result, the general framework of light logics is now full of different systems, and of variants of those systems. It is urgent to get *general, unifying* results to compare the systems and improve on them.

As already mentioned, soundness and completeness are the key results to be obtained when trying to characterize complexity classes by way of logical systems. Completeness is always of an *extensional* nature: one proves that any function in the complexity class under consideration can be represented by a proof in the logical system, without taking care of the *intensional* aspect of the proof itself. Since the same function can be represented by many, distinct proofs (in particular those corresponding to unnatural algorithms, in computer science terminology), proving completeness does not say much about the usefulness of the considered system as a programming language. On the other hand, soundness is a key property: any representable function is in a complexity class. This is often proved by showing that *any* proof in the logical system can be normalized with a bounded amount of resources. This way, soundness proofs give insights on the *reasons* why normalization is a relatively easy computational process. As a consequence, soundness proofs are interesting on their own:

- The intensional expressive power of two or more apparently unrelated systems can be compared through their soundness proofs. This requires the soundness proofs to be phrased in the same framework.
- The possibility of extending sound systems with new rules and features can be made easier if their soundness proofs are designed to be adaptable to extensions in the underlying syntax.

Unfortunately, soundness has been traditionally proved with *ad hoc* techniques which cannot be easily generalized. Different systems has been proved sound with very different methodologies. As an example, Lafont's SLL has been proved to be polytime sound by a simple mathematical argument, while Asperti's LAL requires a complicated argument, in particular if strong soundness is needed [37].

In the last years, much effort has been put in the task of finding powerful, simple and unifying semantic frameworks for implicit computational complexity. In the rest of the section we will give an account of some of these frameworks, with particular emphasis on authors' contribution. What we would like to convey are not the technicalities, of course, but the flexibility of the approaches – simple modifications to a general framework allow for the semantical description (and sometimes also for syntactical results, like soundness) of a wide spectrum of formal systems.

2.1 Context semantics

Context semantics [26] is a powerful framework for the analysis of proof and program dynamics. It can be considered as a model of Girard's geometry of interaction [24], [23] where the underlying algebra consists of *contexts*. As such, it is a dynamic, interactive semantical framework, with many similarities to game semantics [3]. In these semantic frameworks, one manages to identify those proofs interacting in the same way with the environment (*i.e.*, that cannot be distiguished, through normalization, no matter the context in which they are used in a larger proof; technically: observationally equivalent proofs) by proving that the semantic objects which interpret them are the same. In some cases, one can prove that the converse implication holds – this property is called *full-abstraction* in the literature. Context semantics and geometry of interaction have been used to prove the correctness of optimal reduction algorithms [26] and in the design of sequential and parallel interpreters for the λ-calculus [31], [33].

Notice that whenever a proof π is obtained from another proof ρ by cut-elimination (or normalization), π and ρ have the same interactive behaviour and, as a consequence, cannot be distinguished in the semantics. This implies that the "complexity" of normalizing a proof π *cannot* be read out of the interpretation of π itself, in any standard semantical approach. What is appealing in context semantics is that its formulation, differently from the majority of other approaches, allows for easy modifications of the semantics of π (without altering the basic mathematical concepts needed), in such a way that the interpretation of π somehow "reflects" the computational difficulty of normalizing it. A first example of this flexibility of the context semantics, is the well known fact that strongly normalizing proofs are exactly the ones having finitely many so-called regular paths in the geometry of interaction [18]. As a second example, a class of proofs which are not just strongly normalizing but normalizable in elementary time can still be captured in the geometry of interaction framework, as suggested by Baillot and Pedicini [6]. Until recently it was not known, however, whether this correspondence scales down to smaller complexity classes, such as the one of polynomial time computable functions. The usual measure based on the length of regular paths cannot be used, since there are proofs which can be normalized in polynomial time but whose regular paths have exponential length [18].

The solution to the above problem is relatively simple: not *every* (regular) path should be taken into account, but only those having something to do with duplication. This way, one can define the *weight* W_π of any proof as the total number of paths in this restricted class. On one hand, we may prove that W_π is an upper bound to the computational difficulty of normalizing π:

Theorem 2. *There is a polynomial $p : \mathcal{N}^2 \to \mathcal{N}$ such that for every proof π, π normalizes in at most $p(W_\pi, |\pi|)$ steps and the size of any reduct ρ of π is at most $p(W_\pi, |\pi|)$.*

On the other hand, W_π can be proved to be a lower bound to the complexity of normalizing π:

Theorem 3. *For every proof π there is another proof ρ such that π normalizes in n steps to ρ and $W_\pi \geq n + |\rho|$.*

The two results above highlight a strong correspondence between W_π and some crucial quantitative attributes of π, i.e. the (maximal) number of steps to normal form for π and the (maximal) size of reducts of π. Perhaps the most interesting application of this correspondence are new, simple proofs of soundness for elementary, light and soft linear logic. After all, W_π can be defined for any (multiplicative and exponential) proof π, while proving bounds on W_π when some of the rules are forbidden (as in the cited logical systems) turns out to be very simple. Details about the results sketched in this paragraph can be found in [14].

The same technique can be applied to similar, but different, systems. A first example consists in linear lambda calculi with higher-order recursion [13]. There, a parameter influencing the complexity of any term M is the (maximal) size of ground subterms of reducts of M, called the *algebraic potential size* of M. Context semantics is powerful enough to induce bounds on the algebraic potential size of terms, in the same vein as in Theorem 2. Another example is optimal reduction [1], where context semantics allows to transfer known complexity results from global reduction, where proofs are normalized in the usual way, to local reduction, where a slightly different syntax for proofs allows a completely local notion of reduction [5], itself necessary to get optimality for lambda calculus.

2.2 Realizability semantics

Kleene's realizability and its variations have been used for various purposes in modern logic (see [38] for an introduction). In particular, realizability has been used in connection with complexity-constrained computation in several places. The most prominent one is Cook and Urquhart's work [9], where terms of a language called PV^ω are used to realize formulas of bounded arithmetic. There, realizability is used to show polytime soundness of a logic (for a slightly different notion of soundness than the one considered here). Realizers in [9] are typed and very closely related to the logic that is being realized. This implies that any modification to the underlying logical system would require changing as well the language of realizers.

On the other hand, one can use untyped realizers and interpret formulas as partial equivalence relations on those, this way going towards a unifying framework. This has been done by Crossley et al. [10], in the same spirit as Kreisel's untyped realizability model HEO [28]. The proof that the untyped realizers (in this case of formulas of bounded arithmetic) can indeed be normalized in bounded time, has to be performed "externally" (that is, it is not obtained directly from the model). In other words, proving soundess of a logic has been reduced to proving soundness of a certain class of realizers.

Recently, a realizability framework has been introduced by one of the authors, as a joint work with Martin Hofmann [15]. The framework is designed as a refinement of the existing ones. In standard realizability, formulas are usually interpreted as sets of pairs (t, f), where t (a realizer, e.g., a lambda term) realizes f (a denotational object, e.g., a function); in our approach formulas are interpreted as sets of triples (α, t, f), where t realizes f as usual but, moreover, t is *majorized* by α (itself an element of an algebraic structure called a *resource monoid*). The notions of majorization and of resource monoid are relatively simple and designed to guarantee that whenever $(\alpha, t, f) \in A \to B$ and $(\beta, u, g) \in A$, t computes a realizer for $f(g)$ in time bounded by $\mathcal{F}(\alpha + \beta)$ when applied to u ($\mathcal{F}$ is a function from majorizers to natural numbers which is part of the definition of the underlying resource monoid). Proving that $\mathcal{F}$ is bounded (e.g., by something like a polynomial) is usually easy, once a resource monoid is fixed. Therefore, by choosing a proper resource monoid, "intensional" soundness can be obtained as a corollary of "extensional" soundness while giving meaning to a logic. Notice, however, that this way we do not prove anything about the time needed to normalize *proofs*, but we rather prove something about *realizers* corresponding to the intereperted proofs.

Once a resource monoid $\mathcal{M}$ has been fixed, one can present the above construction as a category $\mathscr{L}(\mathcal{M})$: objects are *length spaces* (i.e., the ternary relations described above) while morphisms are realized and majorized functions between them. Noticeably, $\mathscr{L}(\mathcal{M})$ is symmetric monoidal closed for every $\mathcal{M}$. This means that any category $\mathscr{L}(\mathcal{M})$ is a model of (second order) multiplicative linear logic, and that one must only guarantee that the *exponential* structure can be justified when interpreting systems like LAL or SLL in such a category. This has been done indeed [15], obtaining as a byproduct new proofs of soundness for LAL, SAL (the affine version of SLL) and EAL. Noticeably, even systems going beyond the realm of linear logic (e.g., Hofmann's LFPL) can be proved sound.

2.3 Other frameworks

In the last two sections, we focused on semantic frameworks where relevant, quantitative properties of the interpreted proof can be "read out" from the underlying semantic object. This way, semantics can be effectively used as a tool when proving soundness.

Semantics, however, has also a more traditional role (in ICC, as in any other domain), somehow orthogonal to the goals we had so far. A mathematical model of a formal system is, by its very essence, a different description of the syntax, expressed with new concepts, and embedding the syntactical objects in the structures the model is built upon. The more simple, non-syntactical, and mathematically elegant a model is, the more it is interesting and able to give us a fresh look on our formal system. Also in this sense, therefore, semantics is much welcomed in ICC.

Once a mathematical object $\mathscr{M}$ is proved to be a fully complete model of a logical system, itself sound and complete with respect to a complexity class $\mathscr{C}$, the model $\mathscr{M}$ can be understood as a new, presumably simpler, presentation of $\mathscr{C}$. Here, we are not interested in inferring intensional properties of proofs from $\mathscr{M}$ – indeed, not much about the intensional expressive power of a logic can be inferred from any fully complete model for it. We would like, instead, to get global insight on the structure (and properties) of $\mathscr{C}$. As pointed out by Girard [25] denotational models of light logics could thus help in shedding some light on the nature of polytime from an extensional point of view, maybe allowing to attack one of the many open problems in computational complexity. In the rest of this section, we briefly recall three proposals designed around these lines.

First, one should mention Baillot's *stratified coherent spaces* [4], a refinement on Girard's coherent spaces in which semantic stratification mimicks the absence of digging and dereliction in ELL. Indeed the two principles mentioned are not valid in stratified coherent spaces. By a further enrichment, one can obtain a model for LLL through *stratified measured coherence spaces*. In all these cases, however, full completeness does not hold, due to the inherent incompleteness of coherent spaces (already at the level of linear logic).

If one forgets about coherence in coherent spaces, what is left is relational semantics, maybe the simplest model of linear logic. Again, one can define a restricted notion of this model, this time through *obsessional cliques* [30]. Of course, the obtained model is not fully complete with respect to any of the known light logics. However, one can prove relative completeness for ELL: the clique interpreting a MELL proof π is obsessional (in a certain technical sense) *if and only if* π is an ELL proof. A similar result holds for SLL.

The only example of a semantic object which is fully complete with respect to a light logic is the model based on Murawski and Ong's *discreet games* [32]. They can be constructed as appropriate restrictions on usual game structures, already known to be fully complete with respect to MELL.

3 ICC in the small

Giving a *reasonable* cost model for λ-calculus grounded on natural notions is a non trivial challenge. From the logical point of view, the most interesting parameter is the number of β-reductions, but, as we briefly discussed in Sect. 1, it is difficult to give general arguments proving that this simple approach gives rise to a reasonable model. As usual, the problem stems from duplication; in a single step $(\lambda x.M)N \to M[x \leftarrow N]$, the term N may be duplicated as many times as the number of occurrences of x in M. While in general this precludes to use the number of beta steps as the main ingredient in a cost model, there are restrictions of the reduction for which something can be done. In particular, we will concentrate on pure untyped λ-calculus endowed with *weak* (that is, we never reduce under an abstraction) *call-by-value reduction*. The following definitions are standard.

Definition 1.

- *Terms are defined as usual:* $M ::= x \mid \lambda x.M \mid MM$, *where x ranges over a denumerable set. Λ denotes the set of all lambda terms.*
- *Values are defined as* $V ::= x \mid \lambda x.M$.
- *Weak call-by-value reduction is denoted by* $\to_v$ *and is obtained by closing call-by-value reduction under any applicative context:*

$$\overline{(\lambda x.M)V \to_v M[x \leftarrow V]} \qquad \frac{M \to_v N}{ML \to_v NL} \qquad \frac{M \to_v N}{LM \to_v LN}.$$

Here M ranges over terms, while V ranges over values. We denote with $\twoheadrightarrow$ the reflexive and transitive closure of $\to_v$.

- *The length $|M|$ of M is defined as follows, by induction on M: $|x| = 1$, $|\lambda x.M| = |M| + 1$ and $|MN| = |M| + |N| + 1$.*

Weak call-by-value reduction enjoys many nice properties. In particular, the one-step diamond property holds and, as a consequence, the number of beta-steps to normal form (if any) is invariant on the reduction order [16]. These properties enable the definition of interesting cost models for this calculus. We will discuss two different points of view.

First, we consider the case in which λ-terms are represented explicitly. Then, we will discuss the situation where we are allowed to choose more compact (and therefore efficient) representations.

3.1 Explicit representation

With explicit representation, a term M is represented by its explicit, uncompressed description (that is, with a data structure of size $|M|$). In particular, from the representation of a term in normal form, the term is immediately available (a situation which will change for implicit representations). In this case, we define the *difference cost* of a single step of β-reduction $M \to_v N$ as $\max\{1, |N| - |M|\}$. The following definition and results are from [16].

Definition 2 (Difference cost model). *If $M \twoheadrightarrow N$, where N is a normal form, then $Time_d(M)$ is defined as $|M|$ plus the sum of the difference costs of the individual steps of the sequence $M \twoheadrightarrow N$. If M diverges, then $Time_d(M)$ is infinite.*

In view of the invariance of the number of beta-steps to normal form, this is a good definition. As an example, consider the term $\underline{n}\ 2$, where $\underline{n} \equiv \lambda x.\lambda y.x^n y$ is the Church numeral for n. It reduces to normal form in one step, because we do not reduce under the abstraction. To force reduction, consider $E \equiv \underline{n}\ 2\ x$, where x is a (free) variable; E reduces to

$$F \equiv \lambda y_n.(\lambda y_{n-1} \ldots (\lambda y_2.(\lambda y_1.x^2 y_1)^2 y_2)^2 \ldots)^2 y_n$$

in $\Theta(n)$ beta steps. However, $\mathit{Time}_d(E) = \Theta(2^n)$, since at any step the size of the term is duplicated. Indeed, the size of F is exponential in n.

We are left to state that this cost model makes weak call-by-value reduction a reasonable machine. That is, (i) we can simulate Turing machine computations with β-reduction, and (ii) we can implement β-reduction over Turing machines, in both cases with a polynomial overhead on the cost with Turing machines.

Theorem 4. *If $f : \Sigma^* \to \Sigma^*$ is computed by a Turing machine M in time g, then there are a λ-term N_M and a suitable encoding $\ulcorner \cdot \urcorner : \Sigma^* \to \Lambda$ such that $N_M \ulcorner v \urcorner$ normalizes to $\ulcorner f(v) \urcorner$ and $\mathit{Time}_d(N_M \ulcorner v \urcorner) = O(g(|v|))$.*

Theorem 5. *We can construct a multitape Turing machine $\mathcal{R}$ computing the normal form of any term M in $O((\mathit{Time}_d(M))^4)$ steps.*

3.2 Compact representation

If we are allowed to use more efficient representations, we can get more parsimonious cost models. In particular, under certain conditions, we can in fact choose the number of β-steps to normal form as a reasonable cost model.

One the most appealing representations for λ-terms is the graph-based one, known at least since Wadsworth's thesis [40]. A term is represented first with its *syntax tree*, linking then the nodes representing bound variables to the λ-node binding them. The crucial idea, at this point, is to perform reduction without textual substitution, and updating – as much as possible – the arcs of the graph (*i.e.*, performing update of pointers in the implementation). When we have a redex $(\lambda x.M)N \to M[x \leftarrow N]$, just erase the λ and the application nodes of the redex and link the arcs representing x (from the graph of M) to the root of the graph representing N. If there is more than one occurrence of x, N is not duplicated, but simply "shared" by all the arcs represented the different x's. The problem is that, sooner or later, some portion of a redex has to be *actually* duplicated (see [1], pp 14-15). The cost of normalization, thus, has to take into account the size of the duplicated part, which has no relation, in general, with the size of the original term (or with other sensible parameters). In other words, we would be back at the cost model of Sect. 3.1, taking into account the size of the intermediate terms. However, in case of weak (call-by-value and call-by-name) reduction we can drastically improve on Sect. 3.1 and show that the number of β-reduction steps *is indeed* a polynomially invariant cost model. The crucial observation is that, for these reductions, there exist ways to represent λ-terms in which subterms can be efficiently shared. Moreover, for this representation, any subterm needing to be duplicated during the reduction is a subterm *of the original term*. This has been obtained first by Sands, Gustavsson, and Moran [35], by a fine analysis of a λ-calculus implementation based on a stack machine.

We may give a more formal treatment [17], by exploiting techniques as defunctionalization and graph-rewriting. By using defunctionalization, we first encode a λ-term M into a term $[M]_\Phi$ in a (first-order) constructor term rewrite system, where any λ-abstraction is represented by an atomic constructor.

Theorem 6. *Weak (call-by-value and call-by-name) β-reduction can be simulated step by step by first-order rewriting.*

At this point we are left to show that first-order rewriting can be implemented so efficiently that the number of reduction steps can be taken as the actual cost of the reduction (up to a fixed polynomial). Without going into technical details, we may define a further encoding $[\cdot]_\Theta$ that, applied to a first order term, returns a graph. Our original term M would then be translated into $[[M]_\Phi]_\Theta$, composing the two encodings. We prove, further, that graph-rewriting simulates step by step constructor rewriting. We can eventually exploit the crucial observation about the first-order encoding of a λ-term: during the reduction of $[M]_\Phi$ we always manipulate subterms of the original term M.

Definition 3 (Unitary cost model). *For any λ-term M, $Time_u(M)$ is defined as the number of β-steps to normal form, under weak reduction. If M diverges, then $Time_u(M)$ is infinite.*

Some simple combinatorics allows to prove the following theorem.

Theorem 7. *There is a polynomial $p : \mathcal{N}^2 \to \mathcal{N}$ such that for every lambda term M, the normal form of $[[M]_\Phi]_\Theta$ (under weak reduction) can be computed in time at most $p(|M|, Time_u(M))$.*

It is worth observing that $[[N]_\Phi]_\Theta$ is in general very different from N, the former being a *compact* graph representation of the latter. Additional "read-back" work must in general be performed to obtain from $[[N]_\Phi]_\Theta$ its explicit, sequentialized form N. To factor out this time from our cost model should be considered a feature of the model. Indeed, we cannot do miracles. Consider again the term $E \equiv \underline{n}\ \underline{2}\ x$, introduced earlier. Under weak call-by-value reduction this term reduces to normal form in $\Theta(n)$ beta steps, but there is an exponential gap between this quantity and the time just needed to write the normal form, which is of length $\Theta(2^n)$.

4 Conclusions

Proof-theoretical implicit computational complexity is a young research area which has given in the last ten years a fresh look to a number of different subjects. In this short paper we tried to give a (*very*) personal overview of some of its main achievements and of its techniques. Still, we have to admit it is a very fragmented area. Too many systems, too few general results. We have suggested some semantical paths, but it is not sufficient.

An important problem we do not know how to attack is *intensional expressiveness*. We mentioned several complete systems for certain complexity classes. Completeness is an *extensional* notion – for any function in a complexity class, there is a term (proof, program) in our complete system. For the systems we mentioned, the term guaranteed to exists is seldom the most intuitive one, from a programming point of view. Moreover, there are a lot of natural algorithms (terms, proofs, programs) which *do have* the right complexity, but which *are not* expressible in our systems (they are not typable, or get the wrong type). Filling the gap between extensional completeness and the miserable intensional expressiveness of these systems is probably the most interesting challenge of the field.

References

1. A. Asperti and S. Guerrini: *The Optimal Implementation of Functional Programming Languages* (Cambridge University Press, Cambridge 1998)
2. A. Asperti and L. Roversi: Intuitionistic Light Affine Logic. *ACM Transactions on Computational Logic* **3**, 1 (2002) pp 137–175
3. P. Baillot: Approches dynamiques en sémantique de la logique lineaire: jeux et géometrie de l'interaction. (Université Aix-Marseille 2, 1999)
4. P. Baillot: Stratified coherence spaces: a denotational semantics for light linear logic. *TCS* **318**, 1-2 (2004) pp 29-55
5. P. Baillot, P. Coppola and U. Dal Lago: Light Logics and Optimal Reduction: Completeness and Complexity. In: *22nd LICS* (IEEE Comp. Soc., 2007) pp 421-430
6. P. Baillot and M. Pedicini: Elementary Complexity and Geometry of Interaction. *Fundamenta Informaticae* **45**, 1-2 (2001) pp 1-31
7. P. Baillot and K. Terui: Light Types for Polynomial Time Computation in Lambda-Calculus. In: *LICS 2004* (IEEE Comp. Soc., 2004) pp 266–275
8. A. Church: A set of postulates for the foundation of logic. *Ann. of Math. (2)* **33** (1932) pp 346-366
9. S. Cook and A. Urquhart: Functional Interpretations of Feasible Constructive Arithmetic. *Annals of Pure and Applied Logic* **63**, 2 (1993) pp 103-200
10. J. Crossley, G. Mathai and R. Seely: A Logical Calculus for Polynomial-time Realizability. *Journal of Methods of Logic in Computer Science*, **3** (1994) pp 279-298
11. H. B. Curry and R. Feys: *Combinatory Logic*, vol. 1 (North-Holland, Amsterdam 1958)
12. N. Cutland: *Computability: An Introduction to Recursive Function Theory* (Cambridge University Press, Cambridge 1980)
13. U. Dal Lago: The Geometry of Linear Higher-Order Recursion. In: *20th LICS* (IEEE Comp. Soc., 2005) pp 366-375

14. U. Dal Lago: Context Semantics, Linear Logic and Computational Complexity. In: *21st LICS* (IEEE Comp. Soc., 2006) pp 169-178
15. U. Dal Lago and M. Hofmann: Quantitative Models and Implicit Complexity. *Proc. Found. of Software Techn. and Theor. Comp. Sci.* (2005) pp 189-200
16. U. Dal Lago and S. Martini: An Invariant Cost Model for the Lambda Calculus. In: *Logical Approaches to Computational Barriers. LNCS* **3988** (Springer-Verlag, Berlin Heidelberg New York 2006)
17. U. Dal Lago and S. Martini: Beta Reduction is a Cost Model for the Weak Lambda Calculus. Available from the authors (2007)
18. V. Danos and L. Regnier: Proof-nets and Hilbert space. In: *Advances in Linear Logic*, ed by J.-Y. Girard, Y. Lafont and L. Regnier (Cambridge University Press, Cambridge 1995) pp 307–328
19. M. Gaboardi, J.-Y. Marion and S. Ronchi Della Rocca: *Proceedings of 35th ACM POPL. A Logical Account of PSPACE*. To appear
20. G. Gentzen: Untersuchungen über das logische Schließen I and II. *Mathematische Zeitschrift* **39**, 1 (1935) pp 176–210; 405-431
21. J.-Y. Girard: *Proof Theory and Logical Complexity, I* (Bibliopolis, Napoli 1987)
22. J.-Y. Girard: Linear Logic. *TCS* **50** (1987) pp 1–102
23. J.-Y. Girard: Geometry of interaction 2: deadlock-free algorithms. In: *COLOG-88: Proc. of the int. conf. on Computer logic. LNCS.* **417** (Springer-Verlag, Berlin Heidelberg New York 1988) pp 76-93
24. J.-Y. Girard: Geometry of Interaction 1: Interpretation of system F. In: *Proc. Logic Colloquium '88* (North-Holland, Amsterdam 1989) pp 221-260
25. J.-Y Girard: Light Linear Logic. *Inform. and Comp.* **143**, 2 (1998) pp 175–204
26. G. Gonthier, M. Abadi and J.-J. Lévy: The Geometry of Optimal Lambda Reduction. In: *Proc. 12th ACM POPL* (1992) pp 15-26
27. W. A. Howard: The Formulae-as-Types notion of Construction. In: *To H. B. Curry: Essays on Combinatory Logic, Lambda Calculus and Formalism*, ed by J. P. Seldin and J. R. Hindley (Academic Press Inc., New York 1980) pp 479-490
28. G. Kreisel: Interpretation of analysis by means of constructive functions of finite types. In: *Constructivity in Mathematics*, ed by A. Heyting (North-Holland, Amsterdam 1959) pp 101-128
29. Y. Lafont: Soft Linear Logic and Polynomial Time. *TCS* **318** (2004) pp 163–180
30. O. Laurent and L. Tortora de Falco: Obsessional Cliques: A Semantic Characterization of Bounded Time Complexity. In: *LICS* (2006) pp 179-188
31. I. Mackie: The Geometry of Interaction Machine. In: *Proc. 22nd ACM POPL* (1995) pp 198-208
32. A. S. Murawski and C.-H. L. Ong: Discreet Games, Light Affine Logic and PTIME Computation. *CSL* (2000) pp 427-441
33. J. S. Pinto: Parallel Implementation Models for the lambda-Calculus Using the Geometry of Interaction. *Proc. 5th International Conference on Typed Lambda Calculi and Applications* (2001) pp 385-399
34. G. D. Plotkin: Call-by-Name, Call-by-Value and the lambda-Calculus. *TCS* **1**, 2 (1975) pp 125-159
35. D. Sands, J. Gustavsson and A. Moran: Lambda Calculi and Linear Speedups. In: *The Essence of Computation: Complexity, Analysis, Transformation. Essays Dedicated to Neil D. Jones. LNCS* **2566** (Springer-Verlag, Berlin Heidelberg New York 2002)
36. R. Statman: The Typed lambda-Calculus is not Elementary Recursive. *TCS* **9** (1979) pp 73-81

37. K. Terui: Light affine lambda calculus and polynomial time strong normalization. *Arch. Math. Log.* **46**, 3-4 (2007) pp 253-280
38. A. S. Troelstra: Realizability. In: *Handbook of Proof Theory*, ed by S. Buss (Elsevier, 1998) pp 407-473
39. P. van Emde Boas: Machine Models and Simulation. In: *Handbook of Theoretical Computer Science, vol A: Algorithms and Complexity (A)*, (MIT Press, Boston 1990) pp 1-66
40. C. P. Wadsworth: Semantics and pragmatics of the lambda-calculus. Phd Thesis, Chapter 4 (1971)

Quantum Combing

Mario Rasetti

Dipartimento di Fisica e Scuola di Dottorato, Politecnico di Torino (Italy)
mario.rasetti@polito.it

1 Alphabets, sequences, languages and all that

Leading idea of this note is to argue that quantum information manipulation tools may allow us to explore much wider fields than mere computation, reaching beyond its boundaries to touch the very roots of the universal structure of languages. The paper is mostly conjectural and touches just the few technical details necessary to pursue the general argument, because its main aim is simply to show how a complex blend of notions coming from formal language theory, finite group theory, and quantum computation theory can lead to new views. As working study-case the problem of combing finite groups will be dealt with, which bridges language theoretical issues with structural and algorithmic issues.

To begin with, let us recall that the basic ingredient of a language is its alphabet. An alphabet is a finite set. The members of the alphabet are symbols; often characters. A finite sequence over an alphabet $\mathfrak{S}$ is a function from $\{0, 1, ..., N-1\}$ (for some integer $N \geq 1$) to $\mathfrak{S}$. The size of the function's domain is the length of the sequence. One writes sequences in this form: $\mathfrak{aab}$ is a sequence mapping 0 to symbol $\mathfrak{a}$, 1 to $\mathfrak{a}$, and 2 to $\mathfrak{b}$. A sequence of length 1 is written as $\mathfrak{a}$ and the sequence of length 0 as ϵ. A language over an alphabet $\mathfrak{S}$ is a set of finite sequences over that alphabet. For example if we have an alphabet of two distinct symbols, say $\mathfrak{a}$ and $\mathfrak{b}$, then each of the following is a different language over that alphabet

$$\{\} \, , \, \{\epsilon\} \, , \, \{\mathfrak{a}, \mathfrak{b}, \mathfrak{ab}, \mathfrak{ba}\} \, , \, \{\mathfrak{a}, \mathfrak{ab}, \mathfrak{aba}, \mathfrak{abab}, \mathfrak{ababa}, ...\} \, , \, \{\epsilon, \mathfrak{a}, \mathfrak{aa}, \mathfrak{aaa}, \mathfrak{aaaa}, ...\} \, .$$

The first is the empty language, whereas the second is a language containing only the empty sequence; they are different. Notice that while all the sequences in a language are finite, the language itself can be infinite (as in the last two examples).

A family of languages is defined to be any non-empty set of languages over finite alphabets; in particular, the universal set $\mathcal{U}$ denotes the set of all languages with finite alphabet. Many families of formal languages are known, including the four families of the Chomsky Hierarchy [8], [9], [11], [12], (regular sets, context-free languages, context sensitive languages and recursively enumerable sets), recursive sets, and indexed languages. A rigorous formal mathematical setting of context-free languages and of language theory from a group theoretical standpoint has been constructed [29], [21]. Also indexed languages and bounded languages are exhaustively studied in the literature.

A formal language may be reconducted to a machine which recognises it [13], [14], [15], [10], [32]; for example, regular sets are recognised by finite state automata, context-free languages by (non-deterministic) pushdown automata, recursively enumerable and recursive sets by Turing machines and halting Turing machines respectively. Alternatively, a formal language may be generated (defined) by the set of its grammatical rules, as halting Turing machines respectively. Alternatively, a formal and generated by grammars.

It is worth recalling that the philosophy of Chomsky's linguistic theory has changed the long, traditional way of studying language. The nature of knowledge, closely tied to human knowledge in general as it is, made it a logical step for Chomsky to use his theory to analyze the relation between language and the world – in particular, to give life to the study of truth and reference, thus turning linguistics into a science of mind/brain. The theory is based on the assumption that "languages are learnt with limited stimuli" (the so called problem of poverty of evidence), namely the recognition that the input during the acquisition of a natural language is circumscribed and degenerate [13], [14], [15]. Moreover, the output cannot be simply accounted for by learning mechanisms only, such as induction and analogy on the input. Output and input differ both in quantity and quality. A subject may know linguistic facts without instruction or even direct evidence. Knowledge of language is normally attained through exposure, and the character of the acquired knowledge may be largely predetermined [10]. This predetermined knowledge is tied to some "notion of structure" in the mind, which guides the subject in acquiring a natural language of his own.

Too far reaching as it may appear, this is nevertheless a plausible – yet challenging to the very limit of the possible in terms of cognition – motivation for the analysis that will follow.

Listing all the elements of a language necessary to determine its structure is (in principle) possible for finite languages, but of course not for infinite languages and a notation is needed to describe languages. Regular expressions provide one such notation. Several methods support regular expressions using various syntaxes; these syntaxes are not unique.

There are five basic ways to form a regular expression. Any member of the alphabet is itself a regular expression: it describes the language that consists of one sequence which in turn consists of that one symbol exactly once. For example, if the alphabet contains symbols $\mathfrak{a}$ and $\mathfrak{b}$, we have for regular expression $\mathfrak{a}$ the language $\{\mathfrak{a}\}$. Also the symbol ϵ is a regular expression: it describes the language $\{\epsilon\}$. More complex regular expressions are: Alternation, Concatenation, Repetition:

- If E and F are regular expressions, their "alternation", denoted $E|F$, is also a regular expression, that consists of the set of all sequences described by either E or F (or both).
- If E and F are regular expressions, their "concatenation", denoted EF, is also a regular expression, that consists of the set of all sequences one can make by gluing together an ordered sequence described by E and one described by F.
- Using the above notations, one can describe any non-empty finite language, but no infinite language. The key to describing infinite languages is "repetition". If E is a regular expression, then repetition, E^*, is also a regular expression describing sequences that are made up of the catenation of a finite number of sequences, each described by E. If one allowed infinitely large regular expressions (which typically one doesn't), then E^* would describe the same language as

$$\epsilon|E|EE|EEE|\cdots.$$

Given a sequence and a regular expression, one can legitimately ask whether the sequence is described by the regular expression in $\mathcal{O}(N)$ time and $\mathcal{O}(1)$ space, where N is the length of the sequence.

Regular expressions cannot describe all languages. Any language that can be described by a regular expression is a regular language. To go beyond regular expressions, one needs extended context free grammars. An extended context free grammar consists of an alphabet, a nonempty finite set of symbols not in the alphabet (referred to as the set of "non-terminal symbols") and a set of "productions". A production consists of a nonterminal symbol $\mathcal{N}$ and a regular expression E, and is denoted by $\mathcal{N} \mapsto E$. Each nonterminal symbol appears on the left-hand side of exactly one production. The alphabet of the regular expressions is the union of the alphabet, the grammar and the set of nonterminal symbols. One nonterminal symbol is singled out to be the start nonterminal.

This is an example grammar: an alphabet $\{\ell, \mathfrak{a}, \mathfrak{b}\}$, nonterminal symbols $\{\mathcal{S}, \mathcal{B}\}$, where $\mathcal{S}$ is the start nonterminal, and productions, say, $\mathcal{S} \mapsto \ell\mathcal{B}\mathfrak{b}$ and $\mathcal{B} \mapsto \ell\mathcal{B}\mathfrak{b}|\mathfrak{a}$.

Basic idea is that one starts with the start nonterminal and replaces it with any sequence in the language, described by its regular expression. Then one picks up any occurrence of a nonterminal in the resulting sequence, and replaces it with any sequence in the language described by the nonterminal's regular expression. Finally, one repeats the process until there are no nonterminals left. Any sequence one can reach by this procedure is assumed to belong to the language described by the grammar.

One writes $\mathfrak{s} \Rightarrow \mathfrak{t}$ to mean that sequence $\mathfrak{t}$ can be obtained from sequence $\mathfrak{s}$ by picking one occurrence of a nonterminal in $\mathfrak{s}$ are replacing it by one sequence in the language described by that nonterminal's regular expression. Using the example grammar above, one can see that:

$$\begin{aligned} S &: \quad \text{replace } S \text{ with } \ell B\mathfrak{b} \\ \Rightarrow &: \quad \ell B\mathfrak{b} \quad \text{replace } B \text{ with } \ell B\mathfrak{b} \\ \Rightarrow &: \quad \ell\ell B\mathfrak{b}\mathfrak{b} \quad \text{replace } B \text{ with } \mathfrak{a} \\ \Rightarrow &: \quad \ell\ell\mathfrak{a}\mathfrak{b}\mathfrak{b} \,, \end{aligned} \qquad (1)$$

proves that $\ell\ell\mathfrak{a}\mathfrak{b}\mathfrak{b}$ is in the language described by that grammar. The language is

$$\{\ell\mathfrak{a}\mathfrak{b}, \ell\ell\mathfrak{a}\mathfrak{b}\mathfrak{b}, \ell\ell\ell\mathfrak{a}\mathfrak{b}\mathfrak{b}\mathfrak{b}, ...\} \,,$$

and it is not regular. Those languages that can be described by an extended context free grammar are called context free languages.

Acquiring a language in terms of the internal language, corresponds to the change of a subject's mind/brain state: knowing the language $\mathcal{L}$ consists for the subject's mind/brain, initially in a state $\mathfrak{S}_\mathcal{O}$, to be set to another state $\mathfrak{S}_\mathcal{L}$. To explain what this implies for the individual's brain (in particular, its language faculty) that corresponds to its knowing $\mathcal{L}$ is of course task of the brain sciences. Chomsky proposes the crucial hypothesis that there exists a universal grammar ($\mathcal{UG}$). $\mathcal{UG}$ is a characterization of the innate principle of language faculty [13], [14], [15], [10]. The detailed structure of $\mathcal{UG}$ is a system of conditions on grammars; constraints on the form and interpretation of grammar at all levels, from the deep structure of syntax, through the transformational component, to the rules that interpret syntactic structures semantically and morphologically. "Semantics" means indeed the study of the relation between language and the world – in particular, the study of truth and reference. Comprehension of linguistic universals is the study, classified as substantive, of the properties of $\mathcal{UG}$ for a natural language.

Substantive universals concern the vocabulary for the description of language and a formal linguistic universal involves the character of the rules that appear in grammars as well as the ways in which they can be interconnected. The idea is that any language-acquisition device uses primary formal language data as empirical basis for language learning to meet the explanatory adequacy inherent in $\mathcal{UG}$, and to select one of the potential grammars permitted by $\mathcal{UG}$.

The theory of complex systems shows that a vast system such as the human brain, highly flexible and able to process abstract concepts at many different levels as it is, may exhibit features intractable at the algorithmic level. Chomsky's proposal however is indeed a plausible theory of language and when he maintains that the boundary between linguistics and natural sciences will shift or disappear he fully exhibits his far-reaching vision. The theory of mind should aim to determining the properties of the initial state $\mathfrak{S}_\mathcal{O}$ as well as of each attainable state $\mathfrak{S}_\mathcal{L}$ of the language faculty, while the brain sciences should seek to discover the mechanisms of the brain that constitute the physical realizations of such states. Eventually, language theory and the brain sciences will converge in their effort to discover the mechanisms of brain that are the physical realization of the state $\mathfrak{S}_\mathcal{L}$, while $\mathcal{UG}$ is biologically determined.

2 Combing finitely generated groups

A further topic toward the argument of this note is the fact that, as thoroughly and rigorously argued in [29], there exist deep connections between language theory and the properties (structural, combinatorial, topological) of groups. Here – with no substantial loss of generality – the focus will be mainly on finitely presented groups. An automatic group [2] is a finitely generated group equipped with several finite-state automata [32], [1], [40]. These automata can tell whether a given word representation of a group element is in a "canonical form" and whether two elements given by canonical words differ by a generator. A particularly interesting and intriguing structural property of finitely presented automatic groups is "combing". A group is combable if it can be represented by a language of words satisfying a "fellow traveller" property.

More precisely, let G be a group and $\hat{\mathfrak{X}}$ a finite (sub)set of generators of G. An automatic structure of G with respect to $\hat{\mathfrak{X}}$ is a set of finite-state automata: i) the word-acceptor, which accepts for every element of G at least one word in $\hat{\mathfrak{X}}$ representing it; ii) multipliers, one for each $\mathfrak{g} \in \hat{\mathfrak{X}} \cup \{e\}$, ($e$ being the group identity element) which accept a pair (w_ℓ, w_j) of words accepted by the word-acceptor, precisely when $w_\ell \mathfrak{g} = w_j$ in G. The property of being automatic does not depend on the set of generators. In particular, the **braid group** [5] is an automatic group. One can find the properties of groups with combing, as well as the closure properties of the associated classes of groups, in various formal language classes.

The concept of **combing** [6] for **a finitely generated group** has grown out just of the definition of automatic group; evolving to the property that a group is automatic precisely when it possesses a regular synchronous combing. More details will be given below; for the moment, roughly speaking, combing is the process of generation of an orderly set of strands through the Cayley graph of the group, which is regular if it is defined by a finite state automaton. Groups with asynchronous combing – which is an indexed language – can also be defined, by weakening both of the two restrictions on the "language" associated with an automatic group: the geometric "fellow traveller condition" is relaxed from synchronous to asynchronous, and the language theoretic requirement of regularity is replaced by the requirement that the language be indexed (that is, recognised by a one-way nested stack automaton, a type of machine which is more general than a pushdown automaton). An automatic group has a "synchronous" combing which is a regular language.

In order to make the argument more effective, the notions of fellow traveller property and their connection with languages and combings will now first briefly reviewed. Successively their application will be considered to the study of the problem of combing for a specific class o groups: parenthesized (finite) groups, which will be assumed as study case and metaphor.

Let G be a finitely presented group, with identity element e, and let $\mathfrak{X}$ be the finite generating set of G. Assume that $\mathfrak{X}$ is inverse closed, that is, it contains also the inverse of each of its elements. A product of elements in $\mathfrak{X}$ is a "word in G", and one denotes by $\mathfrak{X}^*$ the set of all such words. Let $\Gamma = \Gamma_{G,\mathfrak{X}}$ be the Cayley graph for G over $\mathfrak{X}$, with vertices corresponding to the elements of G, and, for each $x \in \mathfrak{X}$, a directed edge from the vertex g to the vertex $g \cdot x$, labelled by x. Let moreover $d_{G,\mathfrak{X}}$ measure the "chemical" distance between vertices of $\Gamma_{G,\mathfrak{X}}$. For words $w, v \in \mathfrak{X}^*$, one writes $w = v$ if w and v are identical as words, $w =_G v$ if w and v represent the same element of G. Let $\ell(w)$ denote the length of w as a string, and $\ell_G(w)$ the length of the shortest word v with $v =_G w$. It is also possible to extend $d_{G,X}$ to a differentiable metric on the 1-skeleton of Γ [3]. Then each word w can be associated with a differentiable path from e labelled by w such that, for $t < \ell(w)$, the path from e to $w(t)$ has length t, and for $t \geq \ell(w)$, $w(t) = w(\ell(w))$.

Suppose now that v, w are words in $\mathfrak{X}^*$, and that $k \in \mathbb{N}$. Various "fellow traveller" properties can describe the relationship between v and w. We say that v and w "synchronously" k-fellow travel if for all t, $d_{G,\mathfrak{X}}(v(t), w(t)) \leq k$. More generally, one says that v and w "asynchronously" k-fellow travel if there is a strictly increasing, differentiable function $h : \mathbb{R} \to \mathbb{R}$, mapping $[0, \ell(v)]$ onto $[0, \ell(w)]$, with the property that, for all $t > 0$, $d_{G,\mathfrak{X}}(v(t), w(h(t))) \leq k$. v and w asynchronously k-fellow travel with bound M, if for all $t \leq \ell(v)$ the function h satisfies $1/M \leq h'(t) \leq M$. h is the "relative speed function" of v and w, and k the fellow traveller constant. Notice that v and w synchronously fellow travel if and only if they asynchronously fellow travel with relative speed function $h(t) = t$.

A language for G over $\mathfrak{X}$ is a set of words $\mathcal{L}$ in $\mathfrak{X}$ which contains at least one representative for each element of G. It is "bijective" if it contains exactly one representative of each group element; "geodesic" if for each $w \in \mathcal{L}$, $\ell(w) = \ell_G(w)$, and "near geodesic" if, for some arbitrary constant ξ, for all $w \in \mathcal{L}$, $\ell(w) - \ell_G(w) < \xi$.

Suppose further that $\mathcal{L}$ is a language for G. $\mathcal{L}$ has a "synchronous combing" if for some constant k, the k-fellow traveller condition is satisfied by all pairs of words $v, w \in \mathcal{L}$ for which $w =_G vx$ for some $x \in \mathfrak{X} \cup \{e\}$. $\mathcal{L}$ has instead an "asynchronous combing" if for some k the asynchronous k-fellow traveller condition is satisfied by all pairs of words $v, w \in \mathcal{L}$ for which $w =_G vx$ for some $x \in \mathfrak{X} \cup \{e\}$. An asynchronous combing $\mathcal{L}$ is "boundedly asynchronous" if for some constant M, relevant pairs of words asynchronously fellow travel with bound M. Boundedly asynchronous combings are crucial for the study of automatic groups.

2.1 Combing and parenthesized groups

An interesting metaphor and study-case for both the above notions and their interpretation in terms of quantum information theory is a group, Dehornoy group $\mathfrak{B}$ [19], that includes Artin's braid group $\mathcal{B}_\infty$ and Thompson's group $\mathcal{F}$ [45], [38]. The elements of $\mathfrak{B}$ are represented by braids diagrams in which the distances between the strands are not uniform and, besides the usual crossing generators, new rescaling generating operators shrink or stretch the distances between the strands. $\mathfrak{B}$ is a "group of fractions", *i.e.* it is orderable, and admits – besides its group multiplication – a non-trivial self-distributive structure generated by the additional binary operation $(x(yz)) = ((xy)(xz))$ [17]. Moreover $\mathfrak{B}$ embeds in the mapping class group [5] of a sphere with a Cantor set of punctures, and Artin's representation of $\mathcal{B}_\infty$ into the automorphisms of a free group extends to it.

$\mathfrak{B}$ is generated by copies of $\mathcal{B}_\infty$ and $\mathcal{F}$, and its properties are a mixture of those of $\mathcal{B}_\infty$ and $\mathcal{F}$. Indeed the presentation of $\mathfrak{B}$ extends the standard presentations of $\mathcal{B}_\infty$ and $\mathcal{F}$, starting from a geometric approach in terms of parenthesized braid diagrams. Every element of $\mathfrak{B}$ generates a free subsystem with respect to the operation $(x(yz)) = ((xy)(xz))$ – which shows that the self-distributive structure of $\mathfrak{B}$ is non-trivial.

The elements of $\mathfrak{B}$ can be seen as parenthesized braids [30], or braids in which the distances between the strands are not uniform. While an ordinary braid diagram connects an initial sequence of equidistant positions to a similar final sequence, a parenthesized braid diagram connects a parenthesized sequence of positions to another possibly different parenthesized sequence of positions. For this reason, arranging such objects into a group leads to introducing, besides the usual braid generators s_i that create crossings, and s_i^{-1} that generate inverse crossings, new rescaling generators a_i that shrink the distances between the strands in the vicinity of position i, and of course a_i^{-1} that stretch it. The s_i's generate the copy of $\mathcal{B}_\infty$, while the a_i's generate the copy of Thompson's group $\mathcal{F}$. Parenthesized braids and tangles were studied as categories in connection with Vassiliev's invariants of knots: one can therefore expect that they are related to knot polynomials.

Group $\mathfrak{B}$ is constructed using the approach that is standard for braids, namely starting from isotopy classes of braid diagrams. The difference is that diagrams are considered here in which the distances between the endpoints of the strands need not be uniform. Positions are therefore specified using parenthesized expressions, such as $(\bullet((\bullet\bullet)\bullet))$, where grouped positions are to be seen as (arbitrarily) closer than the adjacent ones. This is implemented by considering positions that are indexed not only by integers, but more generally by finite sequence of integers. A braid diagram consists of curves that connect an initial sequence of positions to a similar final sequence of positions. In an ordinary braid diagram, positions are indexed by positive integers and a generic diagram is obtained by stacking one on top of the other elementary diagrams. Here one considers also braid diagrams in which the initial and final positions need not be equidistant. This leads to the feature that, between the positions i and $i+1$, (infinitely) many new positions are possible. Then one has to consider generalized braid diagrams obtained by stacking (finitely many) elementary crossing diagrams ("s_i''") in which all strands near position i cross over all strands near position $i+1$, and rescaling diagrams ("a_i''") in which the strands near position i are shrinked and all strands on their right are translated to fill the gaps. These diagrams form a category, whose objects are the possible sets of positions – which can be specified by parenthesized expressions or, equivalently, finite binary trees – and whose morphisms are the isotopy classes of braid diagrams.

Generating a group out of these objects requires further analysis. In ordinary braid diagrams, the initial and final positions coincide; thus, for each n, concatenating n strand diagrams is always possible, which leads to the braid group $\mathcal{B}_n$. In the extended framework, concatenating two diagrams $\mathcal{D}_1, \mathcal{D}_2$ is possible only when the final set of positions in $\mathcal{D}_1$ coincides with the initial set of positions in $\mathcal{D}_2$, and an everywhere defined product is achievable only when one considers infinite completions. To make a group out of all ordinary diagrams, independently on the number of strands, one embeds recursively $\mathcal{B}_n$ into $\mathcal{B}_{n'}$, for $n < n'$ and the elements of $\mathcal{B}_\infty$ are then represented by diagrams with n arbitrarily large. The set of positions involved in an ordinary braid diagram is an initial interval $\{1, 2, ..., n\}$ of $\mathbb{N}$.

When one turns to parenthesized positions, the role of such intervals is played by finite binary trees. Upon denoting by $\bullet$ the tree consisting of a single vertex and by $t_1 t_2$ the tree with left subtree t_1 and right subtree t_2, every tree has a unique decomposition in terms of $\bullet$, and one can identify trees and parenthesized expressions. Conversely, one associates with every tree a finite set of positions $P(t)$.

Consider now diagrams constructed from two series of elementary diagrams indexed by $\{s_i^{\pm 1}\}$ and $\{a_i^{\pm 1}\}$, namely diagrams specified using a word on these letters. To construct a parenthesized diagram $\mathcal{D}_t(w)$ for w a word and t a tree, one may proceed exactly as for the ordinary diagram $\mathcal{D}_n(w)$ defined for an s-word w and n an integer. For t of size $n+1$, namely defining n positions, $\mathcal{D}_t(w)$ consists of n strands that connect the positions of $P(t)$ to n new positions.

In contrast to the case of $\mathcal{B}_\infty$, the diagrams $\mathcal{D}_t(s_i)$ or $\mathcal{D}_t(a_i)$ require attention when they have to be stacked, since the final positions of the strands need not coincide with the initial ones. However, the changes correspond to a straightforward action on trees: 1) for tree t, the unique sequence of trees $(t_1, ..., t_n)$ such that t factorizes as $(t_1(t_2(...(t_n\bullet)...)))$ is called the (right) decomposition of t, and denoted by $\mathfrak{d}(t)$. For a tree t with $\mathfrak{d}(t) = (t_1, ..., t_n)$ with $n > i$, one defines the trees $t \circ s_i$ and $t \circ a_i$ by:

$$\mathfrak{d}(t \circ s_i) = (t_1, ..., t_{i-1}, t_{i+1}, t_i, t_{i+2}, ..., t_n),$$
$$\mathfrak{d}(t \circ a_i) = (t_1, ..., t_{i-1}, t_i t_{i+1}, t_{i+2}, ..., t_n).$$

One further inductively defines $t \circ w$ for word w in such a way that $t \circ w^{-1} = t'$ is equivalent to $t' \circ w = t$ and $t \circ (w_1 w_2)$ is equal to $(t \circ w_1) \circ w_2$. Such definition implies that the final positions of the strands in $\mathcal{D}_t(s_i)$ and $\mathcal{D}_t(a_i)$ are $P(t \circ s_i)$ and $P(t \circ a_i)$, respectively. Completing the construction of the diagrams $D_t(w)$ is then straightforward; 2) the diagrams $\mathcal{D}_t(s_i^{-1})$ and $\mathcal{D}_t(a_i^{-1})$ are defined to be the mirror images of $\mathcal{D}_{t \circ s_i}(s_i)$ and $\mathcal{D}_{t \circ a_i}(a_i)$, respectively. Then, for w a word and t a binary tree such that $t \circ w$ is defined, the parenthesized braid diagram $\mathcal{D}_t(w)$ is inductively defined by the rule that, if w is xw', where x is one among the $s_i^{\pm 1}$'s and $a_i^{\pm 1}$'s, then $\mathcal{D}_t(w)$ is obtained by stacking $\mathcal{D}_t(x)$ over $\mathcal{D}_{t \circ x}(w')$.

Induction gives, for every tree t and every word w, the diagram $\mathcal{D}_t(w)$, defined if and only if the tree $t \circ w$ is defined, and, in the affirmative, with final positions in $\mathcal{D}_t(w)$ that are $P(t \circ w)$.

One can now finally define the group of parenthesized braids. According to Artin's original construction, braids can be introduced as equivalence classes of braid diagrams. Viewing a diagram as the planar projection of a 3D-figure, one considers the equivalence relation corresponding to ambient isotopy of such 3D figures. This amounts to declaring equivalent those diagrams that can be connected by a finite sequence of appropriate Reidemeister moves. From a topological point of view, parenthesized braid diagrams are just ordinary diagrams, thus they are eligible for the same notion of equivalence.

The task of making a group out of parenthesized braids faces the problem that one cannot compose arbitrary diagrams. Such problem can be solved easily by introducing a completion procedure and defining the group operation on the completed objects. In the case of ordinary braids, where the only parameter is the number of strands, in order to compose two diagrams $\mathcal{D}_{n_1}(w_1)$, $\mathcal{D}_{n_2}(w_2)$ with, say, $n_2 > n_1$, one first completes $\mathcal{D}_{n_1}(w_1)$ into the n_2-diagram $\mathcal{D}_{n_2}(w_1)$ obtained from $\mathcal{D}_{n_1}(w_1)$ by adding $n_2 - n_1$ unbraided strands on the right. Such construction leads to working with infinite diagrams. This is not an obstruction because for each braid word w, the diagrams $\mathcal{D}_n(w)$ make an inductive system when n varies, and, letting $\mathcal{D}_\infty(w)$ to be the limit of such system, one obtains a well-defined product on arbitrary diagrams. Moreover, as completion preserves equivalence, the product so defined induces a group structure, namely just that of $\mathcal{B}_\infty$.

The procedure is similar for parenthesized braid diagrams, the appropriate ordering being the inclusion of trees viewed as sets of nodes: for t, t' trees with $t \subseteq t'$, let $\varGamma_{t,t'}$ denote the completion that maps $\mathcal{D}_t(w)$ to $\mathcal{D}_{t'}(w)$ whenever $\mathcal{D}_t(w)$ exists. The geometric construction of parenthesized braid diagrams makes indeed the completion procedure easy to understand: the diagram $\mathcal{D}_{t'}(w)$ for $t' \supseteq t$ is obtained by keeping the existing strands, and adding new strands in $\mathcal{D}_t(w)$ that lie half-way between their left and right neighbours. The only difference with ordinary diagrams is that there is in general more than one basic extension: the only way to extend the interval $\{1, 2, ..., n\}$ into a bigger interval is to add $n + 1$, while in a tree t, each leaf can be split into a caret with two leaves, so there are $n + 1$ basic extensions when t specifies n positions. As induction shows, splitting the k-th leaf amounts to doubling the k-th strand.

In conclusion, all is needed for mimicking the construction of $\mathcal{B}_\infty$ is: i) for each word w the system $(\mathcal{D}_t(w), \Gamma_{t,t'})$, which is directed, because, for any two trees t, t', there exists a tree t'' that includes both t and t'; ii) diagram concatenation, that induces a well-defined product on direct limits, as the completion $\Gamma_{t,t'}$ is compatible with the product: if $\mathcal{D}_t(w_1)$ and $\mathcal{D}_{tow_1}(w_2)$ exist so that $\mathcal{D}_t(w_1 w_2)$ is defined, then, for each tree t' including t, the diagram $\mathcal{D}_{t'}(w_1 w_2)$ exists and

$$\mathcal{D}_{t'}(w_1 w_2) = \mathcal{D}_{t'}(w_1) \cdot \mathcal{D}_{t' o w_1}(w_2) \ .$$

Notice that completion maps are compatible with diagram equivalence, as the description of completion in terms of strand addition implies.

Consider, for each word w, the union of the inductive system of the $\mathcal{D}_t(w)$. Then concatenation induces an everywhere defined product on such parenthesized braid diagrams, and isotopy induces a well-defined equivalence relation that is compatible with such product.

Finally, the same argument as in the case of ordinary braid diagrams gives the property that "isotopy classes of infinite parenthesized braid diagrams make a group". The group of isotopy classes of such diagrams is the **parenthesized braid group** $\mathfrak{B}$, whose elements are parenthesized braids. By construction, $\mathfrak{B}$ is generated by the elements s_i and a_i. Let $\mathfrak{R}$ denote the list of the following relations with $i \geq 1$ and $j \geq i+2$:

$$s_i s_j = s_j s_i \ , \ s_i a_j = a_j s_i \ , \ a_i a_j = a_{j-1} a_i \ , \ a_i s_j = s_{j-1} a_i \ ,$$
$$s_i s_{i+1} s_i = s_{i+1} s_i s_{i+1} \ , \ s_{i+1} s_i a_{i+1} = a_i s_i \ , \ s_i s_{i+1} a_i = a_{i+1} s_i \ .$$

All relations in $\mathfrak{R}$ induce diagram isotopies, hence equalities in $\mathfrak{B}$. Moreover, $\mathfrak{R}$ includes the standard braid relations, as well as the relations $a_i a_j = a_{j-1} a_i$ for $j \geq i+2$, which provide the standard presentation of Thompson's group $\mathcal{F}$. $\mathfrak{R}$ furnishes thus the full set of relations of $\mathfrak{B}$. In other words, upon denoting by s_* and a_* the families of all s_i's and all a_i's, respectively, one has the presentation [18]:

$$\mathfrak{B} = \langle a_* \, , \, s_* \, ; \, \mathfrak{R} \rangle \ .$$

When one constructs combings out of other combings, it is clear from the construction that properties such as bijectivity, geodesicity or near geodesicity possessed by the original combings would be inherited by the new ones. In general, however the construction itself is independent on those properties. More precisely, 1) any asynchronously combable group is finitely presented; 2) any asynchronously combable group has soluble word problem.

A natural property of combing is the feature that it is a language in one of the families of formal languages, namely, a language recognised by some theoretical model of computation (or, equivalently, defined by a formal grammar). The combing of an automatic group is a regular language. Bijective, asynchronous combings were studied in various formal language families, in particular the families of bounded languages, regular languages, context-free languages and indexed languages, and more generally any full abstract family of languages. This general set includes both the range of fellow traveller conditions (synchronous and boundedly asynchronous, harder to obtain) and the range of language families for which they hold: manifestly such conditions are valid for parenthesized groups in general, and for $\mathfrak{B}$ in particular.

3 Knots, braids, automata and languages

One can now finally address the question of generating combing algorithms in a quantum context. It is by now a generally accepted fact that the laws of quantum theory provide in principle a radically novel and more powerful way of processing information than any classically operated device. In the past few years a great deal of activity has been devoted to devise and to implement schemes for taking actual advantage from such quantum extra power. In particular in quantum computation the states of a quantum system $\mathcal{S}$ are used for encoding information in such a way that the final state, obtained by the appropriate unitary time evolution of $\mathcal{S}$, encodes the solution to a given computational problem. A system $\mathcal{S}$ with state-space $\mathcal{H}$ (the "Quantum Computer" [41]) supports universal quantum computation if any unitary transformation $U \in \mathcal{U}(\mathcal{H})$ can be approximated with arbitrarily high accuracy by a sequence ("network") of simple unitaries ("gates") that the experimenter is supposed to be able to implement. The case in which $\mathcal{S}$ is a multi-partite system is the most relevant, as it allows for entanglement, the crucial quantum feature from which quantum speed-up (polynomial or exponential) is generated.

In the above picture the realization of the quantum network is achieved at the physical level by turning on and off external fields coupled to $\mathcal{S}$ as well as local interactions among the subsystems of $\mathcal{S}$, in other words a basic set of time-dependent Hamiltonians that perform the necessary sequences of quantum logic gates.

At variance with such a standard dynamical view, conceptual schemes [48] based on geometrical and topological approaches to quantum information manipulation exhibit far reaching peculiarities: for example, one can act over the manifold $\mathcal{C}$ of quantum codewords in $\mathcal{H}$ with a trivial Hamiltonian, for example $H\big|_{\mathcal{C}} = 0$, yet obtain nevertheless a non-trivial quantum evolution due to the existence of an underlying geometrical/topological global structure. The quantum gates in this case are realized in terms of operations of purely geometrical/topological nature. Besides being conceptually intriguing on their own, such schemes are noteworthy because they have built-in fault-tolerant features, as certain topological as well as geometrical quantities are inherently stable against local perturbations. This in turn allows for quantum information processing naturally stable against errors.

Most general conceptual schemes of interpretation of quantum computation are indeed based on topological notions [34], [20], [25]; among them, anyonic quantum computation, fermionic quantum computation, localized modular functor quantum field computation, holonomic quantum computation. Such models were shown to be simply different realizations of a unique more general conceptual scheme, that incorporates all of them as particular instances, grounded on their "discretized" counterparts.

The general setting for such universal representation is the "Quantum Spin Network Simulator" [35], [36], [36], [43]. The latter is characterized by dynamical evolution processes, permitting information manipulation, based on the (re)coupling theory of the angular momenta $SU(2)$. The scheme automatically incorporates all the features that make quantum information encoding more efficient than classical: it is fully discrete; it deals with inherently entangled states, naturally endowed with a (non associative) tensor product structure; it allows for generic encoding patterns. Also, it satisfies the whole set of Feynman, R. P.'s requirements [22] for the full characterization of an efficient quantum simulator: i) locality of interactions; ii) complexity capacity growing at most polynomially with the space-time volume of the system; iii) time discreteness (time itself is "simulated" by the number of computational steps). All such basic features are endowed in spin networks. Spin networks are in essence graphs, $\mathfrak{G}$, the node and edge sets of which can be labelled by quantum numbers associated with $SU(2)$ irreducible representations and by $SU(2)$ recoupling coefficients, respectively. A more general version of the spin network resorts to the quantum (deformed) algebra $su(2)_q$, which embodies crucial novel features in its co-algebra structure (in particular, it provides the discretized version of a Chern-Simons topological quantum field theory). Spin networks are the ideal conceptual framework for dealing with tensorial transformations and topological effects in the information coding observables, by modelling the computational space in terms of a set (a graph) endowing all necessary combinatorial and topological rules.

The spin network quantum simulator model can be thought of as a non-Boolean generalization of the quantum circuit model, with unitary gates expressed in terms of the recoupling coefficients $\mathfrak{S}$ ($3nj$ symbols) between inequivalent binary coupling schemes of a finite number, $(n + 1)$, of $SU(2)$-angular momentum variables, and of Wigner rotations, $\mathfrak{W}$, in the eigenspace of the total angular momentum [4]. The basic ingredients of the spin network simulator are then $\mathfrak{G}$, $\mathfrak{S}$, $\mathfrak{W}$, *i.e.*, its computational Hilbert space and admissible elementary gates.

Here a new frontier of quantum information is proposed: the search for algorithms capable of addressing problems in linguistics, mapping them onto problems in low dimensional geometry and topology. The combing of finite groups is the case study addressed. The reference scheme to achieve this aim, is the same already worked out in the literature [26], [27], which plays the role of a metaphor for it: the problem of evaluating Jones polynomials [33]. The latter characterize the topology of knots and links associated with the expectation value of a Wilson loop operator in quantum Chern-Simons field theory in three dimensions. Resorting to the q-deformed spin-network quantum simulator model, which essentially encodes the deformed $su(2)_q$ Racah-Wigner tensor algebra, it was shown that **families of finite-states** and **discrete-time quantum automata** could be implemented, **which accept the language generated by the braid group** as their language and the (deformed) angular momenta (re)coupling scheme as corresponding grammar. The success of the algorithm mentioned is due to the feature that transition amplitudes for such automata are indeed just (colored) Jones polynomials. What is argued here is that the quantum circuit which efficiently simulates the dynamics of these automata, that can be explicitly constructed, if appropriately controlled and sampled with a set of measurements, could approximate not only the knot invariants, but recognize and count the group identity elements necessary to solve the combing problem. As for the complexity of the corresponding circuit, since the time complexity of the spin network automaton is polynomial in the size of the input (depending on the index of the braid group), the algorithm that efficiently simulates the automata is expected to provide an efficient estimation for the latter problem as well.

It is worth noticing, incidentally, that knots and braids, beside being fascinating mathematical objects, are encoded in the foundations of a number of physical theories, either as concrete realizations of natural systems or as conceptual tools. From statistical mechanics of exactly solvable classical lattice models, to string theory, from liquid crystals physics to molecular biology (counting "knotted" configurations of DNA strands), the need for classifying the observed structures in the topological theory of knots and links has generated novel mathematical categories.

On such conceptual side, knot theory reveals a deep, unexpected interaction with quantum field theory, based on its global and not purely local (*i.e.*, topological) features. It was in the seminal work by Witten [46] that it was recognized how knot invariants could be associated with the vacuum expectation value of Wilson loop operators in a three-dimensional non-abelian Chern-Simons quantum field theory with gauge group $SU(2)$.

Braids appear naturally in this conceptual context, since one can always present a knot as the closure of a braid (Alexander theorem). Moreover, braids and parenthesized braids enrich the purely topological nature of the theory, since, as it was shown above, they can be endowed with a group structure. The Artin braid group on n strands indeed encodes all topological information about over- and under-crossings into an algebraic setting and Thompson's group allows to extend the same equivalence-class relations to parenthesized braids. It is the very structure of the spin network quantum simulator, based as it is on recoupling of angular momenta, that allows fully incorporating the extra information connected with parenthesization in natural way: the idea is simply its recursive iteration over different computational graphs $\mathfrak{G}_J$, combinatorially coupling the corresponding total angular momenta **J**, for different **J**'s, to generate the full new computational space needed to host the parenthesized braids.

Problems ubiquitous in many areas of mathematics and physics, which can be cast in the language of braids, that often share the feature of being intractable in the framework of classical information theory, may then instead be possibly dealt with efficiently at quantum level. The arguments presented in previous sections prove that formal symbolic, questions in language theory, and possibly in linguistic, can enrich the family of problems (efficiently) tractable in a quantum framework.

Acknowledgements

Enlightening discussions with Tiziana Bertoletti, Annalisa Marzuoli, Silvano Garnerone, Sorin Solomon, Chris Barrett, Lou Kauffman, Seth Lloyd are gratefully acknowledged.

References

1. A. V. Aho: *J. Assoc. Comp. Math* **15**, 647 (1968)
2. G. Baumslag, S. M. Gersten, M. Shapiro and H. Short: *Journal of Pure and Applied Algebra* **76**, 229 (1991)
3. D. Battaglia and M. Rasetti: Quantumlike Diffusion over Discrete Sets. *Phys. Lett. A* **313**, 8 (2003)
4. L. C. Biedenharn and J. D. Louck: The Racah-Wigner Algebra. In: *Quantum Theory, Encyclopedia of Mathematics and its Applications* textbf9 ed by G. C. Rota (Addison-Wesley Publ. Co., Reading 1981)
5. J. Birman: *Braids, links, and mapping class groups*, Annals of Math. Studies **8** (Princeton University Press, Princeton, 1975)
6. M. R. Bridson: Combings of semidirect products and 3-manifold groups. *Geometric and Functional Analysis* **3**, 263 (1993)
7. J. W. Cannon, W. J. Floyd and W. R. Parry: *L'enseignement mathématique* **42**, 215 (1996)
8. N. Chomsky: Three models for the description of language. *IRE Transactions on Information Theory* **2**, 113 (1956)
9. N. Chomsky: On certain formal properties of grammars. *Information and Control* **2**, 137 (1959)
10. N. Chomsky: *Aspects of the Theory of Syntax* (MIT Press, Cambridge 1965)
11. N. Chomsky: *Language and Mind* (Harcourt, Brace & Jovanovich Inc., New York 1972)
12. N. Chomsky: *Reflections on language* (Pantheon Books, New York 1975)
13. N. Chomsky: *Knowledge of Language* (Praeger, New York 1986)
14. N. Chomsky: *Language and the Problems of Knowledge.* (MIT Press, Cambridge 1988)
15. N. Chomsky and M. P. Schützenberger: The algebraic theory of context free languages. In: *Computer Programming and Formal Languages*, ed by P. Braffort and D. Hirschberg (North Holland Publ. Co., Amsterdam 1963) p 118
16. J. H. Conway, R. T. Curtis, S. P. Norton, R. A. Parker and R. A. Wilson: *Atlas of Finite Groups: Maximal Subgroups and Ordinary Characters for Simple Groups* (Clarendon Press, Oxford 1985)
17. P. Dehornoy: *Braids and Self-Distributivity*, Progress in Math. **192** (Birkhäuser Verlag, Basel 2000)

18. P. Dehornoy: Geometric presentations for Thompson's groups. *Journal of Pure and Applied Algebra* **203**, 1 (2005)
19. P. Dehornoy: The group of parenthesized braids. *Advances in Mathematics* **205**, 354 (2006)
20. E. Dennis, A. Yu. Kitaev, A. Landahl and J. Preskill: Topological quantum memory. *J. Math. Phys.* **43**, 4452 (2002)
21. D. B. A. Epstein, J. W. Cannon, D. F. Holt, S. Levy, M. S. Patterson and W. Thurston: *Word processing in groups* (Jones and Bartlett, Boston 1992)
22. R. P. Feynman: Simulating Physics with Computers. *Int. J. of Theor. Phys.* **21**, 467 (1982)
23. M. H. Freedman, A. Kitaev and Z. Wang: Simulation of topological field theories by quantum computers. *Commun. Math. Phys.* **227**, 587 (2002)
24. M.H. Freedman, A. Kitaev, M. Larsen and Z. Wang: Topological quantum computation. *Bull. Amer. Math. Soc.* **40**, 31 (2002)
25. M.H. Freedman, M. Larsen and Z. Wang: A modular functor which is universal for quantum computation. *Commun. Math. Phys.* **227**, 605 (2002)
26. S. Garnerone, A. Marzuoli and M. Rasetti: Quantum geometry and quantum algorithms. *J. Phys. A: Math. Theor.* **40**, 3047 (2007)
27. S. Garnerone, A. Marzuoli and M. Rasetti: Quantum automata, braid group and link polynomials. *Quantum Information & Computation* **7**, 479 (2007)
28. R. H. Gilman: Formal languages and infinite groups. *Discrete Math. Theoret. Comput. Sci.* **25**, 27 (1996)
29. S. Ginsburg: *The Mathematical Theory of Context-Free Languages* (McGraw-Hill, Inc., New York 1966)
30. S. Ginsburg and M. Harrison: Bracketed Context-Free Languages. *J. Comp. Sci. Soc.* **1**, 1 (1967)
31. N. Goodman: The Emperor's New Ideas. In: *Language and Philosophy*, ed by S. Hook (New York University Press, New York 1969)
32. J. E. Hopcroft, R. Motwani and J. D. Ullman: *Introduction to automata theory, languages and computation* (Addison-Wesley, Boston 1979)
33. V. Jones: A Polynomial Invariant for Knots via von Neumann Algebras. *Bull. Am. Math. Soc.* **12**, 103 (1985)
34. A. Kitaev: Fault-tolerant quantum computation by anyons. *Annals Phys.* **303**, 2 (2003)
35. A. Marzuoli and M. Rasetti: Spin network quantum simulator. *Phys. Lett. A* **306**, 79 (2002)
36. A. Marzuoli and M. Rasetti: Computing spin networks. *Annals of Physics* **318**, 345 (2005)
37. A. Marzuoli and M. Rasetti: Spin network setting of topological quantum computation. *Int. J. Quantum Information* **3**, 65 (2005)
38. R. McKenzie and R. J. Thompson: An elementary construction of unsolvable word problems in group theory. In: *Word Problems*, ed by W. W. Boon, F. B. Cannonito and R.C. Lyndon (North-Holland Publ. Co., Amsterdam 1973)
39. R. McNaughton: Parenthesis grammars. *Journal of the ACM* **14**, 490 (1967)
40. C. Moore and J. P. Crutchfield: Quantum automata and quantum grammars. *Theor. Comput. Sci.* **37**, 275 (2000)
41. M. A. Nielsen and I. L. Chuang: *Quantum Computation and Quantum Information*, (Cambridge University Press, Cambridge 2000)
42. J. Pachos, P. Zanardi and M. Rasetti: Non-Abelian Berry connections for quantum computation. *Phys. Rev. A* **61**, 010305(R) (2000)

43. R. Penrose: Angular Momentum: an approach to combinatorial space-time. In: *Quantum Theory and Beyond*, ed by T. Bastin (Cambridge Univ. Press, Cambridge 1971)
44. W. V. O. Quine: Methodological Reflections on Current Linguistic Theory. *Sinthese* **21**, 386 (1970)
45. R. J. Thompson: Embeddings into finitely generated simple groups which preserve the word problem. In: *Word Problems II*, ed by S. Adian, W. W. Boone, and G. Higman (North-Holland Publ. Co., Amsterdam 1980)
46. E. Witten: Quantum field theory and the Jones polynomial. *Commun. Math. Phys.* **121**, 351 (1989)
47. A. P. Yutsis, I. B. Levinson and V. V. Vanagas: *The Mathematical Apparatus of the Theory of Angular Momentum* (Israel Program for Sci. Transl. Ltd., Jerusalem 1962)
48. P. Zanardi and M. Rasetti: Holonomic quantum computation. *Phys. Lett. A* **264**, 94 (1999)

Proofs instead of Meaning Explanations: Understanding Classical *vs* Intuitionistic Mathematics from the Outside

Dag Westerståhl

Department of Philosophy, University of Göthenburg (Sweden)
Dag.Westerstahl@phil.gu.se

1 Introduction

The conflict between classical and intuitionistic mathematics – henceforth, the *C-I conflict* – has been discussed at length and in depth by a number of famous scholars. Why an outside perspective? Is such a perspective interesting, or even possible?

There are in fact reasons why a somewhat detached account of this conflict might be worthwhile. First, the conflict is *prima facie* very puzzling, and even worrying. Mathematics is a discipline on which much of science, indeed much of our knowledge, rests. Moreover, it is a discipline whose practitioners are supposed to agree among each other more than in other fields about results and methods. Yet here there appears to be a conflict even about basic laws of logic, not to mention specific mathematical claims.

Second, popular accounts of this state of affairs are not very satisfactory. One may be told that the part of mathematics that matters for *practical applications* is unaffected by the C-I conflict. But that leaves the original question even more puzzling: how then can there be a conflict about basic logic? Another idea is that classical and intuitionistic mathematicians simply speak different languages, and only *seem* to contradict each other. There is something to this, of course. But again, if that were the whole explanation, why would the conflict persist?

Specialists in the field haven't paid much attention to explaining what goes on to a wider audience. That's unfortunate, especially since the 'received' view of the matter has undergone significant changes since the days of Brouwer. For example, it is now quite common for intuitionists to see classical mathematics as a *special case* of intuitionistic mathematics. That would have been unthinkable to Brouwer.[1]

But also from the point of view of general epistemology or philosophy of science, this conflict ought to be an ideal object of study. One would be hard put to find other cases of such clear-cut and continued disagreement, about truth and about methods, in the sciences. Discussions on the subject in the philosophy of mathematics abound, but they usually reflect the philosophical aspects of one or the other position in the conflict. What I am after here is the more detached view of the philosopher of science.[2]

For example, it might seem, *prima facie*, that the C-I conflict is a promising case for those who sustain some form of *relativism* about knowledge or truth. Here we have two communities of mathematicians who clearly disagree, but whose disagreement is not easily resolvable by giving one side an advantage over the other. Perhaps *both* are right; perhaps the disagreement is 'faultless'? I am not saying that this is actually the case, but the question could surely be raised, as it recently has been raised for discourse about other things than numbers or functions (e.g. discourse about taste, values, probabilities, knowledge, the future, etc.[3]).

Such an undertaking would not only benefit from an outside perspective, but require one. But is such a perspective really possible?[4] Won't one inevitably be influenced by one's own preferences? Surely there is a such a risk, but it shouldn't make us give up before trying. Being aware of the problem, one can try to avoid falling into the most obvious traps. And if in the end the difficulties become unsurmountable, that too would be a useful insight.

There is, however, a theoretical objection coming from the intuitionist camp, for example in Michael Dummett's version. It stems from the claim that classical mathematics, and more generally the classical notion of truth, is simply incoherent, and therefore ultimately *unintelligible*. Mustn't the lack of intelligibility transfer to any attempt at an 'objective' account of the conflict? This is a serious question. But things are not simple: the unintelligibility claim is not shared by all intuitionists, and it may even be in conflict with

---

[1] The inverse view is also common, that intuitionistic mathematics is just a particular kind of (classical) mathematics. Appearances notwithstanding, these two views are not incompatible; see § 5.2.

[2] [7] focuses on the issue of mutual understanding between the two camps, as I do in this paper. His perspective is that of classical mathematics, however, and a main claim is that the intuitionist cannot state her position clearly without resorting to classical logic. Although that issue is both interesting and relevant, I avoid it here.

[3] See, for example, [10] and references therein.

[4] Traditional relativism denies that this is possible. But the modern versions of relativism referred to in the previous footnote might very well allow it. Relativism applies, it is claimed, to certain discourses, not to all. One could be allowed to be relativist about statements of taste, say, but absolutist about semantics, in particular about the meaning of taste statements.

some other things Dummett says on this subject — Dummett is in fact one of the (few) champions of promoting *mutual understanding* between classical and intuitionistic mathematicians.

A further worry is that an outside view of the C-I conflict will be superficial. Mathematicians are usually (and often justifiably) suspicious of how non-specialists describe what they are up to. They feel that the mathematics should speak for itself. But that would mean that non-specialists should give up any attempt to understand what the conflict is about. And this might even be reasonable if the debate were about number theory, or topology, say. But the debate is (also) about what mathematics *is*. Then it is not enough to just *point* to existing mathematics, especially when the different camps point to different kinds of mathematics.

The structure of the paper is as follows. After some stage-setting in § 2 and § 3, I start with a rather close look in § 4 at the suggestion, mainly due to Dummett, that classical and intuitionist mathematicians should try to achieve mutual understanding by starting from a *common ground*, which is unproblematic in some important sense. My evaluation of this strategy is mostly negative: a basic *asymmetry* as to one side's ability to achieve understanding of what the other is up to will remain. In § 5, I then explore another approach: focusing on *proofs* rather than meaning explanations, and taking account of the avowed intention of most modern intuitionists to make all intuitionistic theorems classical theorems as well, appears to significantly improve the prospects of mutual understanding. Although this indeed promises to eliminate serious *conflict* between the two camps, I make some cautionary remarks at the end of that section, as well as in the concluding § 6.

2 Background

Although there are many variants of intuitionistic as well as classical mathematics, for certain basic issues these differences do not matter much. It is often enough to simply speak (as Dummett does) of *intuitionists* and *platonists*. The principled differences between these two concern the notion of mathematical *truth* and the meaning of the basic *logical constants*. The typical intuitionist takes truth to be what philosophers call an *epistemic* notion: roughly, something is true if it can be *proved*. This puts *computation* at center stage: (intuitionistic) proofs are computations, or directions for finding computations. Accordingly, the meaning of the constants are given as *proof conditions*: some form of the Brouwer-Heyting-Kolmogorov (BHK) conditions for the circumstances under which a statement of a certain form can be asserted.

The typical platonist disagrees. Truth is *not* epistemic: whether something *is* true is unrelated to our ability to *find out* if that is so. A statement for which we will in fact never find a proof might still be true. The meaning of the constants are in terms of the usual (Tarskian) truth conditions. It is a little harder to state the 'typical' platonist way of explaining *why* truth is non-epistemic, but the rough idea is that mathematical statements are *about* some reality or structure suitably independent of us. We need not assume he holds this for all of mathematics; for most of what I say it will be enough to consider *first-order number theory*, *PA*: for any sentence in *PA*, the platonist holds that it is either true or false, regardless of what we know now or will ever know. A *platonist about real numbers* holds the same for statements (in some language that needs to be specified) about the reals, but that doesn't commit him, as these terms are used here, to the same view about the whole of set theory, for example.

It may seem that an obvious weak point for the platonist is the reference to an independent (platonic) reality of abstract objects. The only abstract objects the intuitionist needs are the proofs themselves. But here the platonist counters that this is just appearance: by defining truth as provability you lose the ability to explain the *point* of proofs, which, non-trivially, is precisely to get at the truth. The intuitionist responds that the point is in fact another, having to do with *computability*. And the familiar (philosophical) debate continues. But to begin, at least, we shall ignore the 'why' of proving things, as well as the existence of platonic realities: it is enough to assume that the two parties have the different attitudes towards number-theoretic *statements* indicated above.

3 Setting the stage

A quick glance from the 'outside' seems to indicate that the C-I conflict is very serious indeed. Intuitionists refuse to assert things that platonists find trivially true, and in other cases assert things that classical mathematics outright denies. An example of the former is of course the Law of Excluded Middle, say in the form that for any PA-sentence φ,

(LEM) $\varphi \vee \neg \varphi$

is (logically) true. The intuitionist doesn't *deny* LEM (which to her would mean claiming that it is contradictory), but she certainly doesn't believe we have any reason to assert it. The second kind of conflict is exemplified by Brouwer's theorem

(CONT)   Every function from [0,1] to the real line $\mathbb{R}$ interval is uniformly continuous,

something that every math student learns how to *disprove* at an early stage.

But a common explanation is that only the *words* are the same; the *statements* made are different. The intuitionist means something quite different with words like "or", "not", "real number", etc.[5] This eliminates the immedi-

---

[5] Thus, for example, [3, p. 55]: "The apparent absurdity of this statement is, however, illusory, as is suggested by the following more careful re-statement of it.

ate threat of conflict. Moreover, it seems that *provided* the words are used in the intuitionist way, the platonist too can accept that LEM fails, and perhaps even that CONT holds.[6]

However, the problem doesn't go away so easily:

- Can the respective meaning explanations be provided in a sufficiently clear way, so that *mutual understanding* is achieved?
- Assuming this can be done, and allowing for the meaning differences, can we be sure that no *other* conflicts than those alluded to above, of the stronger or the weaker kind, exist?
- For example, can we be sure, for some principled reason, that the platonist will accept *all* 'translated' intuitionistic claims?
- Even if that were the case, what shall we do with the fact that the converse seems to fail? Intuitionists do *not* accept the classical version of LEM or of the negation of CONT. They might (nowadays) agree that these claims are *consistent*, but they would not assert them, whereas the platonist is happy to admit, for example, that the *intuitionistic version* of LEM fails.
- Thus there seems to be an *asymmetry* as regards mutual understanding, and one would like to know why.

To the specialist, these questions may seem trivial, or misguided. But I will proceed on the assumption that, at least initially, they make sense.

4 Mutual understanding

The intuitionism of Brouwer and Heyting was often presented in rather polemical form. Michael Dummett, however, is a latter-day defender of Brouwer style intuitionism who, in addition to finding support for it from a Wittgenstein-inspired account of how language works, has repeatedly stressed the need for *dialogue* between platonists and intuitionists:

> ...the desire to express the conditions for the intuitionistic truth of a mathematical statement in terms which do not presuppose an understanding of the intuitionistic logical constants as used within mathematical statements is entirely licit. Indeed, if it were impossible to do so, intuitionists would have no way of conveying to platonist mathematicians what it was that they were about: we should have a situation quite different from that which in fact obtains, namely one in which some people found it natural to extend basic computational mathematics in a classical direction, and others found it natural to extend

---

Every intuitionistically definable function from the intuitionistic interval $[0, 1]$ *to the intuitionistic real line is, intuitionistically, uniformly continuous."*

[6] For example, by following the exposition of Brouwer's theory of choice sequences in [17], which takes place in a classical framework.

it in an intuitionistic direction, and neither could gain a glimmering of what the other was at. That we are not in this situation is because intuitionists and platonists can find a common ground, namely statements, both mathematical and non-mathematical, which are, in the view of both, decidable and about whose meaning there is therefore no serious dispute and which both sides agree obey a classical logic. [4, pp. 237-8]

The quote also indicates one road along which Dummett thought mutual understanding could proceed: via the common ground of decidable sentences.

4.1 Decidable sentences as a common ground?

The basic idea seems to be that decidable sentences are *unproblematic*, and therefore mutual understanding can begin with them.

We can avoid any discussion about exactly what *decidable* means here, as follows. First, restrict attention to the language of *PA*. The great advantage of this is that we can assume, without distorting things very much, that

(1) There is no conflict about the meaning of the arithmetical non-logical constants, and therefore no conflict about *atomic* sentences.

In contrast with the case of analysis, the conflict concerns only the logical vocabulary in this case. Now, let D be the set of *PA*-formulas with only *bounded quantification*.[7] Even if D is only a subset of the set of sentences Dummett has in mind, there is no unclarity about the fact that all sentences in D are decidable.

Now, in what sense are sentences in D a *common ground* for the platonist and the intuitionist?

At first sight, it might seem that Dummett holds that these sentences express "basic computational mathematics" and therefore *mean the same* for both. But this cannot be the idea. Sentences in D use the basic logical vocabulary, and Dummett points out time and again that the logical constants have different meanings for the platonist and the intuitionist. Rather, what he means is that the following holds:

(2) For all $\varphi \in D$, the intuitionist asserts φ if and only if the platonist does.

This is the sense in which decidable sentences "obey a classical logic". However, it doesn't *follow* from (2) that they only involve notions concerning which there is no dispute. That would only follow if there were nothing more

---

[7] That is, terms and D-formulas have the following forms:

terms: 0, $S(t)$, $t_1 + t_2$, $t_1 \cdot t_2$
formulas: $t_1 = t_2$, $t_1 < t_2$, $\neg \varphi$, $\varphi \wedge \psi$, $\varphi \vee \psi$, $\varphi \rightarrow \psi$, $\exists x(x < t \wedge \varphi)$, $\forall x(x < t \rightarrow \varphi)$

to the meaning of these sentences than their assertion conditions, so that (2) would *entail* that D-sentences do mean the same to both. But Dummett doesn't favor such a crude behavioristic meaning theory. This is clear from his remarks about the logical constants, and also from his claim that what the intuitionist means can be *explained* in terms which are not in dispute. On the crude meaning theory, there would be nothing further to explain about sentences in D.

It is thus somewhat mysterious how (2) could do the work Dummett wants it to. Consider the following D-sentence:[8]

$$\varphi_0 = prime(2^{10540} + 1) \vee \neg prime(2^{10540} + 1)$$

The intuitionist and the platonist can both assert φ_0, but on very different grounds. For the intuitionist, φ_0 is true since there is an algorithm for determining if a number is prime, which we know in advance will terminate, even if we don't know the outcome for this particular number. The platonist recognizes that this is a ground for asserting φ_0, but he has a much a simpler one: it is a trivial *logical* truth. Surely, this is a strong indication that φ_0 means a different thing for the platonist than for the intuitionist.

So the sense in which decidable sentences constitute a common ground is too weak, it seems. Nor is there a common way they are used in standard explanations of the meaning of the logical constants. For the platonist, decidable sentences play no role at all in that explanation. There is no difference for him between φ_0 and

$$\varphi_1 \vee \neg \varphi_1$$

when φ_1 is undecidable. The intuitionist, on the other hand, might use decidable sentences in a first approximation of the meaning explanations, going beyond them to deal with quantification over infinite domains. No *common* role is played by decidable sentences in these respective explanations.

4.2 A neutral metatheory?

To understand what Dummett is after we must, I think, pay less attention to the class of decidable sentences and the fact that these have the same assertion conditions for everyone. Instead, we should focus on his idea that the respective meaning explanations themselves can be given in terms which are understandable to the opponent. In [4], he is mostly interested in how the platonist can come to understand the intuitionist:

> It is therefore wholly legitimate, and, indeed, essential, to frame the condition for the intuitionistic truth of a mathematical statement in terms which are intelligible to a platonist and do not beg any questions, because they employ only notions which are not in dispute. [4, p. 239]

---

[8] Allowing standard extensions by definition from D-formulas, such as e.g. $prime(x)$.

Dummett goes on to say that this is most naturally done by carefully describing the intuitionistic notion of truth, in terms of the existence of a proof, to the platonist. He comments, concerning the success of such explanations, that although the other side may not accept them as legitimate, "at least the conception of meaning held by each party is not wholly opaque to the other" [4, p. 238]. This remark relates to the fact that the intuitionist insists that mathematical truth *cannot* be explained in the platonist manner. In the other direction no similar problem is mentioned. In fact, the rest of his discussion concerns the very notion of intuitionistic truth: e.g. whether one should require the actual possession of a proof or if it is enough to have the means (in principle) to obtain one. This leads to an intricate analysis of the role of so-called *canonical* proofs, but there is no indication that the platonist should have greater difficulties following these arguments than anyone else.

When Dummett returns to the issue of mutual understanding in [6], his approach is slightly more formal:

> What is needed, if the two participants to the discussion are to achieve an understanding of each other, is a semantic theory as insensitive as possible to the logic of the metalanguage. Some forms of inference must be agreed to hold in the metalanguage ... but they had better be ones that both disputants recognise as valid. ...
> Thus, within sentential logic, the semantics of Kripke trees or Beth trees is insensitive to whether the logic of the metalanguage is classical or intuitionistic: exactly the same forms of inference can be shown valid or invalid on that semantic theory. If both disputants propose semantic theories of this kind, there will be some hope that each can come to understand the other; there is even a possibility that they may find a common basis on which to conduct a discussion of which of them is right. [6, p. 55]

Although Dummett carefully distinguishes formal semantic theories from the 'real thing', i.e. theories of meaning, he apparently thinks that if the language in which such semantic theories are expressed has a logic not in dispute, at least a road towards mutual understanding is open. He is also explicit that 'internal semantics', e.g. a semantics for an intuitionistic theory given in an intuitionistic metalanguage, is of no help here. No technical details are given, but presumably Dummett is referring to intuitionistic proofs of *completeness theorems* for intuitionistic logic. A completeness theorem says precisely that a certain formal semantics captures the notion of validity in a certain logic or theory.[9]

---

[9] It was first believed that completeness theorem for Kripke or Beth semantics for intuitionistic systems could only be proved classically, but Weldman and de Swart realized that if one allows contradictory worlds (worlds in which some sentences are both true and false), completeness with respect to this class of models could be proved intuitionistically. See, for example, [9] for results of this kind.

The point cannot be that the platonist too understands the metalanguage and the logic in which the completeness proof is carried out — if he did there would be no point of the exercise. Rather, the idea must be that there is now a formal characterization of a certain set of intuitionistic validities, whose correctness is accepted by the intuitionist, as well as (*via* the classical completeness proof) by the platonist. One may grant, as Dummett indicates, that this could provide some basis for a discussion between the two on the merits of that system of intuitionistic logic.

Again, this is only understanding in one direction. For truly *mutual* understanding by these means, we would also need an intuitionistically acceptable proof of the completeness of a relevant system of *classical* logic; say, first-order logic. However, it is known that such a proof doesn't exist.[10]

We thus see, following Dummett, that whether one takes the direct route of explaining the intuitionistic meaning of the logical constants, or the more indirect route via completeness theorems, an *asymmetry* appears: it seems fairly clear how the platonist could go about understanding intuitionism, but much less clear how understanding in the opposite direction would work.[11] Indeed, in several other places, Dummett says explicitly that the intuitionist cannot understand or make sense of classical logic or mathematics, because it doesn't make sense: it is *unintelligible*.

4.3 Intelligibility and translation

How seriously should one take Dummett's claims about unintelligibility? On the one hand, he continues Brouwer's antagonistic stance towards classical mathematics, saying that intuitionistic theorems "refute certain classically valid logical laws" [5, p. 84]. One may wonder how a theorem can refute a meaningless statement. On the other hand, he takes the issue of mutual understanding and a common ground very seriously, as we have seen.

Perhaps one should take the unintelligibility claim at face value. Perhaps laws like LEM are refuted in the sense that the only meaningful way to understand them renders them invalid. And perhaps mutual understanding must always be approximate or partial.

At this point, an observer can only note that if Dummett is *right*, the prospects of mutual understanding are bleak indeed. To get any further, he

---

[10] This was shown by Gödel and Kreisel; for stronger versions, see [13]. It should be noted that [8] shows that the fact that every consistent set of sentences (in a countable language) has a model can be proved intuitionistically; see also the exposition in [1]. Classically (but not intuitionistically), the completeness of classical first-order logic follows almost immediately from this fact. So some measure of understanding can perhaps be obtained in this case too.

[11] The first claim is also a standard platonist view: he can follow the intuitionistic explanations of the logical constants, as well as intuitionistic mathematical proofs (given the way the relevant intuitionistic concepts are defined); but he sees no reason to declare that these are the only acceptable proofs.

would have to engage in the philosophical debate, which is not my ambition here. A remaining point, however, would be to account for the fact that classical mathematics *appears* to make sense. After all, it does so to the vast majority of mathematicians.

Intuitionists often explain this via the various *negative translations* that exist from parts of classical mathematics into corresponding constructive theories. The idea is that when the platonist asserts φ, what he really means — or alternatively, all he can be taken to mean — is φ^{neg}, where φ^{neg} is some translation of φ (in the same language) such that, if T_C and T_I are the relevant axiomatic theories, φ and φ^{neg} are equivalent in T_C, and T_C proves φ if and only if T_I proves φ^{neg}.

But there are problems with this view. First, it only concerns certain axiomatized *parts* of mathematics. Second, such translations yield (relative) *consistency* of the classical theories (since they preserve negation), and so the intuitionist can take them to indicate that classical mathematics is at least consistent, but that is a far cry from making sense of it. Of course, an extreme view would be that this is the only sense to be had. But the translation is often taken to show more, namely, that what the platonist mathematician is really after are the translated versions of his theorems. And at this point, the asymmetry in understanding shows up again. For even if φ and φ^{neg} are provably equivalent, if you take a reasonably complex classical theorem φ and tell a platonist that what he *really* means is φ^{neg}, he might just deny that that was what he had in mind when he was thinking about how to prove φ.[12]

In other words, even for these theories (like PA versus its intuitionistic version, Heyting Arithmetic, HA), the platonist and the intuitionist would not *agree* about what the classical mathematician is up to. By contrast, if the platonist 'translates' an intuitionistic statement using the BHK explanations of the logical constants, or further intuitionistic elaborations about meaning as in e.g. [6], they might well agree about the truth or falsity of the statement understood in *this* way.

4.4 Summing up

We started with the need for an outside view, but have so far focused on whether mutual understanding between the two camps is possible. But that's an entirely relevant issue. If we had found, for example, that each party can fully understand what the other is up to, and is willing to admit that both are doing mathematics and that no inconsistencies are likely to arise, then the conflict would only be about which kind of mathematics was most interesting or useful. This is of course highly relevant for matters of research funding or academic appointments, but has little theoretical interest. (It might interest the sociology of science, but hardly the philosophy of science.)

---

[12] The argument hinges on notions of meaning that may themselves be controversial. My point is merely to observe that even if a translation preserves theoremhood, it does not automatically follow that it also preserves meaning.

But that is not what we found. There is a striking asymmetry when it comes to understanding what the other side is up to, however such understanding is supposed to take place. The platonist appears to have no serious difficulties in grasping, at least not in principle, via reinterpretation of the logical vocabulary and other means, the intended content of intuitionistic mathematical claims. This is what many classical mathematicians themselves claim, but we saw that Dummett appears to reason along similar lines.

Problems arise, on the other hand, for how classical mathematics is to be understood. If the intuitionist insists that it is fundamentally flawed, she can try to make sense of at least parts of it *via* negative translations. But it seems unlikely to me that there could be an agreement about *meaning* along these lines. There is likely to be a recognition that what the other side is up to is consistent, but that is a very weak form of agreement.

If we don't want to delve deeply into philosophical questions about meaning, or simply take sides in the conflict, we seem to have reached an impasse.

5 Understanding in terms of proofs

The intuitionist I have so far portrayed is of the original Brouwer style, although in Dummett's version, which differs as to philosophical background but not in mathematical content. But there is a newer brand of intuitionism, that I will simply call *modern intuitionism*,[13] since it is a dominating trend these days. One starting point is [2], whose explicit aim was to do constructive mathematics that looked just like ordinary mathematics, not even apparently contradicting any classical theorems, and not relying on more or less philosophical notions concerning the continuum or other central mathematical objects, but only paying attention to constructivity (to assert that something *exists*, you must provide an algoritm for finding it). More specifically, it proposed to approach the continuum without using Brouwer's choice sequences, or his ideas about the 'creative subject'. An independent effort with similar aims was [11].

This line of work has been carried on by a number of mathematicians, e.g. Per Martin-Löf, Douglas Bridges, Fred Richman, Giovanni Sambin, Thierry Coquand, to mention just a few,[14] and today encompasses an impressive body of mathematics.

Some of the modern intuitionists are still concerned with philosophy and the foundations of mathematics, whereas others prefer to let the mathematics speak for itself. But one thing that separates them from the old style intuitionists is their (explicit or implicit) adherence to the slogan:

---

[13] Some of its practitioners would prefer to avoid the label "intuitionism" altogether, using "constructivism" or "constructive mathematics" instead. But it is just a label here.

[14] Again, I am ignoring the various differences concerning the nature of constructivism/intuitionism and platonism among these scholars.

(*) Every intuitionistic theorem (proof) is a classical theorem (proof).[15]

This appears to provide a way out of the impasse mentioned above.

5.1 Truth and assertability

The impasse stemmed from the radically different notions of truth entertained by the two sides: for one it is a primitive, fundamental, and 'metaphysical' notion; for the other it is a secondary epistemic notion, defined in terms of proof. Although this difference may make mutual understanding impossible at the level of a theory of *meaning*, it is worth pointing out that in one important respect, the differences over truth don't matter. The point is that both parties have essentially the same notion of *assertion*.

Assertions in mathematics are *theorems* (or propositions, lemmas, etc.), and with some simplification (actually a lot) we can say that the main goal of mathematical scientific activity is to deliver theorems. And regardless of any difference over what truth is, both sides agree about the following:[16]

(a) To assert something in mathematics, you need a *proof*.
(b) Provable statements are true.[17]

That is, for the purely mathematical activity, the differences come down what proofs to accept. Certainly, a platonist might claim that there are true statements of arithmetic whose proofs we will never know, or even truths that don't have proofs. But that is not a mathematical claim.

Relying on (*), one may affirm that intuitionistic mathematics is a part of classical mathematics. But the converse affirmation is also popular.

5.2 Classical and intuitionistic mathematics as special cases of each other

The implementation of (*) (in either version) in a specific area of mathematics T often takes roughly the form:

(**) classical version of T = intuitionistic version of $T + AX$

where AX is a particular axiom, like some version of LEM, or the unrestricted axiom of choice, or the power set axiom. For example, HA can be formulated so that one obtains PA simply by adding LEM as an axiom. This has of course been known for a long time, but a result of the work of modern intuitionists has been to extend (**) to ever larger parts of mathematics.

---

[15] For example, Brouwer's CONT is not a theorem of modern intuitionistic mathematics. See also footnote 19.

[16] "...the intuitionist's view is that ...you are not entitled to assert that a theorem is true until it's proved, which sounds much like a realist's view also" [15, p. 124].

[17] At least if we restrict attention to number theory and analysis.

Classical mathematics is then a special case of intuitionistic mathematics in the sense that it allows fewer models (having more axioms); in particular, AX disallows 'computational' models that intuitionists take a special interest in.[18] A different and perhaps clearer way to make the same point is that without AX, many mathematical notions bifurcate. For example, intuitionistic logic distinguishes between a statement's not being true and its leading to contradiction. Or consider *formal topology*, a constructive approach to topology initiated by Martin-Löf and Sambin, where the duality between closed and open sets remains, but a closed set is no longer defined as the complement of an open set; only with classical logic do these two notions collapse into one.[19] For a final example, intuitionistic analysis doesn't have access to the axiom

$$\forall x \in \mathbb{R}(x = 0 \lor x \neq 0)$$

but gets by with slightly weaker principles like

$$a > b \rightarrow \forall x \in \mathbb{R}(a > x \lor b > x)$$

$$\forall x \in \mathbb{R}(\neg(x > 0) \rightarrow x \leq 0)$$

(see [3]). With LEM, one never even thinks of these distinctions.

On the other hand, in another clear sense, intuitionistic mathematics is a special case of classical mathematics, i.e. the special case where one investigates how to get by without certain axioms. For particular axiomatized theories, (**) expresses just that. For mathematics in general, i.e. for (*), this is presumably not something one can prove (see below), but it appears to be a shared conviction. This notion too goes well with the idea that intuitionistic mathematics is the computational part of classical mathematics.[20]

Clearly, from either of these (fully compatible) perspectives, the *conflict* between platonists and (modern) intuitionists becomes less serious. Focus has shifted from what lies behind mathematical truth to what proofs to accept. Indeed, there is no necessity to take a stand, as witnessed by the fact that a number of mathematicians do *both* classical and intuitionistic mathematics. For example, a set theorist can study classical extensions of ZFC, and the

---

[18] See [15] for a forceful statement of this claim.
[19] See e.g. [16]. Thus, (**) should not be taken to entail that both sides use the same language. Roughly, the intuitionistic language extends the classical one, but in such a way that when AX is added, the extra intuitionistic vocabulary can be eliminated.
[20] This statement is imprecise. For some, the computational part of mathematics is essentially recursive function theory. Intuitionists emphasize that recursive functions too must be studied with constructive methods, e.g. without assuming LEM. Also, they reject the idea that intuitionists study subclasses of classical mathematical objects, such as constructive real numbers (as opposed to all real numbers) or recursive functions (as opposed to all function among natural numbers). Instead, they maintain that if you study e.g. number-theoretic functions with constructive methods, these functions will in fact all be computable; see [15] and [3].

status of the Continuum Hypothesis or large cardinal axioms, and at the same time be interested in constructive versions of set theory. At the extremes, there will be platonists who find the abandonment of certain obvious valid methods of proof wholly unmotivated, and intuitionists who see no justification at all in the extra axioms. In between, all kinds of positions are possible. But when the differences have been reduced to whether or not this or that axiom can be used, those interested in philosophical foundations can focus on those axioms, and the others — the majority of mathematicians — can keep studying what follows from what, which proofs are more effective, or more elegant, or more informative, etc. The threat of conflict, in the sense of proving theorems that contradict each other, seems to have disappeared.

End of story? Recall that the peaceful coexistence between classical and intuitionistic mathematics envisaged here wholly builds on (*). I will briefly consider the evidence for (*), and conclude with some remarks indicating that some problems still remain.

5.3 Evidence for (*)

If (*) holds, no inconsistency between classical and intuitionistic mathematics can ever arise. How sure can we be of (*)? As long as we restrict attention to specific theories for which (**) holds, we are safe. But everyone knows that mathematics cannot be fully captured within any formal system, and especially intuitionists have emphasized the open-endedness of the mathematical enterprise: its methods can never be laid down once and for all. This may not matter much to the working mathematician, but it certainly matters for the methodological question of the validity of (*).

How could we *know* (*), *once and for all*? Note that the reformulation of intuitionistic mathematical theories in the form (**) has by no means been an easy matter, but the result of hard mathematical work. The methodological considerations underlying this work are, when they are made explicit,[21] still some form of the BHK explanations of the logical constants. However, these explanations by themselves really don't give full evidence for (*). This observation is not often made, but an exception is Dummett, who notes that the problem lies with intuitionistic *implication*:

> In some very vague intuitive sense one might say that the intuitionistic connective → was stronger than the classical →. This does not mean that the intuitionistic statement $A \to B$ is stronger than the classical $A \to B$, for, intuitively, the antecedent of the intuitionistic conditional is also stronger. The classical antecedent is that A is *true*, irrespective of whether we can recognize it as such or not. Intuitionistically, this is unintelligible: the intuitionistic antecedent is that A is (intuitionistically) *provable*, and this is a stronger assumption. We have to show

---
[21] As in the careful meaning explanations in [12].

that we could prove B on the supposition, not merely that A happens to be the case (an intuitionistically meaningless supposition), but that we have been given a *proof* of A. Hence intuitionistic $A \to B$ and classical $A \to B$ are in principle *incomparable* in respect of strength. We may sometimes have a classical proof of $A \to B$ where we lack an intuitionistic one; but *there is no reason why the converse should not sometimes hold too.* [5, p. 17] (Last italics mine)

To flesh out these remarks, consider the following thought experiment. Suppose $\varphi = \psi \to \theta$ were a sentence – we can even assume it is a numer-theoretic sentence – such that:

(i) there is a construction taking intuitionistic proofs of ψ into intuitionistic proofs of θ;
(ii) there is (in fact) no intuitionistic proof of ψ, but
(iii) there is a classical proof of ψ and a classical proof of $\neg\theta$.

Of course, these claims about existence and non-existence of proofs must be understood relative to some future, not yet discovered, notion of number-theoretic proof. (That's why it is a thought experiment.) Also, assumption (ii) has to be read classically: not in the sense that we can show that ψ's provability would lead to contradiction, but simply that no proof exists. So the thought experiment is only accessible to someone who can make sense of that assumption. But if you cannot do that, probably (*) makes no sense to you either.[22] In any case, these assumptions appear consistent. An instantiation of them would be a counter-example to (*).

The existence of such a counter-example seems very unlikely. For all the known theories which satisfy (**), no such example can exist. Perhaps a more general meta-theorem can be proved, ruling out such examples for a large class of theories. And the issue whether we could give a principled argument that there isn't one, in *all* of mathematics, doesn't look like something that could be proved anyway. My point here is merely that (*) doesn't automatically follow from the standard intuitionistic account of the logical constants.

Incidentally, if there were a counter-example φ, it would not constitute a conflict with classical mathematics, at least from the platonist's viewpoint: he would happily acknowledge that $\neg\varphi$ is true, but also that the intuitionistic reading of φ is true! It would, however, show that the relation between classical and intuitionistic mathematics is not quite what it is usually taken to be.

---

[22] Note that Dummett in the quote above (a) claims that an assumption like (ii) is "intuitionistically meaningless", but (b) uses it to explain the difference between classical and intuitionistic implication.

6 Concluding remarks

6.1 The asymmetry remains

We found that the attempts to achieve mutual understanding between platonists and intuitionists via a common ground of unproblematic statements, or via a meta-theory that was not in dispute stranded — or at least were far from successful — because of the apparent *asymmetry* of understanding that resulted. The platonist could claim he has no principled problem of understanding what the intuitionist is up to. The intuitionist might even agree that this understanding is essentially correct. But if she also insists that classical mathematics is at bottom unintelligible, there can be no corresponding agreement about how to understand classical mathematics. For those who still pursue Brouwer style intuitionistic mathematics, as well as for those who base their adherence to intuitionistic logic on a Wittgenstein-inspired theory of meaning, like Dummett or Prawitz, there is no real possibility of reconciliation. Despite efforts to find a common ground, they must in the end argue that "classical logic contains some invalid forms of reasoning, and consequently has to be rejected" [14, p. 2].

Modern versions of intuitionistic mathematics appear to allow for friendlier relations. We noted that this stance presupposed that every intuitionistic theorem is also a classical theorem, a highly nontrivial claim which does not follow automatically from the standard intuitionistic explanations of what the logical vocabulary means. But the claim has been remarkably borne out in mathematical practice. Let us assume it is true. Does it follow that peaceful coexistence is now unproblematic?

The threat of platonists and intuitionists proving theorems that contradict each other has disappeared. But in an important sense, the asymmetry remains. The platonist still has no problem understanding intuitionistic mathematicians as dealing with the constructive part of mathematics in general. He could even admit that this is a useful and worthwhile enterprise. But nothing similar holds in the other direction. As far as I can see, the intuitionist's only possibility is a *formalist* understanding of classical mathematics: investigating the consequences of certain extra axioms.[23]

The appeal of formalism to mathematicians, of all kinds, should not be underestimated.[24] For one thing, it is a handy retreat position when philosophers or logicians ask too many questions about foundations: I just study

---

[23] The claim that classical mathematics is a special case doesn't really help, if this special case results from ignoring distinctions that one feels should be upheld.

[24] Of course I don't mean Hilbert style formalism, i.e. the idea that the safety of mathematics should be guaranteed by some reduction to a small 'concrete' part of it, about which one is in no way formalist. Formalism here is roughly the view that mathematicians prove theorems in axiom systems, but the choice of axioms is unrelated to questions of truth.

what follows from these axioms. For another, it fits with the *aesthetic* aspects of proofs and theorems, aspects which no mathematician ignores.

> What are the criteria for choosing among axiom systems? Generally there are two opposing criteria: interesting models and beautiful theorems. [15, p. 125]

Presumably, a theorem or a proof is beautiful in much the same way as a game of chess can be beautiful. But, as Richman indicates, beauty has little to do with the truth- or knowledge-seeking aspects of mathematics.

On reflection, formalism is not a solution to the problem but a way to ignore it. Besides, I doubt that there are any formalists about number theory. There is a huge literature on axiom systems for arithmetic, and their models. But this is part of *proof theory* or *model theory*, both established mathematical-logical disciplines. To put it crudely, the object of these investigations is proofs, or models, but not numbers. By contrast, consider the immense efforts mathematicians have spent on long standing number-theoretic claims, such as Fermat's Last Theorem or Goldbach's Conjecture. Clearly, the feeling of mathematicians is that we now know that Fermat's Theorem is *true*, whereas Goldbach's conjecture is still *open*.[25]

To be sure, an intuitionist might not accept this result until she is satisfied about the constructivity of the methods. That is, without a constructive proof she would not think that the *truth* of Fermat's Last Theorem had been established, and would presumably be forced to take a formalist stance on the actual proof. And that would be another illustration of the asymmetry.

6.2 From the outside

What should the outside observer conclude, then, about the C-I conflict? A first impression is that the persistent asymmetry we found might not be that serious after all, at least with modern intuitionism. There is no outright

---

[25] There is an interesting quirk concerning Fermat's Last Theorem, since the actual proof apparently uses methods from category theory not formalizable in ZFC (relying on the existence of inaccessible cardinals; see the discussion in FOM on this issue, for example Harvey Friedman's postings, such as http://cs.nyu.edu/pipermail/fom/1999-April/002992.html), although all specialists are convinced these methods are eliminable and the proof goes through in ZFC. My simple point here is just that virtually everyone agrees that it is the *truth* of Fermat's claim which is at stake here, not which axioms it follows from. The question was unresolved for 350 years, but now it is *settled*. (There are other and perhaps more interesting issues involved, such as why everyone agrees that provability in ZFC, and perhaps even in ZFC + some large cardinal axioms, would guarantee arithmetical truth, and also why no one apparently has found it worthwhile or rewarding to actually perform the elimination of inaccessibles from the proof. But the simple point is sufficient here.)

conflict, and the fact that one participant in the debate has problems understanding what the other is up to doesn't mean that it *cannot* be understood. The other side claims it can.

I am not being ironic here. Without going into the philosophical debate about meaning, I think all our observer can do is to take seriously the claims of the mathematicians involved. If one group of mathematicians insist they have no problems understanding both kinds of mathematics, and another group insist they have serious problems understanding parts of classical mathematics, so be it.

But an equally strong impression is that we haven't really dealt with the heart of the matter. If the differences between platonists and intuitionists eventually boiled down to matters of *taste*, to which kind of mathematics they *liked best* (and therefore should be funded, etc.), the investigation could stop. But more seems to be involved. Consider the question of *why* modern intuitionists have gone to such lengths about asserting only theorems that the classical mathematician can also assert. There is no *a priori* reason to do so. On the contrary, although both insist on using the same logical symbols, the respective meanings they associate with these symbols are manifestly different, so *a priori* one wouldn't be surprised if some *apparent* conflict emerged (as it did with Brouwer style intuitionism). But the tendency has been to avoid even apparent conflicts. Why?

Presumably, part of the answer is that in this way intuitionistic mathematics is will attract more interest among 'traditional' mathematicians. But that can hardly be the main motivation. Surely the main motivation lies in the mathematical work itself, in the fact that it has proved possible to formulate constructive mathematics in this way. This is a striking and non-trivial fact, and it would appear to merit some principled explanation. Then, the asymmetry might come to look natural, rather than problematic. It seems to me that such an explanation has not yet been given.[26]

That much can perhaps be gleaned from the outside. Providing an explanation, however, most likely would require inside work.

Finally, what about relativism? I think that question too must await an explanation of the kind just asked for. Consider the statement

(3) The real numbers can be well-ordered.

This is a claim students learn to prove during a first set theory course, but which intuitionists (modern or traditional) refuse to believe in. The platonist may argue, as we have seen, that the sentence (3) can express two different claims, the second entailing that we can somehow *compute* such a well-ordering, and he may agree with the intuitionist that we have have no grounds for asserting *that*. A relativist take on this, however, is different. The relativist

---

[26] As noted, I don't think explanations via negative translations are adequate in the required sense.

must argue that there is in fact only one claim, but that the *context of assessment* determines its truth value.[27] In the standard classical set theory context of assessment, (3) is true; in the intuitionist context, the very same claim or proposition is not true.

There is the issue of whether such a relativist stance is internally coherent. Many philosophers doubt that. But setting that issue aside, isn't there some plausibility in the (vague) idea that platonists and intuitionists do talk about the same things, but assess them in different ways? If they only talked about different things, or said different things that only appear similar because the same words are used, their disagreement would be somewhat trivial. But there is a strong impression that it is not trivial in that way. An explanation of the 'real' relation between classical and constructive mathematics, and of the way platonists and intuitionists understand each other, should clarify this situation too. Whether some form of relativism is involved is, I think, anybody's guess.

Acknowledgements

I am grateful to Per Martin-Löf, Peter Pagin, Dag Prawitz, Giovanni Sambin, Jouko Väänänen, and an anonymous referee, for helpful comments or advice, but none of them is to be held responsible for anything I say here. The paper develops some points made in [18]. Work on it has benefited from grants by the Swedish Research Council and the Bank of Sweden Tercentenary Foundation.

References

1. S. Berardi and S. Valentini: Krivine's intuitionistic proof of classical completeness (for countable languages). *Annals of Pure and Applied Logic* **129** (2004) pp 93–106
2. E. Bishop: *Foundations of Constructive Analysis* (McGraw-Hill, New York 1967)
3. D. S. Bridges: Constructive truth in practice. In: *Truth in Mathematics*, ed by H. G. Dales and G. Oliveri (Clarendon Press, Oxford 1998) pp 53–69
4. M. Dummett: The philosophical basis of intuitionistic logic. Reprinted in: *Truth and other Enigmas* (BLA, London 1978) pp 215–247
5. M. Dummett: *Elements of Intuitionism* (Oxford University Press, Oxford 1977) (Revised and reprinted in 2000)
6. M. Dummett: *The Logical Basis of Metaphysics* (Harvard University Press, Harvard 1991)
7. G. Hellman: Never say "Never"! On the communication problem between intuitionism and classicism. *Philosophical Topics* **17** (1989) pp 47–67
8. J.-L. Krivine: Une preuve formelle et intuitionistique du Theoreme de Completude de la Logique Classique. *Bulletin of Symbolic Logic* **2** (1996) pp 405–21
9. J. Lipton: Kripke semantics for dependent type theory and realizability interpretations. In: *Constructivity in Computer Science, LNCS* **613** (Springer, Berlin 1992) pp 22–32

---

[27] This would be relativism in the sense of [10].

10. J. MacFarlane: Making sense of relative truth. *Proceedings of the Aristotelian Society* **105** (2005) pp 321–39
11. P. Martin-Löf: *Notes on Constructive Mathematics* (Almquist & Wiksell, Stockholm 1970)
12. P. Martin-Löf: *Intuitionistic Type Theory* (Notes by G. Sambin) (Bibliopolis, Napoli 1984)
13. D. C. McCarty: Undecidability and intuitionistic incompleteness. *Journal of Philosophical Logic* **25** (1996) pp 559–565
14. D. Prawitz: Meaning and proofs: on the conflict between classical and intuitionistic logic. *Theoria* **43** (1977) pp 2–40
15. F. Richman: Intuitionism as generalization. *Philosophia Mathematica* **5** (1990) pp 124–128
16. G. Sambin: Some points in formal topology. *Theoretical Computer Science* **305** (2003) pp 347–408
17. A. Troelstra and D. van Dalen: *Constructivism in Mathematics; An Introduction. Vols. I–II* (North-Holland, Amsterdam 1988)
18. D. Westerståhl: Perspectives on the dispute between intuitionistic and classical mathematics. In: *Ursus Philosophicus, Philosophical Communications*, web series no. 32, Gothenburg University (2004)

Proof as a Path of Light

Rossella Lupacchini

Dipartimento di Filosofia, Università di Bologna (Italy)
rossella.lupacchini@unibo.it

According to certain medieval philosophers, perspective is a "demonstrative science" as it reveals the connection between sensible and intelligible visions by means of the mathematical rules of geometric optics. Carrying over concepts and methods of the medieval *perspectiva naturalis* into a plane surface, the Renaissance *perspectiva artificialis* unfurls a new "pictorial" space. To appreciate the impact of quantum theory on determinism and computation issues this paper will adopt a "perspectival" approach: the architecture of the theory, first captured in the 'real' three dimensional space, will lead us into a new 'imaginary' space. Here the bilateral symmetry coupling any possibility with its negation, sized by "complex probability amplitudes", may dissolve the 'ignorance' of classical probabilities as well as the 'blindness' of finite mechanical procedures.

1 In the light of light

To understand the rules underlying the structure of the world, nature must be examined through the mechanism of vision. In the thirteenth century, Robert Grosseteste wrote that "Physicists know what [*quid*], whereas experts of optics [*perspectivi*] know why [*propter quid*]".[1] The *quid* is the "substantial cause", whereas the *propter quid* is the cause derived from the definition of the thing, since a definition is the beginning and the end of the proof. Light is the first beginning, the form, or *quid*, or essence, common to all natural beings, that allows the construction of a "demonstrative science".[2] By connecting the Platonic doctrines of light to the Aristotelian conception of a demonstrative science, Grosseteste sought a causal explanation of the generation of forms through the analysis of the behaviour of creatures according to the different

---

[1] *De iride* 1230-33 [25, p. 145].
[2] *Commentarius in Posteriorum Analyticorum libros* 1228-30 [24].

'modes' in which light acts. The idea of an active luminous matter, capable of generating forms, was further developed in the *De multiplicatione specierum*[3] by Roger Bacon. Species is the effect of natural agency and presupposes both an action and a substratum of matter inclined to receive it and be modified by it. The laws of the multiplication of species follow the rules of natural perspective, namely of optics.

The demonstrative science of the medieval *perspectivi* is of Aristotelian character and pursues substantial causes, rather than functional links; nevertheless the conception of the science of light as a demonstrative science hands over a mathematical language to visual phenomena. Thus the problem of vision leads to search for an adequate reading of the spatiality and of the geometry of things revealed by the light radiating through the space. However the problem of the objectivization of the visual impression, as we shall see later on, demands a new "perspectival" space. According to Panofsky, the antique conception of an inhomogeneous space of aggregates is abandoned when the Romanesque, fully matured by the middle of the twelfth century, sheds all vestiges of antiquity and renounces all spatial illusionism.

> For if Romanesque painting reduced bodies and space to surface, in the same way and with the same decisiveness, by these very means it also managed for the first time to confirm and establish the homogeneity of bodies and space. It did this by transforming their loose, optical unity into a solid and substantial unity [38, p. 51].

The way in which the Romanesque resolves the antithesis between the bodies and the space in a new plastic composition ought to be regarded as a concrete application of the theoretical principles which the medieval perspective derived from the Arabic philosopher Al-Kindi.

Although both Grosseteste and Bacon, regarded the spiritual enlightenment coming from faith above the human and philisophical experience of vision, there is the spiritual enlightenment which comes from faith, their ideal of a contemplative science (in the Greek tradition) is dialectically related to an 'active' conception of nature derived from *De radiis* (the theory of rays)[4] by Al-Kindi. In the worldview of Al-Kindi, things act and leave signs or impressions of their action in different ways, according to certain rules. All actions occur in line with rays and the laws of radial propagation, namely in accordance with geometric optics. However the Euclidean geometry cannot correctly explain what rays are, since the rays are not empty and abstract mathematical entities but they must have the power of operating. Hence geometry must be 'materialized'; if the rays were really one-dimensional, like the lines of the Euclidean geometry, they would not be physically determined, or perceivable. Vision is a sensory phenomenon and the certainty of the truth of demonstrative and geometrical sciences ought to rest on sensory evidence.

---

[3] Written before 1267.

[4] The Arabic text (IX cen.) was translated in Latin in the XII century.

Light is not an abstraction, it is a material quid which generates sensible and intellectual experience through its action. Matter, the first essence,[5] anticipates form. Specification or, in general, differentiation is bound to the capability of recognizing forms: such a capability that allows 'identifing' something through sight and, as a result, 'knowing' something.

Keeping in mind this image of light as a (material) *quid* giving rise to our knowledge through its action, as well as the conception of a demonstrative science, based on on the capability of recognizing forms and on the rules of light propagation, we are going to explore what quantum physics tells us about the effectiveness of proof.

2 Photons, observability, and limitation of size

Consider a photon which impinges on a beam splitter – that is a randomising device reflecting and transmitting the light with the same probability – and propagates via two different paths (Fig. 1A). After its encounter with the beam splitter the photon will be registered with the same probability either in the detector *A* or in the detector *B*. Now let the photon reach, via two symmetrical alternative paths, another beam splitter and again propagates *via* one of two paths (as shown in Fig. 1B).

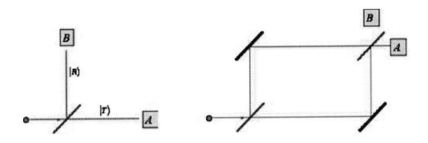

Fig. 1. A: Beam splitter   B: Quantum interference effect

According to the probability laws, we would expect the photon to be registered with the same probability either in the detector *A* or in the detector *B*. However, according to quantum physics, if the two paths are exactly symmetrical, and therefore "indiscernible", then there is a 100 per cent probability that the photon reaches one of the two detectors, call it *A* the *light detector*,

---

[5] The five essences or principles of reality are: matter, form, place, motion, time. Mater is the first essence or the principle from which the other four essences are derived.

and 0 per cent probability that the photon reaches the detector B, the *darkness detector*. It appears as if some action took place between the two beam splitters that prevents the photon from reaching B, an action which involves an invisible *quid* travelling at the speed of light exactly as a photon would. This action of invisible counterparts affecting the motion of photons that we observe is due to *quantum interference* and applies not only to photons but to all quantum particles.

Quantum interference has a peculiar character: as if it were a creature of darkness, when observed it may suddenly disappear. If a measurement device is placed after the first beam splitter of our "Mach-Zehnder interferometer", as such an experimental apparatus is called, to show which path is taken by the photon, then the probability that the photon reaches one of the two detectors is balanced (it becomes 0.5 for each); by observing the path of the photon, the darkness detector lights up and the interference vanishes. The question then arises as to which evidence one can have of quantum interference. A possible answer is that this evidence is to be obtained *a posteriori*, as a kind of 'global' property of an ensemble of identical particles. The series of pictures in Fig. 2 shows a sequence of independent events in a "double-slit experiment" performed with electrons. In this case, the quantum particles travel through a screen in which there are two slits and then reach a photographic plate. The particles which initially appear to draw up randomly in the plate, end up in a typical interference pattern where light and darkness zones follow one another.[6] Interference emerges *step by step*. By considering each particle as a discrete unit, the probability distribution of the electrons in the various points of the plate is derived by the ratios between the number of electrons counted up in each point and the total number in the ensemble. But what is the relationship between probability and interference? How to recognize harmonics in the coarse score of statistics? What enables one particle to conform to the rules of such harmonics?

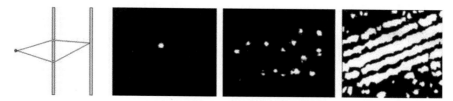

Fig. 2. Two slit interference pattern

To connect probabilities and interference, quantum theory introduces the notion of the *complex* probability *amplitude* and defines the probability of

---

[6] The experiment was first performed at the Bologna University [37].

an event as the squared modulo of its probability amplitude given by a complex number. When an event can occur in several mutually exclusive ways, its probability amplitude is obtained by summing up the probability amplitudes for each alternative. Thus interference is brought out as a consequence of complex numbers having a *phase*, that is to say an angle θ which sizes the distance between any complex number and the 'real' axis in the *Argand plane*.[7] If there are two alternatives, like in the double-slit experiment or in the Mach-Zehnder interferometer, whose amplitudes are α_1 and α_2, the probability amplitude α of the event is $\alpha = \alpha_1 + \alpha_2$ and the corresponding probability p is not $p = p_1 + p_2$ but rather

$$p = |\alpha|^2 = |\alpha_1 + \alpha_2|^2 = |a_1|^2 + |a_2|^2 + a_1^* \alpha_2 + a_1 \alpha_2^*$$
$$= p_1 + p_2 + (a_1^* \alpha_2 + a_1 \alpha_2^*).$$

It is evident that the last term $(a_1^* \alpha_2 + a_1 \alpha_2^*)$ is to answer for the fail of classical probability laws, or rather to be praised for quantum interference. How does it change probability?

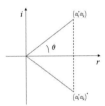

Fig. 3. Probability amplitudes in the Argand plane

As the two addenda are complex conjugate to each other, there is *bilateral symmetry*[8] between them in the real axis (see Fig. 3); accordingly their sum can be expressed as the product of their (same) modulo times $2\cos\theta$, where θ gives their symmetrical *angular separation* from the real axis:

$$a_1^* \alpha_2 + (a_1^* \alpha_2)^* = |a_1^* \alpha_2| \, 2\cos\theta = 2\cos\theta \sqrt{p_1 p_2}.$$

Thus, when there are several mutually exclusive ways in which an event can occur, the probability is the sum of the probabilities for each single way sharpened by an additional term which marks the '*symmetrical distance*' between

---

[7] The Argand plane is a two-dimensional plane where we can visualize any complex number c as a point and locate it by means of Cartesian coordinates (x, y) such that: $c = (x + iy)$ or in polar form as $c = |c| \, e^{i\theta} = |c| \, (\cos\theta + i\sin\theta)$.

[8] More about this in § 4.

any pair of them. In this way quantum interference moulds physical phenomena, as in Fig. 2, by guiding mutually exclusive events to match in a mosaic where each tessera is labelled by a complex number.

The question then arises as to what prevents the effects of quantum interference when the particle path is observed. Is there really no way of observing the path without cancelling out the interference? Perhaps watching the particle in a very weak light may help.[9] According to the de Broglie's relation $p = \hbar/\lambda$, to have weaker light means to increase its wavelength, namely to have light of a redder colour. However, to be distinguishable two objects must be separated by a minimal distance not shorter than the wavelength of the light illuminating them. With reference to the double slit experiment, this implies that, when the wavelength of light is longer than the distance between the two holes, the distinguishability of the two paths is lost; then, being 'aware' of a multiplicity of indiscernible options, each particle is 'free' to choose its place in the interference pattern. The conclusion, as it is stated by the *uncertainty principle*, is that performing an experiment to determine 'which path' is taken by a quantum system entails disturbing its state so much as to cancel any interference effect.

Since measurement involves an interaction of the system with a measuring device, it has a 'perturbative' character. Determining the value of a physical quantity, an *observable* after quantum theory, is an *effective* process, not only for the 'measurer', as it changes its 'state of knowledge', but also for the system as, in getting in one "pure" state of the measured observable, it needs to get out of some other *pure* state of an "incompatible" observable. In classical mechanics, instead, measurement has no effect on the system; though the accuracy of observation can be infinitely sharpened (in principle) by increasing precision and power of the (optical) instruments, these instruments - as powerful as they can be – are undetectable to the system. To discharge this view quantum theory "gives an *absolute meaning to size*". As Dirac writes:

> We have to assume that there is a limit to the finiteness of our powers of observation and the smallness of the accompanying disturbance - a limit which is inherent in the nature of things and can never be surpassed by improved technique or increased skill on the part of the observer [12, § 1].

Turning into mathematics a requirement of observability is also crucial in Turing's conceptual analysis of effective calculability. The so-called *Church-Turing thesis* asserting that "what is effectively calculable is calculable by a Turing machine" presupposes two assumptions: (1) the interpretation of effectively calculable as calculable *by* a 'computer'; (2) the *Turing thesis* that a human computer is *computationally* equivalent to a Turing machine. Investigating calculations as symbolic processes carried out by a computer, Turing

---

[9] Cf. [21, § I.6].

is able to impose restrictions on the steps permitted in computations and justify them through an analysis of the idealized capacities of the computer available for its execution, *i.e.* a person. In other words, (2) is needed to motivate boundedness conditions on the computer involved in (1). Thus Turing portrays a *human* computer as a finite state machine. Considering the actions of a person performing a calculus and searching the easiest way to mimic them, he speculates that a machine could play the role of the computer. The operations of computing are viewed as modifications of a mechanical device consisting of a scanner moving back and forth along a tape. It suffices to consider two kinds of steps: a shift of one cell along the tape, and a change of symbol (0 to 1 or 1 to 0) on the scanned cell. This is all is needed for an abstract machine to reproduce the effectiveness of the computer. Now the calculability of a function can be identified with the possibility of printing its values as sequences of 0s and 1s on the tape of a "Turing machine".

Turing's strategy, in arguing for the adequacy of his notion, is to point out the essential human capacity involved in computing, *i.e. distinguishing* symbols, and to formulate it in terms of finiteness conditions on the symbols scanned by the machine. The process of computation rests on the process of *observation*. If it is assumed that the computation is performed by reading and printing (or cancelling) symbols on a potentially infinite tape divided into squares, then there is a *lower bound* on the distance between symbols, to rule out that some of them get "arbitrarily close", and an *upper bound* on the number of symbols, to ensure that all symbols (which take part in a computational step) can be observed at one glance.[10] In accordance with experience, if we admitted an infinity of symbols, some of them would be confused [44, § 9]. The states relevant to the computation must be "immediately recognizable." As a Turing machine is 'computationally' equivalent to a person, what Turing conceives is a kind of simulation which renders the machine functionally indistinguishable from a human computer. In so far as what happens in "human computing activity" can be described in terms of a local, deterministic, finite procedure, Turing's conceptual analysis echoes Hilbert's words:

> The fundamental idea of my proof theory is none other than to describe the activity of our understanding, to make a protocol of the rules according to which our thinking actually proceeds. Thinking, it so happens, parallels speaking and writing: we form statements and place them one behind another. If any totality of observations and phenomena deserves to be made the object of a serious and thorough investigation, it is this one [27, p. 475].

---

[10] Refining Turing's conceptual argument as to the boundedness constraints on computability, Gandy [22] replaces the sensory limitations of a human computer with physical limitations: a lower bound on the size of atomic components and an upper bound on the speed of signal propagation (a locality condition set by relativity theory). Notice that to set an upper bound on the velocity of light is equivalent to set a lower bound on its wavelength.

The *deterministic* character of Turing machines also satisfies Gödel's requirement for a limited freedom in the activity of the mathematician. "If anything like creation exists at all in mathematics, then what any theorem does is exactly to restrict the freedom of creation" [23, p. 314]. Nevertheless, while the (universal) Turing machine, as derivative of mathematical experience is anchored in space and time, for Gödel mathematical knowledge involves an objective reality out of space and time.

> It is correct that a mathematical proposition says nothing about the physical or psychical reality existing in space and time, because it is true already owing to the meaning of the terms occurring in it, irrespectively of the world of real things. What is wrong, however, is that the meaning of the terms (that is, the concepts they denote) is asserted to be something man-made and consisting merely in semantical conventions. The truth, I believe, is that these concepts form an objective reality of their own, which we cannot create or change, but *only perceive and describe* [23, p. 320].

What does such an objective reality mean? If knowledge comes in with experience, beyond the limits of experience, it has no meaning. As lucidly emphasized by Cassirer, if a finite intellect is limited, this limitation holds not only a negative mark, but a positive one as well. It does not show an accidental and exterior limit of the intellect, it is rather a necessary condition for its activity and fruitfulness. It is not licit to conceive a limit as a simple obstacle to avoid: it rather delimits the domain of our thought and our knowledge, the domain in which they find their concrete meaning [8, chap. II]. Thus the restriction of the intellect to the conditions and limits set by experience is its only possibility of realization.

Although from different perspectives, both Turing computability and quantum physics pose two fundamental questions. Would it be possible to conceive computation or measurement without assuming a limit to the finiteness of our means of observation? Could an adequate description of any effective procedure dispense with the *medium* of the agent (computer or measurer) working out the operations involved?

3 Quantum observables

Observables are physical quantities which can be measured on a system. Any physical theory is about observables, but the classical presupposition that observables are *made* out of objective properties – *beables*[11] – is not tenable in quantum theory because the observables it considers are *incompatible*. We have already met incompatible observables of a photon, that is path and interference. As we have seen, an experimental arrangement which allows to

---
[11] Cf. Bell [5].

answer the question 'which path?', gives no answer to the question 'which detector?'. But what incompatible observables are is written in the Hilbert spaces. And according to the mathematical theory of Hilbert spaces, incompatible observables of a two-state quantum system are represented within the *same* two-dimensional *complex* Hilbert space by operators which are *mutually transformable*[12] and *do not commute*.

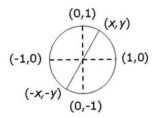

Fig. 4. States of incompatible observables on the unit circle

Consider three incompatible observables like the components of the polarization of a photon selected by means of three filters: the first for vertical polarization 'V', the second for linear polarization at 45° 'L' and the third for right-hand circular polarization 'C'. Each of these observables is assumed to have two values, '+' and '-': if the photon goes through the polarizing filter the outcome is '+', if it does not the outcome will be '-'. We can try to represent the polarization states of the photon in a two-dimensional space $\mathbb{R}^2$. In Fig. 4, the "eigenvectors"[13] of the operator **V** corresponding to the observable V can be associated with the vectors $|v_+\rangle = (0, 1)$ and $|v_-\rangle = (0, -1)$ while the eigenvectors of **L** can be the vectors $|l_+\rangle = (1, 0)$ and $|l_-\rangle = (-1, 0)$. Although in doing so the orthogonal pure states of one observable are separated by an angle π rather than $\frac{\pi}{2}$,[14] this picture is a convenient way to visualize the *symmetry* and *continuity* constraints on probabilities associated with incompatible observables. Any pure state of the photon, say $|v_+\rangle$, assigns probability 1 to exactly one value of one observable, namely $(V, +)$, and probability 0 to the opposite value $(V, -)$, and the same probability 0.5 to the values of the incompatible observables $(L, +)$ and $(L, -)$. Accordingly, any pure state of one observable is equidistant from the pure states of the other observable. The point at issue is that no pure state of one observable can coincide with a pure state of the other, for they must be *distinguishable* (as the operators

---

[12] That is to say, for any pair of operators - say **A** and **B** - there exists a unitary operator **U** such that $\mathbf{A} = \mathbf{UBU}^{-1}$.
[13] The vectors which represent the pure states of a physical quantity.
[14] In quantum theory, any pair of pure states is represented by a pair of orthogonal rays on the Hilbert space.

corresponding to such observables do not commute); still we can 'rotate' axes to transform the diagram of V-results into the diagram of L-results (as the operators are mutually transformable). Thus pure states of the same observable are mutually orthogonal, while pure states of incompatible observables are mutually "oblique".

To translate these ideas into experiment, one can let a photon impinge on a polarizing filter and observe if it passes through it or not. Imagine a photon gets through a vertical polarizing filter, then the state of the photon is $|v_+\rangle$ and its probability to pass through a linear filter at 45° or a right-hand circular polarizing filter is 0.5. By performing a measurement, we can observe the photon going through a second filter, say the linear one, and getting the state $|l_+\rangle$. Knowing its vertical polarization state $|v_+\rangle$ after the former measurement, can we predict that it will confirm such a state passing again through the vertical polarizing filter? No, we cannot as the uncertainty principle teaches us that a polarization state is not an objective property *of* the photon: selecting a polarization state by means of the appropriate filter, any previous incompatible polarization state of the photon is lost. As the photon, after the second measurement, is in the pure state $|l_+\rangle$, its position in the unit circle, given by the point (1,0), is equidistant from the points corresponding to the eigenvectors of **V**; hence the probability of each vertical polarization state of the photon is 0.5. That is how probabilities are assigned to quantum observables according to the uncertainty principle.

Reflecting about the significance of quantum theory, John Bell underlines that quantum theory is fundamentally about the results of 'measurements', and therefore presupposes a 'measurer' (or subject) in addition to the 'system' (or object). But a theory about 'measurement' implies incompleteness of the system and unanalyzed interventions from outside. Here is why the subject-object distinction is viewed as an issue "at the very root of the unease that many people still feel in connection with quantum mechanics." Bell raises the question as to how it can again become possible "to say of a system not that such and such may be *observed* to be so but that such and such be so"[5, p. 41]. In other words, could completeness and determinism be restored?

The unease which Bell connects to the distinction system-measurer in quantum mechanics has its analogues in the unease roused by the incompleteness theorems in mathematics. In both cases the crucial issue originates from the difficulty of describing an 'inside-outside' relationship ignoring one part of the relation, as if a relationship was an 'objective property' of the system. By connecting the 'effectiveness' of the formal system to the 'resolution power' of the computer, as we have seen above, a Turing machine takes Hilbert's proof theory to its limits.[15] Those limits overlap with the classical physics'. Hermann Weyl states the question as it follows:

> The 'physical process' undisturbed by observation is represented by a mathematical formalism without intuitive [*anschauliche*] interpre-

---

[15] Proving the unsolvability of the *Entscheidungsproblem*.

tation; only the concrete experiment, the measurement by means of a grating, can be described in intuitive terms. This contrast of physical process and measurement has its analogue in the contrast of formalism and meaningful thinking in Hilbert's system of mathematics. As it is possible to formalize an intuitive mathematical argument, so it is true that measurement by a grating G may be interpreted as a physical process. In doing so one has to extend the original system S to a system S^* by inclusion of the grating G. But as soon as we want to learn something about S^* that can be told in concrete terms, then the undisturbed course of events [...] must again be disrupted by subjecting S^* to the test of a grating outside S^* [48, p. 261].

However, as we learn from quantum theory, the limits of Hilbert's "metamathematics" are not the limits of Hilbert's mathematics!

In classical mechanics, a complete description of how things are now is given by a point in the phase space and the Hamiltonian function determines how things change with time. However, to make classical physics deterministic, an additional assumption is needed, namely that the physical system, is *closed*. In quantum mechanics, a deterministic description of how the state of a system evolves with time is given by the Schrödinger equation; however quantum theory is inherently probabilistic as a quantum state cannot assign values to all physical quantities associated with the system. A quantum state does not describe how things are but their probability relations.

In determining the probability of a certain value for a physical quantity, quantum theory differs from classical physics as to the impossibility of performing certain measurements simultaneously. A measurement is not a 'formal' process, it is an *action* which involves two actors – a system to be observed and an observer – and establishes a kind of 'exclusive' relationship between them. Once the action is performed the 'pair' is modified through an exchange of information. Given this, any measurement is 'perturbative' and demands to make a choice. The observer has to choose, among incompatible aspects of a physical system, which one to probe. As far as measurement is viewed as an interaction, with the twin requirement of freedom in choosing the observable to be questioned and capability of recognizing the answer, it demands to sharpen the probability relations associated with its results and, consequently, to sharpen their mathematical representation.[16] This leads to locate the mathematical space on *complex* field, hence to Hilbert spaces.

Rephrasing Bell, we may rather say that quantum theory is fundamentally about the 'probabilities' of measurements and, as it presupposes in addition to the system (or object) a 'measurer' (or subject), the distinction between 'objective' probability and 'subjective' probability disappears. Indeed, quantum observables reveal a meaningful *symmetry* between system-object and observer-subject. In the light of this symmetry, let us look at the significance of complex probability amplitudes.

---

[16] Cf. Wheeler [50].

4 The reality and its double

The unease described by John Bell is not felt neither by the hero Perseus nor by the Renaissance artists. They master the object-subject symmetry which rules the observation or the visual phenomenon in a superb way. The myth tells us that Perseus can accomplish his task and kill the monstrous Medusa, thanks to a highly-polished bronze shield given to him by Athena. The polished shield, acting like a mirror, allows him to approach the Gorgon and cut off her head, without exposing himself to her petrifying gaze. The success of Perseus' exploit depends on letting Medusa's reflection guide him.

4.1 The geometry of vision and the colour of numbers

The Renaissance art presents us with the "linear" or "artificial" perspective. The invention is attributed to the architect Filippo Brunelleschi. It provides an effective rational procedure for transferring the visual 'reality' into an artificial space:

> first is sight, that is to say the eye; second is the form of the thing seen; third is the distance from the eye to the thing seen; fourth are the lines which leave the boundaries of the object and come to the eye; fifth is the intersection, which comes between the eye and the thing seen, and on which it is intended to record the object...[17]

The result is a painted or drawn scene somehow indistinguishable from the image captured over a transparent window or in a mirror. To accomplish this task the three-dimensional visual 'appearance' is *projected* on a plane surface. When the natural perspective set up the view of the space, the linear perspective intervenes on that view and transforms the empirical space into a purely mathematical space, *i.e.* infinite, unchanging, and homogeneous [38]. A remarkable shift from this abstract space to a more 'synthetic' system is due to those artists who find the appropriate means to 'substantiate' the *perspectival* view of the space.

While the geometry of light frames the image of the reality in two dimensions,[18] lines and colours drive a new 'pictorial' space to emerge from that image. As a case in point, we could consider Piero della Francesca's "perspective synthesis of form and colour"[35]. We may also translate this synthesis in mathematical language and see Piero's *perspectiva pingendi* as a coherent system which combines the geometry of light with the algebra of colour. In fact an analogous attempt, though not as successful, is that of Descartes who conflates algebra and geometry in his vision of the world. To distinguish and

---
[17] Piero della Francesca 1474 [40, p. 64]
[18] According to Alberti, the geometry of light is the *medium* through which the eye could measure the properties of the forms and geometrical construction of the space as a prerequisite for proper painting [1].

recognize the points of an infinite and homogeneous space, Descartes attaches numbers to them, instead of colours. The numbers give 'voice' to the points in his algebraic geometry, whereas the values of the physical quantities become points of the space, and the matter, by penetrating through the space, becomes *res extensa*. In this way Descartes tries to rewrite the 'geometrizate' physics in the language of algebra[19] and to depict an 'algebraic' method of proof which would aspire to "operate by means of the numbers *what the ancients made with the figures*".[20]

Descartes shares the Renaissance belief in a mathematical harmony of the universe and the view of geometry as the most eloquent expression of the architecture regulating that harmony. However he does not share the Leonardo da Vinci's conviction that arithmetic is essentially inferior to geometry. Leonardo rests his claim on the fact that arithmetic deals with "discontinuous quantities" while geometry deals with "continuous quantities". The number does not possess the magic of the geometric forms. Accordingly he finds most disturbing the rupture of harmony in the ratio of side and diagonal of a square; there is no 'symmetry' or proportion between them as they are *incommensurable* quantities. For Leonardo the beauty and perfection of geometry is revealed, above all, by the *divina proportione* of the five regular polyhedra,[21] the so-called 'Platonic' solids (Fig. 5).

Art and science of the Renaissance seek the harmony of the nature in static forms, not in dynamical laws.[22] Indeed, if located in infinite and homogeneous space, motion and rest are indistinguishable. Koyré [31] observes that if 'classical' science replaces a world of qualities with a world of quantities and the world of the becoming with the world of the being this is because numbers and figures are motionless. The very concept of motion, shaped in geometric terms, does not describe a process but a *status*. However, when nature takes a mathematical form, mathematics in turn ends up by taking in the character of natural science; mathematical entities, 'naturalized', are to be considered not so much in their 'being' as in their 'becoming'.[23] Rather than the Cartesian *res extensa*, a new 'naturalized' mathematics, that of the differential calculus, allows classical physics to be written. It is Newton[24] then who succeeds in constructing the modern physical world by recreating distance between matter and space and, above all, by transforming mathematics in the

---

[19] Descartes also tries to convert geometrical problem into geometric ones known as to be solved by Arabic and Renaissance Italian algebraists.

[20] *Regulae ad Directionem ingenii* 1628 [13].

[21] In the first printed edition of Luca Pacioli's treatise *De Divina proportione* (Florence 1509), Leonardo's illustrations provide the five regular polyhedra with a 'visible' spacial configuration.

[22] Even Kepler made an attempt to reduce the distances in the planetary system to regular bodies which are alternatively inscribed and circumscribed to spheres (*Mysterium cosmographicum* 1595).

[23] Cf. Hadamard [26].

[24] *Philosophiae naturalis principia mathematica* 1687.

guise of physics [31]. And yet the conceptual aporia between substance and form, being and becoming, or discrete and continuum, ignored or neglected, remains. Physics, indifferent to the aporia, runs its course and achieves, with the Hamilton-Jacoby's analytical mechanics, its more coherent and rigorous expression.

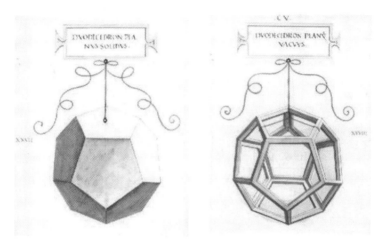

Fig. 5. Pentagondodecahedron as designed by Leonardo for *De Divina proportione*

Whether the question concerning the form, either mathematical or artistic, of the world is solved or dismissed, that form rouses a problem of meaning. Which expressive value is to be attributed to the different intellectual elaborations of experience? On this matter, Panofsky's reading of the perspective as one of the "symbolic forms" fastening 'spiritual' contents to concrete sensible signs, stands out. The different perspectival views of space[25] are determined by the subject; nevertheless, this very variety, "as paradoxical as it may sound", purifies the space of all subjective admixtures. The arbitrariness of direction and distance within modern pictorial space, for Panofsky, bespeaks and confirms the indifference to direction and distance of modern intellectual space; it corresponds to the stage when perspective replaces the simple Euclidean 'visual cone' with the universal 'geometrical beam' of Desargues' general projective geometry; it paves the way, we may add, for the stage when physics would replace the deceiving objectivity of the modern worldview with a new quantum 'inter-objectivity'. Thus perspective provides art and science with a method for recognizing the 'symbolic' value, *i.e.* significant-expressive, of any

---

[25] The "high space" of Italians perspectival constructions is to confront with the "near space" of the North's ones (Netherland in particular) and the "oblique space" such like Altdorfer's [38, pp. 69-70].

artistic or scientific representation, from the Euclidean one to that meaningfully abstract of Hilbert spaces. This method is "ambivalent" as it states the distance between subject and object and in turn dissolves this distance in an autonomous equidistant space:

> Perspective subjects the artistic phenomenon to stable and even mathematically exact rules, but on the other hand, makes that phenomenon contingent upon [...] the individual: for these rules refer to the psychological and physical conditions of the visual impression, and the way they take effect is determined by the freely chosen position of a subjective 'point of view' [38, pp. 67].

Panofsky's emphasis is on the "directional indifference of space" which, as it allows the *objectification of subjectivity*, seals off religious art from the realm of the magical as well as from the realm of the dogmatic, but then it opens it to the realm of the "visionary"[38, p. 72]. In a somehow similar way, the peculiar arbitrariness of orientation of quantum theory in Hilbert spaces seals off science from the realm of the metaphysical determinism. As Hermann Weyl writes:

> The concepts with which it [quantum theory] deals are not qualities or attributes which can be obtained from the objective world by direct cognition. They can only be determined by an indirect methodology, by observing their reaction with other bodies, and their implicit definition is consequently conditioned by definite laws of nature governing reactions. [...] But scientists have long held the opinion that *such constructive concepts were nevertheless intrinsic attributes of the* 'Ding an sich' [...] *In quantum theory we are confronted with a fundamental limitation to this metaphysical standpoint* [47, p. 76].

Quantum world contrasts so much the Cartesian world as the Leonardo's one. Whereas Descartes' attempt of pursuing an algebraic construction of physics is bound to fail as, in Koyré's words, "there is no motion in numbers", and Leonardo diminishes arithmetic as the number does not possess "the magic of the geometric forms", complex numbers open mathematics to the realm of the forms and motion as they open up an entirely new dimension. In the complex space any number, to begin with the 'one' moulded by the Euler's formula $1 = e^{i\pi}$, brings to mind the circle of which is ray and the rotation drawing that form. Form and motion of complex numbers hold the architecture of quantum theory in Hilbert spaces.

4.2 From a pespectival point of view

We have learned from the uncertainty principle that, as soon as a polarization state of a photon is selected by means of the appropriate filter, any previous incompatible polarization state of the photon is lost. However the structure of the probabilities relations is preserved and now we can also appreciate its beauty with 'Leonardo's eye'. To see how the probability function connects quantum states we can use a perspectival procedure: first we try to capture the probabilities relations on the 'visual' space, then we let those relations lead us into a new 'pictorial' space.[26]

There is a one-to-one correspondence between pure states of a two-state quantum system and points of the unit sphere in $\mathbb{R}^3$. Consider a triad of *mutually transformable* observables,[27] each of which has two possible values $i = [+, -]$ corresponding to two *distinguishable* (namely, orthogonal) pure states $|\sigma_+\rangle$ and $|\sigma_-\rangle$, and locate a pure state $|v\rangle$ of the system in one point $\sigma = (\phi, \theta)$ on the unit sphere, where the azimuthal angle ϕ can vary as $-\pi < \phi \leq \pi$ and the longitude θ as $-\frac{\pi}{2} < \theta \leq \frac{\pi}{2}$. The state $|v\rangle$ assigns probabilities $p_v(\sigma_i)$ to the values of all observables of the family. How? As mentioned in connection with the polarization states of light, the *obliquity* of any two states,[28] like $|\sigma_+\rangle$ and $|\vartheta_+\rangle$, tells us that they correspond to observables that do not commute or, as we may say, that are not 'commensurable'. However, in contrast with the antique view of mathematics, in quantum physics 'incommensurability' does not mean 'lack of proportion', *i.e.* lack of symmetry, quite the opposite. Indeed the set of pure states as well as the probabilities grid display a 'spherical' symmetry, and the "arbitrariness of orientation" within the representational space seals off the probability function from the particular point in which the state $|v\rangle$ is located. The probability is a *symmetrical* function δ of the distance, *i.e.* the *angular separation* α, between any pair of states $|v\rangle$ and $|w\rangle$ corresponding to the points σ and ϑ on the sphere: $p_\sigma(\vartheta) = \delta(\alpha_{\sigma,\vartheta}) = p_\vartheta(\sigma)$; hence $|v\rangle$ assigns probability 1 to exactly one point, that that coincides with its own 'point of view' σ, namely when $\alpha = 0$. The probabilities relations are also *symmetrical* about the axis, somehow 'North-South', defined by the points σ and $\sigma^* = (\pi \pm \phi, \theta)$; hence $|v\rangle$ assigns the same probability $p_v(\sigma)$ to all points on the same 'latitude' as σ. That is how any pure state $|v\rangle$ spreads probabilities *continuously* over the sphere.

Now the group of symmetries of the unit sphere S in $\mathbb{R}^3$ is the set of all rotations of S about its centre, that is orthogonal transformations of three variables that leave invariant the angular separation of all pairs of points on the sphere and the 2-norm: $|v|^2 = (\sigma_x^2 + \sigma_y^2 + \sigma_z^2)$. Since our observables are mutually transformable for each pair of them S_σ and S_ϑ there must be a rotation by $\pm\left(\frac{\pi}{2}\right)$ which takes S_σ and S_ϑ to coincide (Fig. 6). In fact, if we confine the system in the subspace $\mathbb{R}^2$ where the two observables can be represented by the two matrices

---

[26] For a more rigorous and detailed treatment of the representational capacity of Hilbert spaces, see Hughes [28, Part I].
[27] See footnote 14 above.
[28] Mind that the 'toy model' in $\mathbb{R}^3$ represents two *oblique* states - namely, pure states of incompatible observables - as orthogonal, whereas it represents *orthogonal* pure states of the same observable as 'antipodes'.

$$S_\sigma = \tfrac{1}{2}\begin{pmatrix} 1 & 0 \\ 0 & -1 \end{pmatrix} \text{ and } S_\vartheta = \tfrac{1}{2}\begin{pmatrix} 0 & 1 \\ 1 & 0 \end{pmatrix},$$

we can verify that $R_{\frac{\pi}{2}} S_\sigma = S_\vartheta$ and $R_{-\left(\frac{\pi}{2}\right)} S_\vartheta = S_\sigma$;[29] the rotation $R_{\frac{\pi}{2}}$ which takes S_σ into S_ϑ has its inverse $R_{-\left(\frac{\pi}{2}\right)}$ which takes S_ϑ into S_σ. But we can also keep on rotating our system and verify the symmetry of the rotations group of the sphere; by repeating the rotation which takes S_σ into S_ϑ we 'overturn' S_σ into its 'mirror image' $-S_\sigma$:

$$R_{\frac{\pi}{2}}\left(R_{\frac{\pi}{2}} S_\sigma\right) = (-1) S_\sigma.$$

No surprise, given that a rotation by $\pm(\pi)$ is nothing but a 'reflection' (-1). Then two rotations by $\pm(\pi)$ take the observable back to the original form, $(R_\pi)^2 S_\sigma = S_\sigma$. We may recall that reflection of the horizontal plane in a line L, i.e. *bilateral symmetry*, is an 'improper' rotation in two-dimensions as it interchanges left and right. To carry any point σ into its mirror-image σ^* by a 'proper' rotation, one more dimension is needed; so it can be brought about in space through a rotation around L by π, an *Umklappung* (overturning) as Weyl calls it. The rotation symmetries of the space can be condensed in the so-called "four-group" $4G$ which consists of the identity and the *Umklappung* around three mutually perpendicular axes.

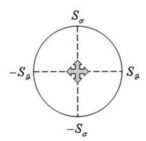

Fig. 6. Mutually transformable observables in $\mathbb{R}^2$

To overview, one rotation by $\pm\left(\frac{\pi}{2}\right)$ interchanges incompatible observables, four rotations by $\pm\left(\frac{\pi}{2}\right)$ or, if you wish to atomize more a continuous process, $2n$ rotations by $\pm\left(\frac{\pi}{n}\right)$ take the observable back into its original form, $\left(R_{\pm\left(\frac{\pi}{n}\right)}\right)^{2n} = 1$, and finally two rotations by $\pm\left(\frac{\pi}{2}\right)$ take the observable into

---

[29] Any rotation by θ in $\mathbb{R}^2$ is represented by the matrix $R_\theta = \begin{pmatrix} \cos\theta & -\sin\theta \\ \sin\theta & \cos\theta \end{pmatrix}$.

its 'mirror image' $\left(R_{\pm\left(\frac{\pi}{2}\right)}\right)^2 = (-1)$. All this is fine as much as two incompatible observables are considered. But given that our mutually transformable observables are three, when the system goes from one incompatible state into another, how to distinguish which of the other two? It is clear that the system can go from a pure state of one observable, say $|\sigma_+\rangle$, into the alternative state $|\sigma_-\rangle$ following (at least) four different symmetrical paths, each of which passes through one of the two alternative states of the other two incompatible observables. All these paths are equally plausible as *no asymmetry* pushes the system to make a choice; and yet they must be distinguishable. How to distinguish in $\mathbb{R}^3$ the 'symmetrical' rotations carrying one state of the system into its mirror-image *via* two incompatible alternative states? This 'doubling' of possibilities, which flows from symmetry, suggests to 'square' the space of our representation as to create the appropriate distance between all of them.

Can we think of some improper rotation which could present each possibility with one more dimension where to reflect its image? The reflection caught in '(-1)' is a good hunch. Even though reflection can be brought about in the 'real' space by proper rotations, it can draw a new "visionary" space. Which roads alternative to proper rotations are feasible from (-1)? Let complex numbers open up them. For continuity allows 'decomposing' any linear transformation, we can take the $2n$-root of any 'rotation' by $\pm\left(\frac{\pi}{n}\right)$, in particular we can take the square root of (-1)! The following table shows how complex numbers entry into matrixes and how these matrixes are 'specularly' related with those of identity and negation:

$$1^{\frac{1}{2}} = \pm \begin{pmatrix} 1 & 0 \\ 0 & 1 \end{pmatrix} \qquad N^{\frac{1}{2}} = \frac{1}{2} \begin{pmatrix} 1 \pm i & 1 \mp i \\ 1 \mp i & 1 \pm i \end{pmatrix}$$

$$(-1)^{\frac{1}{2}} = \begin{cases} i = \begin{pmatrix} i & 0 \\ 0 & i \end{pmatrix} \\ iN = \begin{pmatrix} 0 & i \\ i & 0 \end{pmatrix} \end{cases} \qquad (-N)^{\frac{1}{2}} = \begin{cases} -i = \begin{pmatrix} -i & 0 \\ 0 & -i \end{pmatrix} \\ -iN = \begin{pmatrix} 0 & -i \\ -i & 0 \end{pmatrix} \end{cases}$$

According to Weyl the requirement for improper rotations in $\mathbb{R}^3$ leads to introduce the reflection in the origin which carries any point σ into its 'antipode' σ^*. This operation R^* commutes with all rotations R of the sphere, $R^*R = RR^*$. By including improper rotations of the form R^*R in the four-group $4G$ we obtain the group $4G^* = 4G + R^*4G$ which 'doubles' $4G$.

From the perspectival view in $\mathbb{C}^2$ one can see how the third "Pauli matrix" comes into existence:

$$\mathbf{1} = \begin{pmatrix} 1 & 0 \\ 0 & 1 \end{pmatrix} \qquad \begin{pmatrix} 0 & 1 \\ 1 & 0 \end{pmatrix} = 2S_{\sigma_x}$$

$$2S_{\sigma_z} = \begin{pmatrix} 1 & 0 \\ 0 & -1 \end{pmatrix} \quad i \begin{pmatrix} 0 & -1 \\ 1 & 0 \end{pmatrix} = 2S_{\sigma_y}$$

To transfer the group of rotations of the space $\mathbb{R}^2$ into the Hilbert space $\mathbb{C}^2$ the relevant result is due to Felix Klein: to any rotation which leaves invariant the angular separation between points of the unit sphere, there corresponds two unitary operators $\mathbf{U}$ and $-\mathbf{U}$ on the set of rays of $\mathbb{C}^2$, which leave invariant the angular separation between rays. A mapping of rotation operators on $\mathbb{R}^3$ onto unitary operators of $\mathbb{C}^2$ is consistent with the mapping which takes the point $\sigma = (\phi, \theta) \in S$ of $\mathbb{R}^3$ into the ray L_v of $\mathbb{C}^2$ whose projector $\mathbf{P}_v$ is given by:

$$\mathbf{P}_v = \begin{pmatrix} \cos^2\left(\frac{\phi}{2}\right) & \cos\frac{\phi}{2}\sin\frac{\phi}{2}e^{-i\theta} \\ \cos\frac{\phi}{2}\sin\frac{\phi}{2}e^{i\theta} & \sin^2\left(\frac{\phi}{2}\right) \end{pmatrix}.$$

Now suppose the state $|v\rangle$ coincides with $|v_+\rangle$:

$$|v\rangle = \begin{pmatrix} \cos\frac{\phi}{2}e^{-i\theta} \\ \sin\frac{\phi}{2}e^{i\theta} \end{pmatrix}.$$

Assume that the system is in the state $|\sigma_z\rangle = \begin{pmatrix} 1 \\ 0 \end{pmatrix} \in \mathbb{C}^2$ which corresponds to the point $\sigma_z = (0,0)_{\phi,\theta} = (0,0,1)_{x,y,z}$ on the unit sphere S. The probability which the state $|\sigma_z\rangle$ assigns to $|v\rangle$ is:

$$p_{\sigma_z}(v) = \langle \sigma_z | \mathbf{P}_v \sigma_z \rangle = \cos^2\left(\frac{\phi}{2}\right) = \delta(\alpha_{v,\sigma_z})$$

where α_{v,σ_z} is the 'angular separation' between $|\sigma_z\rangle$ and $|v\rangle$.

If we take as possible values of each observable S_σ, $+1$ and -1, then we have $\mathbf{S}_\sigma = \mathbf{P}_{v_+} - \mathbf{P}_{v_-}$. By expressing the points on the unit sphere in the Cartesian coordinate $\sigma = (\sigma_x, \sigma_y, \sigma_z)$, we can see the Pauli matrixes S_{σ_x}, S_{σ_y} and S_{σ_z} as special cases of

$$S_\sigma = \begin{pmatrix} \sigma_z & \sigma_x - i\sigma_y \\ \sigma_x + i\sigma_y & -\sigma_z \end{pmatrix}$$

and verify that $S_\sigma = \sigma_x S_{\sigma_x} + \sigma_y S_{\sigma_y} + \sigma_z S_{\sigma_z}$. Moreover, we can derive the probability function δ from the *expectation value* of S_σ:

$$\langle S_\sigma \rangle_v = \sigma_x \langle S_{\sigma_x} \rangle_v + \sigma_y \langle S_{\sigma_y} \rangle_v + \sigma_z \langle S_{\sigma_z} \rangle_v$$

and the assumption:

(*) $\quad \langle S_\sigma \rangle_v = p_v(\sigma_+) - p_v(\sigma_-) = 2p_v(\sigma_+) - 1.$

Since the system is in the state $|\sigma_z\rangle$ whose angular separation from $|v\rangle$ is $(\alpha_{v,\sigma}) = \phi$, a convenient Cartesian coordinate system is such that $\sigma(\phi, 0) = (\sin \phi, 0, \cos \phi)$ and $\sigma_z = (0,0) = (0,0,1)$. The expectation value of the two observables S_{σ_x} and S_{σ_y} vanishes $\langle S_{\sigma_x} \rangle_v = \langle S_{\sigma_y} \rangle_v = 0$ whereas the expectation value of the third one is $\langle S_{\sigma_z} \rangle_v = 1$. Thus:

$$\langle S_\sigma \rangle_v = \sigma_z \langle S_{\sigma_z} \rangle_v = \cos\phi \langle S_{\sigma_z} \rangle_v = \cos\phi$$

and confronting this equation with the (*), we obtain the probability function:

$$p_v(\sigma_+) = \frac{1+\cos\phi}{2} = \cos^2\left(\frac{\phi}{2}\right)$$

which depends on the angular separation α as

$$\delta(\alpha_{v,\sigma}) = \cos^2 \tfrac{1}{2}(\alpha_{v,\sigma}).$$

Thus the symmetrical character of the quantum probability function and the key role of complex amplitudes can be captured in the formula:

$$p_v(\sigma) = \langle v|\sigma\rangle \langle \sigma|v\rangle = p_\sigma(v).$$

5 In the light of darkness

The hero Perseus who guided us to the realm of quantum amplitudes with his resplendent shield is also endowed with the helmet of invisibility borrowed from Hades. This last section will explore how one can takes advantage of invisibility.

We have seen that, according to quantum physics, the probability that a photon going through a Mach-Zehnder interferometer reaches the light detector A is 1. This outcome is to be explained as determined by quantum interference (Fig. 1B). By contrast, when a measurement device is placed between the two beam-splitters, the interference vanishes and the probability that the photon reaches the darkness detector B becomes 0.5 (Fig. 7A). Thus, when the darkness detector B is on, the presence of a measuring device between the two beam-splitters is 'brought to light'. It is worth noticing that in this case, namely when the darkness detector is on, there is a 50 per cent probability that no interaction takes place between the measuring device and the photon (Fig. 7B). If the photon does not interact with the measuring device, where is that light coming from? Again, it looks like an invisible *quid* is in charge of some 'spooky action'.

But quantum theory also asserts that the probability that no interaction takes place between the measuring device and the photon (reaching the darkness detector) can be stretched until approaching 1. What can explain this stretching is the so-called "quantum Zeno effect", after Zeno from Elea. This effect is actually an eloquent picture of an argument of the Eleatic philosophy questioning the atomistic determinism: motion is not attainable by summing up motionless states. To figure out what is at issue, we can consider as incompatible observables two components of the polarization of the light. Each photon can be observed in two alternative orthogonal states of polarization, that we can now connect to its possible paths through the interferometer as follows. Imagine that the first beam-splitter reflects vertical polarized photons and transmits horizontal polarized photons, while the second beam-splitter reflects photons with linear polarization by 45° and transmits photons with linear polarization by 135°. To fix the picture, when no measurement is performed between the two beam-splitters, again the probability that a photon going through the interferometer reaches the darkness detector is 0; however, when a measurement reveals the path taken by the photon between the two beam-splitters, that probability becomes 0.5. How to perform such a measurement?

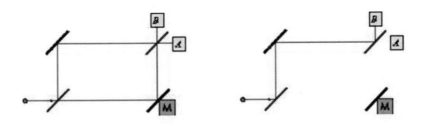

Fig. 7. A: Measuring the photon path   B: Measurement without interaction

As we know (from § 4.2), the probability that an horizontal polarized photon passes through an horizontal polarizing filter is 1, that it passes through a vertical filter - namely a filter rotated by 90°- is 0, that it passes through a filter rotated by θ is $\cos^2(\theta)$. Of course, if the polarization state of the photon is rotated by 90° before impinging on the horizontal filter, then its probability of going through becomes 0 (Fig. 8). Thus if a rotation operator by 90° and a horizontal filter are placed along the lower path in the interferometer, then the photon transmitted by the first beam splitter will be absorbed by the filter. We also know that the action of a rotation operator by $(\pi/2)$ can be decomposed in a series of n steps each of which rotates the state of the photon by $(\pi/2n)$. If $n = 6$, in six steps the polarization state of the photon turns from horizontal into vertical and the photon will be absorbed by the horizontal filter. Can we follow the evolution of the photon step by step, from one polarization state into the other? Yes, we can by inserting a horizontal filter after each polarization rotator (Fig. 9). What happens? Passing through the first rotator, the state of the photon is turned by 15° and the probability of passing through the horizontal filter is $\cos^2(15)$. Now if the photon goes through, its polarization state is again horizontal. By repeating this process five times, the probability that the photon goes through all the six filters is $\left[\cos^2(15)\right]^6$. Moreover, increasing the number n of steps, that is to say decreasing the angle (by which any rotator turns the polarization state of the photon), the probability that the photon goes through the measuring device increases accordingly.[30] As the number n of rotators by $(\pi/2n)$ approaches 'infinite' $n \to \infty$, the probability that the photon is not absorbed, *i.e.* observed, by the filter approaches 1. In other words, watching how the polarization state of the photon is supposed to evolve step by step, in fact it prevents its evolution. It is then possible to conceive a suitable modification of the arrangement above so that when a photon reaches the darkness detector and signals the presence of a measuring device, it is almost certain that this device did not measure (absorb) the photon [32].

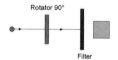

Fig. 8. Orthogonal rotator and horizontal filter

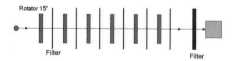

Fig. 9. Action of an orthogonal rotator decomposed in 6 steps

---

[30] Recall $\cos\theta \to 1$ as $\theta \to 0$.

From a 'classical' point of view, no doubt all this certainly echoes the Eleatic critics to the Democritus' atomism. On the other hand, in the symmetrical and non-separable system-measurer perspective, one can see how the photon and the filter are not invisible to each other as a result of a series of system-measurer entanglements: first, the photon gets entangled with the vertical-horizontal beam-splitter, second this composite system gets entangled with the series of rotators and horizontal filters (in the lower arm of the interferometer), thirdly this further composite system gets entangles with the second beam-splitter, and finally the darkness detector lights up. It is thanks to subsequent 'meta-entanglements' due to quantum interference that the darkness detector is able to reveal the presence of a measurement device in the interferometer even though the photon and this device remain somehow invisible to each other. The quantum Zeno effect, first noticed by von Neumann [46], is described as a consequence of *dynamical* features introduced by a series of measurements: the evolution of a quantum state can be slowed down (or even halted in some limit) when measurements are frequently performed on the system, in order to check if it is still in its initial state [17]. However, even Democritus, who first creates the universal schema of the 'atomistic determinism', is aware of the difficulty: motion, or evolution, cannot be described as a series of states. It requires the *void* (empty space) for its representation.

But the void is no sensuous 'given', no thing-like reality; nevertheless, without this concept, a consistent physical representation is not to be attained. As lucidly emphasized by Cassirer, the thought of 'non-being' is no dialectical construction; but, on the contrary, it is taken as the sole means of protecting physics from the extravagances of a speculative idealism. In the conception of the real, this sensuous 'nothing' has the same place and the same inviolable validity as the 'something'.

> It is impossible to relate scientific thought merely to being, as the Eleatic idealism had attempted; non-being is just as necessary and unavoidable a concept. The Eleatics in their denial of non-being not only robbed thought of one of its fundamental instruments, but *they destroyed the phenomena themselves by giving up the possibility of understanding them in their multiplicity and mutability* [7, p. 167].

Through an eloquent interference of the 'visible' with the 'invisible' as well as of the system to be observed with the observer, quantum theory protects physics from metaphysical determinism insofar as it gives back to the physical phenomena their multiplicity and mutability. Quantum theory defines its notions through a double opposition: on the one hand, to metaphysical speculation, and, on the other, to 'non-perturbative' - that is to say 'non-perspectival' - observation. As it connects all the multiple and 'symmetrical' outcomes opened to the system by the observer, quantum interference is not to be understood as an objective property of the system but rather as a mathematical functional *relation*. It provides the *whole system-measurer objectivity* with a structure made out of complex probability amplitudes.

6 Final remarks

In his *Symmetry*, Hermann Weyl credits Leonardo for making up "a *complete* list of *orthogonally inequivalent* finite groups of orthogonal transformations." Actually Leonardo has learnt all he knows about geometric forms and their symmetries from fra Luca Pacioli[31] who had turned to him for illustrating his treatise *De Divina Proportione*. So it is Pacioli who knows why in three dimensions there are only five regular polyhedra; he explains to Leonardo - let the story go loosely - that while the regular polygons are connected with the finite groups of proper and improper rotations in the plane, the polyhedra are connected with the finite groups of proper rotations around the centre or around an axis through the centre. (Of course he could not concern about implausible improper rotations in the space.) Leonardo is conquered and draws the geometrical bodies both in their solid form and in a skeletal manner [29, p. 62]. Leonardo's eye is so much enchanted by the beauty of the geometric forms as to be afraid of loosing it in details. On the other hand, by tackling unsolvable problems such as squaring the circle or diagonalazing the square's side, all canons about *divina proportione* fail.

Besides these 'Renaissance perspective' canons, complex numbers present us with one more: the bilateral symmetry between 'beables' and its mirror-image, 'not-beables', as they both result by taking the squared root of being. And it is most ironic that the Italian Renaissance thought also yields complex numbers. The invention of such numbers is attributed to the algebraist Girolamo Cardano, who works out a general algebraic solution to the cubic and quadratic equations. The irony of the history lies in the fact that Cardano and Leonardo know each other and are tackling what we can now see as 'quantum incompatible observables'. If Leonardo does not see the magic of the forms in numbers, Cardano does not see the magical form of his numbers. They both miss the bilateral symmetry of the issue. Their visions conflate, as it were, in one dimensional space.

Hilbert's proof theory is an attempt to 'secure' what even Leonardo fears to lose, the magic of the continuum; it is pursued by 'projecting' it into 'discreteness', and it fails. No consistency proof for analysis and set theory can be attained within finitist mathematics [42, p. 259]. The perspectival view from Hilbert spaces, as it reveals the *effectiveness* of the bilateral symmetry of complex numbers, it allows mathematical description to radiate in two dimensions. In complex space the 'distinguishability' between numbers or points, consequently between Turing's symbols, does not vanish, it is rather emphasized by

---

[31] Franciscan Friar from Sansepolcro strongly influenced in his work by Piero della Francesca his fellow countryman.

their complex amplitudes. It also applies to a single number as it is to be distinguished from its complex conjugate. Quantum theory focus on probability relations and runs into complex probability amplitudes. Squaring their modulo they become probabilities. But the complex character of quantum amplitudes cannot be locked in 'computation'. On the contrary, as it breaks into a new dimension, it also opens up entirely new computational paths. Quantum theory, as the previous pages have tried to show, has been guided to complex spaces by a search for 'more effective' distinguishability; once in the new perspectival space, the *bilateral symmetry involved in the very notion of distinguishability* emerges out of quantum probability amplitudes and proves the irreducibility of quantum probabilities to classical ones. Can the effectiveness of proof as well benefit from a '*double*-distinguishability' in complex space?

Acknowledgements

Many thanks to Annarita Angelini, Artur Ekert, Stefano Mancini, Saverio Pascazio, Giorgio Sandri, Wilfried Sieg and my sister. Each of them contributed in very different ways to outline the picture presented in this paper.

References

1. L. B. Alberti: *De Pictura* (1435), Basilea 1540. Critical edition by L. Mallé (Florence 1950)
2. Al-Kindi: *De radiis* (Arabic text IX century; Latin trans. XII century). French trans. *De radiis* (Éditions Allia, Paris 2003)
3. R. Bacon: *The Opus Majus*, ed by J. H. Bridges (Minerva, Frankfurt 1964)
4. F. Bailly and G. Longo: Randomness and Determination in the interplay between the Continuum and the Discrete. *Mathematical Structures in Computer Science* **17**, 2 (2007)
5. J. S. Bell: Subject and Object. In: *The Physicist's Conception of Nature* (Reidel, Dordrect 1973). Reprinted in Bell [6]
6. J. S. Bell: *Speakable and Unspeakable in Quantum Mechanics* (Cambridge Univ. Press, Cambridge 1993)
7. E. Cassirer: *Substanzbegriff und Funktionsbegriff*, Berlin 1910. English trans. *Substance and Function* (Open Court, Chicago-London 1923)
8. E. Cassirer: *Determinismus und Indeterminismus in der modernen Physik* (Erlanders Boktryckeri Aktiebolag, Göteborg 1937)
9. C. Calude and M. Stay: From Heisenberg to Gödel via Chaitin. *International Journal of Theoretical Physics* **44**, 7 (2005)
10. C. Cellucci: *Le ragioni della logica* (Laterza, Bari 1998)
11. C. Cellucci: Mathematical discourse *vs* mathematical intuition. In: *Mathematical reasoning and heuristics*, ed by C. Cellucci and D. Gillies (College Publications, London 2005)
12. P. Dirac: *The Principles of Quantum Mechanics* (Clarendon, Oxford 1930)
13. R. Descartes: *Oeuvres*, ed by C. Adam and P. Tannery (Vrin, Paris 1996)

14. D. Deutsch: Quantum theory of probability and decisions. *Proceedings of the Royal Society of London* **A455** (1999)
15. J. Earman: *A Primer on Determinism* (Reidel, Dordrecht 1986)
16. A. Ekert: Complex und unpredictable Cardano (*e-print* arXiv:0806.0485v1)
17. P. Facchi and S. Pascazio: Quantum Zeno and inverse quantum Zeno effects. In: *Progress in Optics 42*, ed by E. Wolf (Elsevier Science BV 2001)
18. D. Flament: *Histoire des nombres complexes* (CNRS Éd., Paris 2003)
19. G. Fano: *Mathematical Methods of Quantum Mechanics* (McGraw Hill, New York 1971)
20. S. Feferman: *In the Light of Logic* (Oxford Univ. Press, Oxford 1998)
21. R. P. Feynman, R. B. Leighton and M. Sand: *The Feynman Lectures on Physics*, vol. III (Addison-Wesley, Reading 1970)
22. R. Gandy: Church's thesis and principles for mechanisms. In: *The Kleene Symposium*, ed by J. Barwise, H. J. Keisler and K. Kunen (North-Holland, Amsterdam 1980)
23. K. Gödel: Some basic theorems on the foundations of mathematics and their implications (Gibbs Lecture 1951). In: *Collected Works III* (Oxford Univ. Press, Oxford 1995)
24. R. Grosseteste: *Commentarius in Posteriorum analyticorum libros*, ed by P. Rossi (Olschki, Florence 1981)
25. R. Grosseteste: *Metafisica della luce. Opuscoli filosofici*, ed by P. Rossi (Rusconi, Milan 1986)
26. J. S. Hadamard: Newton and the Infinitesimal Calculus. In: *Newton Tercentenary Celebration of the Royal Society of London* (Cambridge Univ. Press, Cambridge 1947)
27. D. Hilbert: Die Grundlagen der Mathematik (1927). English trans. The foundations of mathematics. In: *From Frege to Gödel, a source book in mathematical logic, 1879-1931*, ed by J. van Heijenoort (Harvard Univ. Press, Cambridge 1967)
28. R. I. G. Hughes: *The Structure and Interpretation of Quantum Mechanics* (Harvard Univ. Press, Cambridge Mass. 1989)
29. M. Kemp: *The Science of Art* (Yale Univ. Press, New Haven 1990)
30. M. Kemp: *Leonardo* (Oxford Univ. Press, Oxford 2004)
31. A. Koyré: La synthèse newtonienne. In: *Archives Internationales d'Histoire des Sciences*, 3 (1950); reprinted in *Études newtoniennes* (Gallimard, Paris 1968)
32. P. Kwiat, H. Weinfurter and A. Zeilinger: *Scientific American* Nov. (1996).
33. D. C. Lindberg: *Roger Bacon and the origins of Perspectiva in the Middle Ages* (critical edition and English translation of Bacon's Perspectiva) (Clarendon, Oxford 1996)
34. G. Lolli: *QED. Fenomenologia della dimostrazione* (Bollati Boringhieri, Turin 2005)
35. R. Longhi: *Piero della Francesca* (Rome 1927). Reprinted in: *Opere complete* (Sansoni, Florence 1963)
36. G. Longo: Laplace, Turing and the "imitation game" impossible geometry: randomness, determinism and programs in Turing's test. In: *The Turing Test Sourcebook*, ed by R. Epstein, G. Roberts and G. Beber (Kluwer, Dordrecht 2007)
37. R. Merli, G. F. Missiroli and G. Pozzi: On the statistical aspect of electron interference phenomena. *American Journal of Physics* **44** (1985)

38. E. Panofsky: *Die Perspektive als "symbolische Form"* (B. G. Teubner, Leipzig-Berlin 1927). English trans. *Perspective as Symbolic Form* (Zone Books, New York 1997)
39. R. Penrose: *Shadows of the Mind* (Oxford Univ. Press, Oxford 1994)
40. Piero della Francesca: *De Prospectiva pingendi* (1474). Critical edition by G. N. Fasola (Florence 1942)
41. V. Scarani : *Initiation à la physique quantique* (Vuibert, Paris 2003) English trans. *Quantum Physics. A First Encounter* (Oxford Univ. Press, Oxford 2006)
42. W. Sieg: Relative consistency and accessible domains. *Synthese* **84** (1990)
43. W. Sieg: Hilbert's proof theory. In: *Handbook of the History of Logic* Volume 5: *Logic from Russell to Church* ed by D. M. Gabbay and J. Woods (Elsevier 2008)
44. A. M. Turing: On computable numbers with an application to the *Entscheidungsproblem*. *Proceedings of the London Mathematical Society*, Serie 2, **43** (1937). Reprinted in: *The Essential Turing*, ed by B. J. Copeland (Clarendon, Oxford 2004)
45. B. C. van Fraassen: *Quantum Mechanics* (Clarendon, Oxford 1991)
46. J. von Neumann: *Mathematische Grundlagen der Quantenmechanik*, Springer, Berlin 1932. English trans. *The Mathematical Foundations of Quantum Mechanics* (Princeton Univ. Press, Princeton 1955)
47. H. Weyl: *Gruppentheorie und Quantenmechanik* (Methuen and Company, Ltd. 1931). English trans. *The Theory of Groups and Quantum Mechanics* (Dover, New York 1950)
48. H. Weyl: *Philosophy of Mathematics and Natural Sciences* (Princeton Univ. Press, Princeton 1949)
49. H. Weyl: *Symmetry* (Princeton Univ. Press, Princeton 1952)
50. J. A. Wheeler: Bits, quanta, meaning. In: *Problems in Theoretical Physics*, ed by A. Giovannini, F. Mancini and M. Marinaro (Salerno 1984)
51. J. A. Wheeler and W. H. Zureck (eds.): *Quantum Theory and Measurement* (Princeton Univ. Press, Princeton 1983)
52. W. K. Wootters: Statistical distance and Hilbert space. *Physical Review D*, vol. 23, n. 2 (1981)
53. W. K. Wootters: Local accessibility of quantum states. In: *Complexity, Entropy, and the Physics of Information*, ed by W. H. Zurek (Addison-Wesley, Redwood City CA 1990)

Computability and Incomputability of Differential Equations

Guido Gherardi

Dipartimento di Filosofia, Università di Bologna (Italy)
guido.gherardi@unibo.it

In the following discussion we are going to deal with the problem of computability of differential equations, and we will outline some of the most important results achieved in this area, mainly due to K. Weihrauch and N. Zhong. In particular, a large part of the paper will concern the debate about the computability of the wave equation.

First of all, we have to specify what we mean by the term "computability". More precisely, we have to characterize a specific theory which provides a notion of computability that we consider reliable. It is well known indeed that for real numbers we do not have a single generally accepted theory of computation. The situation is therefore different from the case of natural numbers, where the Turing machines approach has become standard.

The main philosophical question we investigate throughout the whole paper is then essentially the following: computers are used every day in ordinary life and in scientific research, and the results they produce are welcomed as trustworthy. Nowadays computers of different capacities run in every scientific department. In many cases their programmers and users employ them (implicitly or explicitly) to perform computations on real numbers or at least to execute calculations which are supposed to approximate, as accurately as possible, some real number values. This happens when one computes the decimal expansions of numbers like π or e, or when physicians aim to find solutions for differential equation systems (as in the case of the wave equation).

Nevertheless, the situation at the theoretical level is rather more complicated. Currently, we still cannot rely on a unique theoretical model of computability for real numbers, and accordingly there is no general agreement on the effective power of *real computing* (again, the case of the wave equation provides a perfect example).

There comes then the problem of making a choice among the different approaches, or even more, of justifying the need for such models, since most of the users may not understand the necessity for a theoretical foundation of real computing.

1 The continuum and the discrete

The question of the existence of real numbers is a long-term problem, open since the time of the Pythagorians, who had to admit them necessarily even if this was opposite to their philosophical interpretation of the world. The opinion of Dedekind is well known: he believed that only natural numbers exist in nature, whereas real numbers are a human invention. Many other authors have discussed this, and not only within the community of the philosophers of science. This is, in fact, the philosophical problem of more general interest, that of the continuum/discrete dichotomy. It has been argued by some authors, like René Guénon (see for example [10]), that the assumption of the existence of real numbers is a consequence of the wrong claim of measuring the spatial continuum using numerical, and thus discrete, tools. According to Guénon, space and numerical quantity have different natures (the earlier is a continuum, the latter is discrete), therefore any attempt to reduce the spatial extension to a numerical measure cannot be completely satisfying:

> There is a basic essence difference among these two modalities of the quantity, so that a perfect correspondence cannot be established; in order to find some remedy, as long as this is possible, one tries somehow to reduce the intervals which exist in the discontinuity of the numerical sequence by introducing new numbers, first of all fractions [...] Nevertheless, there is always necessarily something in the discontinuous nature of numbers that precludes a perfect equivalence with the continuum [...] in this it appears the insufficiency of fractional numbers, and we can say of any other possible kind of number [...]

We will not investigate the problem of the continuum/discrete opposition in its entirety, but we point out how the formulation of theories for real computing has re-introduced the challenge of "squaring" the continuum by the discrete. This reduction leads to several problems, so that some computer scientists have developed a skeptical attitude towards the concrete existence of real numbers in nature. This opinion is described for example in chapters 5 and 6 of a recent book [4] by G. Chaitin, one of the eminent fathers of randomness theory. He wonders whether real numbers should be considered a cheat and states that a number of *infinite precision* is actually something unreal! He is convinced that digital information theory cannot be applied satisfyingly to a physical or mathematical world made of real numbers, since such numbers contain an infinite quantity of information, and a human mind, or a computer, cannot actually have access to infinite information. Currently, there are a lot of expectations, today very fashionable, that information theory and computer science may provide a new interpretation (even a re-formulation) of physical laws. And since the nature of computation is digital, thus discrete, some may be tempted to reject real numbers, and even more, very radically, classical mathematical physics as based on differential equations. Quite surprisingly, this opinion has been expressed also by one of the most important physicians of the twentieth century, R. Feynman [6].

This is not the philosophical perspective we consider in this paper. The approach we present accepts the existence of real numbers, and is interested in developing a computability theory suitable to handle them. This is the goal of what we mean by "*real computing*".

On a first level we can roughly classify the different approaches to real computing in two groups. The first group founds real numbers computation on the digital and discrete nature of Turing machines (and contemporary computers). The second group believes that a satisfying theory of computability for real numbers should somehow subordinate computation laws to the peculiar nature of these numbers. However, both groups assume (or at least they do not argue about) the existence of the continuum. We will not present this *struggle* in detail but we point out again how a generally accepted theory of computation for real numbers is not yet available. We refer the reader interested in this discussion to the last chapter of [14] and to the introduction of [1], or also to [9].

We simply outline some essential theoretical divergence between the two approaches. The classical method of dealing with real computing sees Turing machines as capable of operating on infinite bit sequences; the basic idea is that one needs an infinite amount of information to codify the exact value of a real number (for example, through digital expansions). Accordingly, one needs an infinite amount of time to execute ordinary operations over the reals, like sum or product, and also to execute the *identity test* between two real values. In concrete, thus in a finite quantity of time, one can only expect to obtain reliable rational approximations of the exact operation results.

L. Blum, F. Cucker, M. Schub and S. Smale do not accept this point of view. In [1] they say:

> Computer science is oriented by the digital nature of machines and by discrete foundations given by Turing machines. For numerical analysis, systems of equations and of differential equations are central and this discipline depends heavily on the continuous nature of the real numbers. [...] We believe that the Turing machine as a foundation for real number algorithms can only obscure concepts.

In reality, even the Turing machine approach is paying increasing attention to differential equations, and this is indeed the topic of the present paper.

The theoretical model of L. Blum, F. Cucker, M. Schub and S. Smale postulates that one can operate on real numbers as if they were atomic objects. Therefore they assume it is possible to execute the basic operations of sum, difference, multiplication and quotient, as well as to compare the exact values of two real numbers, in a finite amount of time (even in a single computation step). This paradigm has met with strong criticism from K. Weihrauch, who in [14] states:

In a finite amount of time every physical information channel can transfer only finitely many bits of information and every physical memory is finite. Since there are uncountably many real numbers,[1] it is impossible to identify an arbitrary real number by a finite amount of information. Therefore, it is impossible to transfer an arbitrary real number to or from a computer in a finite amount of time or to store a real number in a computer.

To avoid this concrete limit, in the real-RAM model supported by Blum, Cucker, Schub, and Smale the problem of the "discretization" of the continuum must necessarily re-appear: rather than using infinite sequences of bits to codify real numbers, the alternative is to compute functions on a highly dense set of rational numbers distributed uniformly in some real interval, in order to approximate real function graphs as accurately as possible.

Maybe a computability theory for a mathematical branch is not satisfying unless it proves that the fundamental operations in that field are computable. For example, in classical recursion theory one easily proves that the sum and the product, which are the fundamental operations over the natural numbers, are computable: it is straightforward to write algorithms for Turing machines to compute such operations or to define them through Kleene recursive numerical operations.

In the model we adopt, the so called *"Type-Two Theory of Effectivity"* (*TTE*), or *"Theory of Representations"*, which is a very natural extension of Turing machine computability theory, one proves that the fundamental operations over the reals are computable. Beside the basic four operations, we mention the roots of any degree n (we recall indeed that the notion of real closed fields is actually defined in terms of existence of solutions for odd degree polynomials). This is not true for the real-RAM machine approach to real computing, where, for example, the square root operation is not computable.

Obviously, one may postulate the computability of certain operations by introducing *ad hoc* axioms, but this may sound too arbitrary. Remarkably, in classical recursion theory the two different methods are equivalent: one can define each recursive operation in terms of sum and product, or, conversely, prove recursivity of sum and product by elaborating simple programs for Turing machines. For the reals, these two approaches no longer coincide, thus leading to the two different groups of theories we have mentioned. Of course, an axiomatic approach requires the formulation of particular axioms depending on the mathematical theory under study. This is not exactly the case of the Turing machine approach. For example, in set theory the basic operations are union and intersection. These operations can be easily proved to be computable by Turing machines with respect to closed sets: one needs only to find a satisfactory method to codify the mathematical objects involved

---

[1] Notice the expression *"there are* uncountably many real numbers": the existence of (all) real numbers is openly accepted.

(in this case, closed sets), rather than postulating the computability of such operations by introducing new axioms [8], [3].

2 Computable analysis

Before giving a technical presentation of computable analysis as modelled in the TTE computational paradigm, we discuss its philosophical interest. We would like to call it a "constructive platonism". The underlying idea of computable analysis is that the classical mathematical world exists in all its amplitude. The role of the computable analyst is then that of elaborating algorithms in the language of Turing machines to simulate specific mathematical functions, and this is the constructive part of the subject. An extreme application of this principle would be that of accepting within the mathematical world only objects and functions that can be effectively generated by some computer program. This is typical, indeed, of some constructive schools, like Markov's. But this is not the attitude of TTE computable analysis. Nevertheless, an immediate correspondence with intuitionistic mathematics can be found in the proof that all computable functions (following the TTE definition) are continuous. Notwithstanding, computable analysis is far from accepting only the existence of continuous functions, as intuitionism does. Indeed, even discontinuous functions are objects of investigation in computable analysis, in terms of "degrees of incomputability". Summing up, there is a classical mathematical world, and then a constructive job which consists in codifying the objects and in elaborating algorithms to simulate mathematical operations by machines, or in classifying their degrees of incomputability.

As we have mentioned above, TTE employs Turing machines which can handle infinite sequences of bits. The necessity for this is the *"infinitary"* nature of real numbers (and many other mathematical objects): one needs to codify an infinite quantity of information to characterize a single real number. Typical and well known infinitary codings of real numbers are obtained through decimal expansions and Cauchy sequences of rational numbers.

The Turing machines used in this model are said to be of *Type-2* (therefore we say "Type-2 Theory of Effectivity"). These machines are conceived as having:

- possibly many one-way read-only input tapes,
- several two-ways reading-writing working tapes,
- a one-way write-only output tape.

Such machines transform infinite sequences of natural numbers. For the input tape heads the only instruction that is allowed is to read digits while moving rightwards. On the output tape the machine can only write digits and move rightwards, and therefore no correction can be made. In this way, at any stage of the computation, the result, even if it is *incomplete*, is anyway *reliable*. All ordinary Turing machine instructions are allowed for working tape heads: to move rightwards, to move leftwards, to read, to write, and to erase digits.

The prohibition of erasing digits on the output tape has a fundamental topological consequence: all computable functions are continuous (with respect to the space $\mathbb{N}^\mathbb{N}$ of infinite digit sequences, the so called "Baire space"). Roughly speaking, a function over this space is continuous when any pair of infinite sequences sharing finite initial segments of increasing length yields ever greater coincidence between corresponding images.

A basic result by C. Kreitz and K. Weihrauch (sometimes referred to as the *Representation Theorem*) shows a fundamental correspondence for certain mathematical structures that admit reasonable codification systems: a mathematical function defined on these structures is continuous if and only if it admits a continuous simulation with respect to opportunely selected codifications. We now explain this result in detail for second countable topological T_0-spaces. We recall that a space is second countable if it is generated by a countable sub-base (we may call "*atomic properties*" the open sets in this sub-base). A space is T_0 when given any two different elements x, y in this space, there is an open set U such that $x \in U$ if and only if $y \notin U$: in other words, there cannot be two different objects sharing the same atomic properties. An object is then uniquely determined by the (countable) list of all its atomic properties. This way of denoting objects by means of atomic properties is said to be the *standard representation* of that space (with respect to any chosen sub-base): once an enumeration of the countable sub-base is fixed, one can identify objects using sequences of natural numbers. The term "*representation*" is the technical expression used in the literature to mean any possible way of codifying the elements of a space through infinite numerical sequences. A representation of a space is then said to be "*admissible*" if it is equivalent to the standard representation, *i.e.* it can be reduced through a continuous translation into the standard representation and vice versa. A more general definition characterizes a representation as admissible when it is a continuous function and any other continuous representation of the space can be reduced by a continuous translation into such representation.

The typical standard representation (denoted in the literature by ρ) of $\mathbb{R}$ is given by the set of all maximal sequences of open intervals with rational end points whose intersections are singletons. Any single real number is then uniquely determined by the list of all the rational open intervals it belongs to. A straightforward generalization defines ρ^n for the Euclidean spaces $\mathbb{R}^n$, with $n \geq 0$.

A well known equivalent representation (thus, admissible) of $\mathbb{R}$ is obtained by using Cauchy sequences of rational numbers with a fixed computable modulus of convergence: one can decide uniformly when a rational number in any such sequences approximates the respective limit within a margin of 2^{-n} for any $n \in \mathbb{N}$. The definition of Cauchy representation can be immediately extended to all other *separable* metric spaces (metric spaces including a countable dense subset).

The technical word used to denote any simulation in the Baire space of a (partial) mathematical function $f :\subseteq X \to Y$ is *"realization"*. Any realization of f maps then each name of any $x \in \mathrm{dom}(f)$ (with respect to a given representation δ_X of X) to a name of $f(x)$ (with respect to a given representation δ_Y of Y.). The (partial) function f is then said to be *"computable"* (with respect to δ_X, δ_Y) if it has a computable realization (with respect to the same representations), in other words, if there is a Type-2 Turing machine which transforms any name of any given element $x \in \mathrm{dom}(f)$ into a name of $f(x)$ (where such names are of course infinite sequences of natural numbers which depend on the chosen representations δ_X, δ_Y of X and Y, respectively).

We would like to point out this fundamental fact: in the theory of representations there is no absolute notion of computability: a function may be computable with respect to some representations, but non computable with respect to some others.

It is then important to find reasonable, or even privileged, representations for a given space. The Representation Theorem helps us in the case of T_0 second countable spaces. This theorem says indeed that a (partial) function $f :\subseteq X \to Y$, for X and Y second countable T_0-spaces, is continuous if it has some continuous realization with respect to admissible representations of X and Y. This is the reason why admissible representations are so important (and they are also very natural, because of their equivalence to standard representations).

As we have mentioned before, the theory of representations not only suggests a valid definition of computability, it also provides suitable tools for a detailed analysis of the notion of incomputability. After some preliminary intuition, the theory of incomputability has been elaborated systematically by V. Brattka in [2]. More precisely, Brattka has characterized infinite levels of incomputability: the more the level increases, the more the functions are incomputable. This hierarchy is generated by the functions $C_k : \mathbb{N}^{\mathbb{N}} \to \mathbb{N}^{\mathbb{N}}$:

$$C_k(p)(n) = \begin{cases} 0 \text{ if } \exists n_k \forall n_{k-1} \exists n_{k-2} ... Q n_1 : p(\langle n, n_k, n_{k-1}..., n_1 \rangle) \neq 0 \\ 1 \text{ otherwise} \end{cases}$$

for any $k \in \mathbb{N}$, where

$$Q n_1 = \begin{cases} \exists n_1 \text{ if } k \text{ is odd} \\ \forall n_1 \text{ if } k \text{ is even.} \end{cases}$$

The incomputability levels of these functions are strictly increasing: C_0 is computable and C_n is "more complicated" than C_m, for $n > m$.

A partial function $F :\subseteq \mathbb{N}^{\mathbb{N}} \to \mathbb{N}^{\mathbb{N}}$ on the Baire space $\mathbb{N}^{\mathbb{N}}$ is said to be Σ^0_{k+1}-*computable* if it can be computably reduced to C_k. This means that we could compute F if only we had a method to compute C_k (via uniform translation of any input of F into an input of C_k, and then of the output of C_k in the corresponding output of F). Vice versa, F is Σ^0_{k+1}-*hard* if C_k is reducible to F in a similar way. A function which is Σ^0_{k+1}-hard, for $k \geq 1$, maps some computable object to an incomputable object, therefore it is not a computable function. When both reductions hold, we say that F is Σ^0_{k+1}-*complete*: in this case the function F is exactly complicated as C_k.

The class of the computable functions coincides with the class of the Σ^0_1-computable functions.

This hierarchy turns out to be an effective version of the ordinary Borel hierarchy of degrees of discontinuity, as some results by Brattka show. Such results generalize the relation between computability and continuity at higher levels, as we are about to explain.

We recall that in classical topology a (partial) function $f :\subseteq X \to Y$, for X and Y metric spaces, is said to be Σ^0_k-measurable when for any open set U, the pre-image $f^{-1}(U)$ is a Σ^0_k-set in the Borel structure induced by the relative topology on $\mathrm{dom}(f)$. Therefore the Σ^0_1-measurable functions are exactly the continuous maps. If a function $F :\subseteq \mathbb{N}^\mathbb{N} \to \mathbb{N}^\mathbb{N}$ is Σ^0_k-computable, then it is Σ^0_k-measurable with respect to the relative topology of its domain as a subset of the Baire space.

An immediate extension of the notion of computability for functions between separable metric spaces can be obtained in the following way: a function $f :\subseteq X \to Y$ is said to be Σ^0_k-computable if it has a Σ^0_k-computable realization (with respect to the admissible Cauchy representations of X and Y).

A generalization of the Representation Theorem proved by V. Brattka asserts that for f total, f is Σ^0_k-measurable if and only if it has some Σ^0_k-measurable realization. Thus, for separable metric spaces the concept of Σ^0_k-measurability defined in terms of realizations and the concept of Σ^0_k-measurability through Borel sets simply coincide.

3 The Cauchy problem for ordinary differential equations

Having introduced the previous preliminary notions, we can now investigate the topic of our paper: the computability properties of the Cauchy problem, thus the possibility of computing solutions for differential equation systems.

This problem has been addressed by several authors. A simple case is given by ordinary differential equations of the first order:

$$\frac{d\varphi}{dx} = f(x, \varphi(x))$$

G. Peano has proved that if $f(x,t)$ is a continuous real-valued function on the rectangle $-a \leq x \leq a$, $-b \leq t \leq b$, where $a, b > 0$, then the Cauchy problem $\frac{dt}{dx} = f(x,t)$, with initial condition $t(0) = 0$, has a continuously differentiable solution $t = \varphi(x)$ on the interval $-\alpha \leq x \leq \alpha$, for $\alpha = \min(a, b/M)$, where $M = \max\{|f(x,t)| : -a \leq x \leq a, -b \leq t \leq b\}$. Nevertheless, this theorem does not hold *computably*: in [11] M. B. Pour-El and J. I. Richards have proved that there is a function $f(x,y)$ computable in the square $[-1,1]^2$ such that no solution is computable on any interval $[0,\delta]$, for $\delta > 0$.

S. G. Simpson has provided a further contribution to this problem in the context of reverse mathematics. We may consider reverse mathematics as a sort of "foundation without dogmas": its scope is not to decide which mathematical results are admissible in mathematics (like constructivism does), but to classify them on the basis of the *weight* of the different ontological assumptions needed to prove them. Climbing up the ontological hierarchy, the ontological depth increases. Since reverse mathematics (usually) employs classical logic, one may argue that a very strong ontological assumption is tacitly accepted within the inner foundations of the theory; nevertheless the aim of reverse mathematics is that of measuring the ontological weights of mathematical theorems with *impartiality*. The question to answer should not be "Are you allowed to prove A?", but rather "What must you admit in order to prove A?"

Simpson has proved (see [13]) that the Peano Theorem is *logically* equivalent to the weak Kőnig's Lemma WKL_0: each infinite binary tree has an infinite path. Therefore a solution satisfying the conditions of Peano's Theorem can exist if and only if one accepts the existence of a path in any infinite binary tree.

This result can be translated in the framework of computable analysis, and this proves that the construction of the solution suggested by Simpson is *almost computable*, i.e. it is not computable but nevertheless *simpler* than C_1. This characterization refines the result by Pour-El and Richards. We will see another refinement of the same result in the next section.

4 The incomputability result on the wave equation and its discussion

The two authors that have probably spent more time studying the computability aspects of the Cauchy problem are K. Weihrauch and N. Zhong. In particular, they have written several papers about partial differential equations, and some of them concern the discussion of the computability of the wave equation (see for example [15], [17]).

The wave propagation is a fundamental example of partial differential equations system in mathematical physics:

$$\frac{\partial^2 u}{\partial x^2} + \frac{\partial^2 u}{\partial y^2} + \frac{\partial^2 u}{\partial z^2} - \frac{\partial^2 u}{\partial t^2} = 0$$

$$u(0,x,y,z) = f(x,y,z)$$

$$\frac{\partial u}{\partial t}(0,x,y,z) = 0$$

for $x,y,z,t \in \mathbb{R}$.

Classically the wave equation has a unique solution depending on the initial condition f and on the initial velocity (that we assume here to be 0). Nevertheless, Pour-El and Richards [12] have proved the existence of a computable initial condition f such that the unique corresponding solution at time 1 is not a computable function:

Theorem 1 (Pour-El-Richards). *Consider the wave equation with the initial conditions* $u(t, x, y, z) = f(x, y, z)$, $\frac{\partial u}{\partial t}(t, x, y, z) = 0$ *at time* $t = 0$. *Let* D_1 *and* D_2 *be the following two cubes in* $\mathbb{R}^3$:

$$D_1 = \{(x, y, z) : |x| \leq 1, |y| \leq 1, |z| \leq 1\}$$

$$D_2 = \{(x, y, z) : |x| \leq 3, |y| \leq 3, |z| \leq 3\}.$$

There exists a computable function $f(x, y, z)$ *in* $C(D_2)$ *such that the solution* $u(t, x, y, z)$ *at time* $t = 1$ *is continuous, but is not a computable function in* $C(D_1)$.

By commenting this result, Weihrauch and Zhong [17] say:

> These examples have considerably disconcerted logicians and computer scientists, as well as physicists, most of whom accept the Church-Turing Thesis or at least believe that wave propagation can be predicted (arbitrarily precisely) by means of digital computers.

The say even more:

> These results bother physicists, in particular, who are convinced that wave propagation is computable and in fact write computer programs which predict the future behavior of waves from initial conditions.

In Weihrauch and Zhong's opinion, this belief concerns the whole sector of mathematical physics [17]:

> [...] most physicists believe that for processes which can be described by well-established theories (finitely many point masses interacting gravitationally, electromagnetic waves, quantum systems etc.) the future behavior can be computed with arbitrary precision, at least in principle, from sufficiently precisely given initial conditions, where the computations can be performed on digital computers, and hence on Turing machines.

This is a typical case of what we mentioned in the introduction. Computers are programmed and then run everyday, and often they should simulate, somehow, computations over the reals, as in the case of mathematical physics. Nevertheless, there is a lack of knowledge of the theoretical foundations for real computing. Therefore, when a result like that of Pour-El and Richards is formulated, it can be hard for people to give an appropriate interpretation of these results.

4.1 Weihrauch's and Zhong's solution

In [15] and [17], K. Weihrauch and N. Zhong have developed a detailed analysis of the incomputability result of Pour-El and Richards.

The two authors have dealt with a more general version of the wave propagation, with initial conditions f, g where possibly $g \neq 0$:

$$\frac{\partial^2 u}{\partial x^2} + \frac{\partial^2 u}{\partial y^2} + \frac{\partial^2 u}{\partial z^2} - \frac{\partial^2 u}{\partial t^2} = 0$$

$$u(0, x, y, z) = f(x, y, z)$$

$$\frac{\partial u}{\partial t}(0, x, y, z) = g(x, y, z)$$

for $x, y, z, t \in \mathbb{R}$.

The TTE-approach to computable theory is strongly topological, and therefore Weihrauch and Zhong have constructed some topologies on the space of continuous functions.

First of all take the topology on $C(\mathbb{R}^3)$ generated by the "*rational amplitude boxes*" as atomic properties: an open set in the sub-base consists of all the functions which map a certain closed ball $\overline{B}(a, r))$, for $a \in \mathbb{Q}^3$, within a certain interval (c, d), for $r, c, d \in \mathbb{Q}$. The use of rational numbers only assures the enumerability of all the atomic properties. More precisely, we let:

$$\sigma = \{R_{arcd} : a \in \mathbb{Q}^3, r, c, d \in \mathbb{Q}, r > 0, c < d\}$$

where
$$R_{arcd} = \{f \in C(\mathbb{R}^3) : f(\overline{B}(a, r)) \subseteq (c, d)\}.$$

Let τ be the topology generated by the sub-base σ and let $\zeta :\subseteq \mathbb{N}^{\mathbb{N}} \to C(\mathbb{R}^3)$ be the corresponding standard representation of $C(\mathbb{R}^3)$.

Therefore a ζ-name of $f \in C(\mathbb{R}^3)$ is a list of (natural numbers coding) all its rational amplitude boxes; thus it is a list of all the 4-tuples:

$$(a, r, c, d) \qquad a \in \mathbb{Q}^3, r \in \mathbb{Q}^+, c, d \in \mathbb{Q} \qquad c < d$$

such that
$$f(\overline{B}(a, r)) \subseteq (c, d).$$

The authors have then considered a topology for the set of continuous functions with continuous partial derivatives:

$$C^1(\mathbb{R}^3) = \{f \in C(\mathbb{R}^3) : \partial_{x_i} f \in C(\mathbb{R}^3), 1 \leq i \leq 3\}.$$

The topology is defined so that, roughly speaking, its corresponding standard representation denotes any function $f \in C^1(\mathbb{R}^3)$ through a combination of ζ-names of f and of its partial derivatives $\partial_{x_1} f, \partial_{x_2} f, \partial_{x_3} f$ as continuous functions. More precisely, on the set $C^1(\mathbb{R}^3)$ define the topology τ^1 generated by the following set of atomic properties:

$$\sigma^1 = \{R^i_{arcd} : 0 \leq i \leq 3, a \in \mathbb{Q}^3, r, c, d \in \mathbb{Q}, r > 0, c < d\}$$

with $R^0_{arcd} = R_{arcd} \cap C^1(\mathbb{R}^3)$ and $R^i_{arcd} = \{f \in C^1(\mathbb{R}^3) : \partial_{x_i} f \in R_{arcd}\}$.
Let τ^1 be the topology generated by the sub-base σ^1 and let $\zeta^1 :\subseteq \mathbb{N}^{\mathbb{N}} \to C^1(\mathbb{R}^3)$ be the corresponding standard representation of $C^1(\mathbb{R}^3)$.

Surprisingly, Weihrauch and Zhong have proved that the Cauchy problem of the wave propagation is computable (in some sense):

Theorem 2 (Weihrauch-Zhong). *The solution operator* $S : (f, g, t) \mapsto u(t, .)$ *of the wave equation problem mapping* $f \in C^1(\mathbb{R}^3)$, $g \in C(\mathbb{R}^3)$ *and* $t \in \mathbb{R}$ *to the solution* $u(t, .) \in C(\mathbb{R}^3)$ *is* $(\zeta^1, \zeta, \rho, \zeta)$-*computable.*

The reason for this positive result lies in the choice of the codification methods. Please observe, in fact, that f has continuous partial derivatives, while this condition is not required for g. Hence, the corresponding names of f and g depend on the two different topologies on $C^1(\mathbb{R}^3)$ and $C(\mathbb{R}^3)$, respectively, i.e. the name of f must also contain the information about its partial derivatives, but this is not the case for g. The key of the proof is that for $t \in \mathbb{R}, x \in \mathbb{R}^3$, the operation

$$(f, g, t, x) \mapsto u(t, x) =$$

$$= \int_{S^2} [tg(x + tn) + f(x + tn) + t\nabla f(x + tn) \cdot n] d\sigma(n)$$

is $(\zeta^1, \zeta, \rho, \rho^3, \rho)$-computable (here $\nabla = (\frac{\partial f}{\partial x} + \frac{\partial f}{\partial y} + \frac{\partial f}{\partial z})$) and observe that $f \mapsto \frac{\partial f}{\partial x_i}$ is trivially (ζ^1, ζ)-computable).

As an immediate corollary we have:

Corollary 1 (Weihrauch-Zhong). *The special solution operator* $f \mapsto S(f, 0, 1)$ *is* (ζ^1, ζ)-*computable.*

Therefore, if f is a computable function as an object of $C^1(\mathbb{R})$ (thus, if it has a computable codification of itself and its partial derivatives), then $u(1, .)$ cannot fail to be computable.

Nevertheless, the incomputability result by Pour-El and Richards must necessarily re-appear somehow. In fact, Weihrauch and Zhong show first of all that in some other sense the Cauchy problem of the wave propagation is not computable: this comes from a discontinuity property which arises in the change of the source topology (and coherently of the information given in input):

Theorem 3 (Weihrauch-Zhong). *For any* $t \in \mathbb{R} \setminus \{0\}$ *the wave propagator* $S_t : f \mapsto S(f, 0, t)$, *which sends the initial conditions* $f \in C^1(\mathbb{R})$ *and* $g = 0$ *to the solution at time* t, *is not* (τ, τ)-*continuous.*

Please observe that in this case the name of the function f does not include any information about its partial derivatives, and this lack of information, which corresponds to a change in the source topology, makes the difference. We give the fundamental hint of this fact. Let $f_n(x) = n^{-1} \sin n ||x||^2$ for $x \in \mathbb{R}^3$, $n > 0$, and let $f(x) = 0$. Then $f_n \to_\tau f$ (but observe that $\frac{\partial f_n}{\partial x_i} \not\to_\tau \frac{\partial f}{\partial x_i}$, hence $f_n \not\to_{\tau^1} f$!)

We have that $S_t(f) = 0$ and one can prove that:

$$S_t(f_n)(0) = u_n(t,0) =$$
$$= 2t^2 \cos(nt^2) + n^{-1} \sin(nt^2),$$

in particular $|S_t(f_n)(0)| > d$ infinitely often for some $d > 0$. Take a positive $e < d$. Then $S_t(f) = 0 \in R_{0,1,-e,e}$. Suppose S_t is continuous. There is then an open set $U \in \tau$ such that $f_n, f \in U$ for any sufficiently large n and $S_t(U) \subseteq R_{0,1,-e,e}$. But infinitely often $S_t(f_n) \notin R_{0,1,-e,e}$, which is a contradiction.

If the operator S_t is not continuous with respect to the topologies chosen in Theorem 3, then by Representation Theorem it cannot be computable with respect to the corresponding representations.

Moreover Weihrauch and Zhong prove that the solution operator S_1 is not only incomputable for topological reasons, but also for pure computability theoretical motivations. In fact S_1 maps some computable f to a continuous but incomputable $u(1,.)$. A result by Myhill constructs a computable function $F \in C^1(\mathbb{R})$ whose derivative F' is continuous but not computable, since $F'(1)$ is not a computable real number. Weihrauch and Zhong show in [17] a particular case in which the initial conditions are $f = F(||x||)$, $g = 0$ and:

$$u(t,0) = \frac{\partial}{\partial t}[F(t) \cdot t] = F(t) + F'(t) \cdot t.$$

The function $u(t,.)$ is continuous for all $t \in \mathbb{R}$, but for $t = 1$ and $x = 0$ we obtain $u(1,0) = F(1) + F'(1)$, a number which is not computable. Since the input $(1,0)$ is computable, but its image is not, then $u(1,.)$ cannot be a computable function, even if continuous. In this way the result by Pour-El and Richards is re-obtained with respect the topology τ: indeed F is a computable input with respect to ζ, but not to ζ', since F' is not computable.

Recall that, for $k \geq 2$, a Σ_k^0-hard function maps some computable objects to incomputable ones and notice that it is discontinuous, since C_{k-1} is so. For the wave equation solution operator (with respect to the source topology τ) we have an important example of this kind, and indeed it is a discontinuous function which maps some computable objects to incomputable ones. Hence its incomputability is due to reasons of double nature: both topological and computational. Weihrauch and Zhong say:

> According to a general theorem by Brattka any function of sufficiently high degree of discontinuity with some weak computability properties

maps some computable elements to non-computable ones. For the instances we have considered, the solution operator of the Cauchy problem is computable if it is continuous, and has computable irregularities if it not continuous.

To conclude, we can say, at least theoretically, that a Turing machine can describe correctly the propagation of the wave, provided that suitable information is given in the input. This kind of information is suggested by Theorem 2. The result of Pour-El and Richards apparently denies the possibility of describing the wave propagation process by using Turing machines. But this may simply be a consequence of a wrong approach to the problem. In fact, the incomputability result is obtained when a considerably relevant part of information is missing. If all relevant information is given in the input, the system is perfectly predictable by Turing machines.

Obviously this opens an interesting philosophical question: in the framework of the theory of representations, the computability property is strictly connected to the chosen codification method. Hence it seems questionable whether one can actually speak of the possibility of describing a physical process using Turing machines, if one assigns to the term "possibility" an "absolute" meaning. Some processes could be classified as "non computable" only because the appropriate type of information has not been identified yet. A fundamental problem is then determining what information should be considered relevant in a given context. One possible method can be that of selecting the kind of information which allows the formulation of computable versions of the main results of the theory, if such information exists. But this choice method may sound in general a bit artificial, and maybe one should then evaluate, case by case, if the selected type of information is "natural". In the case of the wave equation, we can accept that the codification system which assures the computability of wave propagator is, theoretically, very natural: it is indeed a fundamental property, and not simply a redundant detail, the fact that the initial condition f has continuous derivatives; hence it is reasonable that one has to codify these derivative in the input. We conjecture that the kind of information needed to make the description of a physical process computable should always sounds at the end very natural. This should hold at least at the level of pure mathematics. Nevertheless, at the practical level there could be some difficulties. The required information could be in fact very hard to obtain. For example, we have mentioned the result by Myhill regarding the existence of computable functions which are continuously differentiable, but whose derivatives are not computable.[2] Therefore it is impossible to give always in input all relevant information required. When f is computable, one could consider this function as being concretely constructible in a laboratory (at least, through approximations) and worthy of consideration. Unfortunately, this could not be the case of some $\partial_{x_i} f$, which could be

---

[2] One can prove easily that the map $f \mapsto \partial_{x_i} f$ is not computable by comparing the positive Theorem 2 with the negative Theorem 3.

not computable. Therefore, through this incomplete information, no Turing machine could in general output the amplitude boxes of the wave at a given computable time.

4.2 Cenzer and Remmel's approach

The incomputability results found by Pour-El and Richards regarding Peano's Theorem and the wave propagation have been analyzed and generalized by other authors, following different approaches. In [5] D. Cenzer and J. B. Remmel have used the classical arithmetical hierarchy studied in recursion theory to evaluate the complexity of index sets of continuous functions which satisfy certain properties. The arithmetical hierarchy is an instrument to classify the level of incomputability of sets of natural numbers. A Σ_k^0-set[3] is the collection of all the $x \in \mathbb{N}$ that satisfy a formula of the kind $\exists x_1 \forall x_2 ... Q x_k \varphi(x_1, ..., x_k, x)$, where $\varphi(x_1, ..., x_k, x)$ is a recursive predicate and $Q := \forall$ if k is even, otherwise $Q := \exists$. The complement of a Σ_k^0-set is said to be a Π_k^0-set, and a set is Δ_k^0 if it is both Σ_k^0 and Π_k^0. The recursive sets are exactly the Δ_1^0-sets, whereas the r.e.-sets coincide with the Σ_1^0-sets. The hierarchy is conservative, but the Σ_k^0-complete sets do not belong to levels smaller than k, and the same is true for Π_k^0 and Δ_k^0-sets. Let $(V_i)_{i \in \mathbb{N}}$ and $(U_i)_{i \in \mathbb{N}}$ be enumerations of the open rational balls (open balls with rational coordinates centers and rational radii) in $\mathbb{R}$ and in $\mathbb{R}^n$, respectively (it is preferable when the open balls are enumerated so as to respect certain trivial inclusion properties, but it is not necessary to insist here on these details). Cenzer and Remmel consider that a natural number e is a code for a (computable) continuous function $f : \mathbb{R}^n \to \mathbb{R}$ if the e-th Turing machine (as defined in classical recursion theory) computes a total function $g : \mathbb{N} \to \mathbb{N}$ such that:

- $\forall m, n \, (U_m \subset U_n \to U_{g(m)} \subset U_{g(n)})$
- $\forall k, m \, \exists r \, \forall t \, (U_t \subseteq [-k, k]^n \wedge \text{diam}(U_t) < 2^{-r} \to U_{g(t)} < 2^{-m})$.

Denote the function f by F_e when $e \in \mathbb{N}$ is a code for f. Cenzer and Remmel have proved then that:

Theorem 4 (Cenzer-Remmel). *The set A of all the numbers $a \in \mathbb{N}$ for which there exists $\delta > 0$ such that the ordinary differential equation $\frac{d\varphi}{dx} = F_a(x, \varphi(x))$ has a computable solution φ within some interval $[-\delta, \delta]$ with $\varphi(0) = 0$ is Σ_3^0-complete.*

This result in some sense improves the incomputability result obtained by Pour-El and Richards regarding Peano's Theorem, *i.e.* about the existence of computable functions f such that $\frac{d\varphi}{dx} = f(x, \varphi(x))$ has no computable solution φ within the usual rectangles. Indeed one can see that the set of indexes for computable functions $f(x, y)$ is Π_2^0-complete, while by Theorem 4 the set of

---

[3] Please notice that we now use light-face letters, since this hierarchy must not be confused with the previous bold-face classification.

those which admit some computable solution is Σ_3^0-complete, thus the two sets do not coincide (the second is more complicated).

We now come back to the wave equation problem. Cenzer and Remmel have in fact proved a result similar to Theorem 4 for the case of the wave propagation:

Theorem 5 (Cenzer-Remmel). *The set A, whose members are the numbers $a \in \mathbb{N}$ such that the wave equation $\frac{\partial^2 u}{\partial x^2} + \frac{\partial^2 u}{\partial y^2} + \frac{\partial^2 u}{\partial z^2} - \frac{\partial^2 u}{\partial t^2} = 0$ with initial condition $u(0, x, y, z) = F_a(x, y, z)$ and $\frac{\partial u}{\partial t}(0, x, y, z) = 0$ has computable solutions, is Σ_3-complete.*

Following the comments to Theorem 4, one sees immediately that Theorem 5 is another way to prove the existence of computable initial conditions f for which the wave equation has no computable solution.

5 Partial differential equations: computability for a general case

Partial differential equations are quite complex objects, classified in different categories, and therefore it is not easy to elaborate a general theory of solvability of the Cauchy problem. This is true for classical mathematics, and coherently, for computable analysis, too. In [19] Weihrauch and Zhong have searched a general method to computably solve partial differential equations. In this paper the two authors write:

> There have been many studies on computability of solutions of partial differential equations (PDEs). The majority of results obtained so far deals with inidividual equations, for example, linear heat, wave, or Schrödinger equation, and the KdV equation. This may have to be the case for non linear equations, for different non linear equations generally have little in common and they may have to be dealt with on a case-by-case basis. But how about linear PDEs? Is there any Turing algorithm computing solutions of a class of linear PDEs?

They have found a possible general solution to the question through the notion of infinitesimal generator, that we now explain according to [19].

A parameter family $W(t)$, $0 \leq t \leq \infty$, of bounded linear maps $A : X \to X$, for X a Banach space, is called a "*semigroup of bounded linear operators on X*" if:
(i) $W(0) = I$, where I is the identity operator on X;
(ii) $W(t + s) = W(t)W(s)$ for every $t, s \geq 0$.
The linear operator A defined by $\mathrm{dom}(A) = \{x \in X : \lim_{t \to 0^+} \frac{W(t)x - x}{t} \text{ exists}\}$ and
$$Ax = \lim_{t \to 0^+} \frac{W(t)x - x}{t} = \frac{d^+}{dt} W(t)x|_{t=0}$$

is called the "*infinitesimal generator of the semigroup* $W(t)$".

A semigroup $W(t)$, $t \geq 0$ of bounded linear operators on X is "*uniformly continuous*" if $\lim_{t \to 0^+} ||W(t) - I|| = 0$.

By showing how to compute a uniformly continuous semigroup $W(t)$ from its infinitesimal generator, and conversely, Weihrauch and Zhong can prove that the solution operator $S(A, x) = u$ of the Cauchy problem

$$\frac{du(t)}{dt} = Au(t) \qquad u(0) = x,$$

for $t > 0$, $x \in X$, and $A : X \to X$ a *bounded* linear operator, is computable.

The case for unbounded operators requires instead further concepts.

The resolvent $\rho(A)$ of a linear operator A is the set of all complex numbers β for which $\beta I - A$ is invertible, thus $R(\beta, A) := (\beta I - A)^{-1} : X \to X$ is a bounded linear operator (complex numbers can be coded as pairs of reals).

A triple (A, θ, M), for $A :\subseteq X \to X$ an *unbounded* linear operator, $\theta \geq 0$ and $M \geq 1$, is a *piece of type-IG* (=infinitesimal generator) *information for* A if:

1. A is a closed map and $\overline{\text{dom}(A)} = X$;
2. the resolvent set $\rho(A)$ of A contains the interval (θ, ∞) and $||R(\lambda, A)^n|| \leq \frac{M}{(\lambda - \theta)^n}$ for $\lambda > \theta$, $n \in \mathbb{N}$.

We codify (A, θ, M) through some standard name of θ and M, and a name of A which consists of a code for a countable dense subset in $graph(A)$. Then the Cauchy problem $\frac{du(t)}{dt} = Au(t)$ with $u(0) = x$ for $t > 0$, $x \in X$, is computable.

Another relevant result in computational mathematical physics is then achieved: the computability of the heat equation

$$\frac{\partial^2 u}{\partial x^2} + \frac{\partial^2 u}{\partial y^2} + \frac{\partial^2 u}{\partial z^2} - \frac{\partial u}{\partial t} = 0$$

$$u(0, x, y, z) = f(x, y, z)$$

for $f \in L^2(\mathbb{R}^3)$, that means that $f \mapsto u \in C([0, T], L^2(\mathbb{R}^3))$ for $0 \leq t \leq T$ where T is any positive computable real number, is computable with respect to admissible representations of the corresponding spaces involved.

6 Further relevant results in computable physics

The case of the Schrödinger equation is more complicated and can be considered similar to that of the wave equation. As Weihrauch and Zhong point out in [16]:

> Like the wave propagator the computability of the solution operator of the Schrödinger equation is closely related to the space of the initial conditions.

More precisely, with respect to the inhomogeneous linear Schrödinger equation:

$$u_t = i\left(\frac{\partial^2 u}{\partial x_1^2} + \frac{\partial^2 u}{\partial x_2^2} + \ldots + \frac{\partial^2 u}{\partial x_d^2}\right) + \phi$$

for $t \in \mathbb{R}$, $i = \sqrt{-1}$, with the forcing term $\phi(t, x_1, \ldots, x_d)$ and the initial condition $u(0, x_1, \ldots, x_d) = f(x_1, \ldots, x_d)$, they have considered as possible initial conditions spaces $L^p(\mathbb{R}^d)$ and the space of Sobolev functions.

We will not give many technical details, but we point out the crucial point of the result. If we consider the initial condition as an element of a Sobolev space, and therefore we codify it through a very natural representation of such space, then the Schrödinger propagator solution operator $S : (t, \phi, f) \mapsto u(t, .)$ is computable. If we consider instead the Schrödinger propagator as operating in a $L^p(\mathbb{R}^d)$ space, then this propagator is computable only in one case, precisely when $p = 2$.

Again, the question about the possibility to describe a physical system is strictly dependent on the chosen representation.

The last topic concerning mathematical physics we mention is the Korteweg-de Vries equation. The question about its computability is one of the open problems listed by Pour-El and Richards in [12]. Some authors have provided partial solutions to this question. In [7] Gay, Zhang and Zhong have analyzed the periodic case

$$\frac{\partial u}{\partial t} + u\frac{\partial u}{\partial x} + \frac{\partial^3 u}{\partial x^3} = 0 \qquad t \geq 0, x \in (0, 2\pi)$$

$$u(0, x) = f(x)$$

$$u(t, 0) = u(t, 2\pi) \qquad \frac{\partial u}{\partial x}(t, 0) = \frac{\partial u}{\partial x}(t, 2\pi) \qquad \frac{\partial^2 u}{\partial x^2}(t, 0) = \frac{\partial^2 u}{\partial x^2}(t, 2\pi).$$

Some mathematicians have proved the appropriateness of Sobolev spaces $H^s([0, 2\pi])$ for $s > \frac{3}{2}$ for the initial value function. This defines $f \mapsto u$ as a non linear operator from $H^s([0, 2\pi])$ to $C([0, T], H^s([0, 2\pi]))$. The authors have then proved that for computable numbers $s > \frac{3}{2}$ and $T > 0$, and for a computable $f \in H^s([0, 2\pi])$, the corresponding unique solution u is computable.

In [20], N. Zhong alone has re-considered the Korteweg-de Vries equation with other initial conditions, and has provided another positive answer to the question by Pour-El and Richards, showing the computability of the solution operator under certain conditions.

Weihrauch and Zhong have considered in [18] the simpler case

$$\frac{\partial u}{\partial t} + u\frac{\partial u}{\partial x} + \frac{\partial^3 u}{\partial x^3} = 0 \qquad t, x \in \mathbb{R}$$

$$u(0, x) = f(x).$$

This problem has been classically proved to be well-posed for all classical Sobolev spaces $H^s(\mathbb{R})$, for $s \geq 0$. Weihrauch and Zhong have proved a partial (but quite general!) effective version of this classical result, by showing that the corresponding solution operator $S : H^s(\mathbb{R}) \to C(\mathbb{R}, H^s(\mathbb{R}))$ is computable for $s \geq 3$.

Different philosophic conclusions may be deduced from this very concise outline of results concerning computable physics. In [17], Weihrauch and Zhong write:

> Apart from the wave equation, two other partial differential equations with applications in physics have been analyzed similarly, the linear and a non linear Schrödinger equation. In both cases the propagator is computable.

We think that one can agree with their opinion, provided that one accepts a precise meaning of the term "computable". This assertion should be interpreted as: "They are Turing computable, in the sense that there is *at least some natural way* to codify the inputs for which the corresponding solution operators are Turing computable". For each case there is indeed at least some reasonable representation that allows a computable simulation of the physical system. In the case of the Schrödinger equation we have even just seen how different mathematical structures, which have been used successfully in pure mathematics in the study of a physical system, can have different physical relevance (we say "physical", since we consider the Turing machine to be a physical model). Physically, some structures are more appropriate than others for an accurate description of a certain reality. This is evident in the case of the Schrödinger equation, where the Sobolev space permits a more general computational treatment than the space $L^p(\mathbb{R}^3)$.

The situation for the Korteweg-de Vries equation is a little more complicated, but we think that the answers to the question of Pour-El and Richards which are nowadays available are satisfying, since the results obtained are quite general. In any case, Weihrauch and Zhong leave us with the general problem still open, when they conclude:

> However, computability of many other physical processes, for example, boundary value problems, has not yet been investigated. Therefore, there still might be a physical device which beats the Turing machine.

References

1. L. Blum, F. Cucker, M. Schub and S. Smale: *Complexity and Real Computation* (Springer-Verlag, Berlin Heidelberg New York 1998)
2. V. Brattka: Effective Borel measurability and reducibility of functions. *Mathematical Logic Quarterly* **51** (2005) pp 19-44
3. V. Brattka and G. Gherardi: Borel Complexity of Topological Operations on Computable Metric Spaces. To appear in *Journal of Logic and Computation*

4. G. Chaitin: *Meta Math! The Quest for Omega* (Vintage Books 2006)
5. D. Cenzer and J. B. Remmel: Index sets for computable differential equations. *Mathematical Logic Quarterly* **50** (2004) pp 329-344
6. R. Feynman: *The Character of Physical Law* (The MIT Press, Boston 1965)
7. W. Gay, B. Y. Zhang and N. Zhong: Computability of solutions of the Korteweg-de Vries equation. *Mathematical Logic Quarterly* **47** (2001) pp 93-110
8. G. Gherardi: Effective Borel Degrees of some Topological Functions. *Mathematical Logic Quarterly* **52**, 6 (2006) pp 625-642
9. G. Gherardi: Paradigmi di computazione per i numeri reali. *Bollettino dell'Unione Matematica Italiana* To appear
10. R. Guénon: *Les principes du calcul infinitésimal* (Gallimard, Paris 1946)
11. M. B. Pour-El and J. I. Richards: A computable ordinary differential equation which possesses no computable solution. *Annals Math. Logic* **17** (1979) pp 61-90
12. M. B. Pour-El and J. I. Richards: *Computability in Analysis and Physics* (Springer-Verlag, Berlin Heidelberg New York 1989)
13. S. G. Simpson: *Subsystems of Second Order Arithmetic* (Springer-Verlag, Berlin Heidelberg New York 1999)
14. K. Weihrauch: *Computable Analysis* (Springer-Verlag, Berlin Heidelberg New York 2000)
15. K. Weihrauch and N. Zhong: The wave propagator is Turing computable. *Lecture Notes in Computer Science* **1644** (1999) pp 697-706
16. K. Weihrauch and N.Zhong: Is the linear Schrödinger propagator Turing computable? *Lecture Notes in Computer Science* **2064** (2001) pp 369-377
17. K. Weihrauch and N. Zhong: Is wave propagation computable or can wave computers beat the Turing machine? *Proceedings of the London Mathematical Society* **85** (2002) pp 312-332
18. K. Weihrauch and N. Zhong: Computing the solution of the Korteweg-de Vries equation with arbitrary precision on Turing machines. *Theoretical Computer Science* **332** (2005) pp 337-366
19. K. Weihrauch and N. Zhong: Computable analysis of the abstract Cauchy problem in a Banach space and its applications I. *Mathematical Logic Quarterly* **53** (2007) pp 511-531
20. N. Zhong: Computable analysis of a boundary-value problem for the Korteweg-de Vires equation. *Theory of Computing Systems* **41** (2007) pp 155-175

Phenomenology of Incompleteness: From Formal Deductions to Mathematics and Physics

Francis Bailly and Giuseppe Longo

Physique, CNRS, Meudon & LIENS, CNRS–ENS et CREA; Paris (France)
http://www.di.ens.fr/users/longo

This paper is divided into two parts. The first proposes a philosophical frame and it "uses" for this a recent book on a phenomenological approach to the foundations of mathematics. Gödel's 1931 theorem and his subsequent philosophical reflections have a major role in discussing this perspective and we will develop our views along the lines of the book (and further on). The first part will also hint to the connections with some results in Mathematical physics, in particular with Poincaré's unpredictability (three-body) theorem, as an opening towards the rest of the paper. As a matter of fact, the second part deals with the "incompleteness" phenomenon in Quantum physics, a wording due to Einstein in a famous joint paper of 1935, still now an issue under discussion for many. Similarities and differences w.r. to the logical notion of incompleteness will be highlighted. A constructivist approach to knowledge, both in mathematics and in physics, underlies our attempted "unified" understanding of these apparently unrelated theoretical issues.

Part I
Revisiting *Phenomenology, Logic, and the Philosophy of Mathematics*[1]

Constructivism is the most common philosophical attitude in the mathematics (and practice) of Computing and this in contrast with the prevailing debate in mathematical circles still ranging from Platonism to Formalism. But, what do we mean, today, by "conceptual construction", in the broadest sense? Phenomenology may provide one possible answer to this, by a deeply renewed understanding of Weyl's (and Brouwer's) ideas, in a perspective close to Husserl's

---

[1] Part I is a largely expanded version of a review of [47], which appeared in *Metascience*, a review journal in Philosophy of Science, 15.3, 615-619, Springer, 2006.

philosophy. Tieszen's book proposes a critical account of modern views in the foundations of mathematics, which is of direct concern for the logician and the theoretician in natural sciences who wants to reflect on the constructive principles of the mathematical intelligibility of the world. We will refer to this book to go further and motivate a broadening of the notion of "construction" as given in formal deductions and arithmetical computations, in either classical or intuitionistic frames. By this broadening, we will understand the incompleteness phenomenon as a "gap" between mathematical construction principles and formal proof principles, following and further devellopping some ideas hinted in [2].

Tieszen's perspective is original, as the Philosophy of mathematics has been largely dominated by a contraposition between Ontologism and Nominalism, as recalled above. This separated the foundational analysis of mathematics both from our lifeworld and from other scientific domains, including physics where mathematics has a constitutive role. By correlating foundational issues in mathematics and physics, along the lines of [2], we will try to recompose the foundational break, at least as for the issue of incompleteness.

Part I.1

The first part of Tieszen'book is dedicated to an introduction to a Husserlian perspective in the foundations of mathematics. It is interesting per se, as a broad survey of Husserl's phenomenology. This is made possible by the relevance that Husserl himself gave to Logic and mathematics in his philosophy of knowledge: writings on Logic and Arithmetic are among the earliest of Husserl's and the related issues accompany his lifelong work.

The constitution of ideal objects, in Husserlian terms, is based on a clear distinction between the transcendental perspective and psychologism. It is the human subject who makes science possible, yet the common endeavour of the historical community should not be confused with the individual analysis: epistemology is a genetic analysis, provided that history is not understood in the usual limited sense, explains Husserl in the "Origin of Geometry" (1933). There are different types and levels of consciousness, which allow the historical dynamics of knowledge: science is built up from the lifeworld experience of human subjects on the basis of active abstraction, idealization, reflection, formalization. The objectivity of knowledge is a constructed one, a result of the interaction by an active subject, beginning with "kinaesthesia", in a living body, in everyday world of life. Meaning is not the passive interpretation of independent signs, but it is constructed in this interaction, it is the result of a "friction" and of structuring of this very world by our attempts to give sense to it; meaning is the result of an action. Of course, we dare to add, this must be understood in a broad sense: Quantum Mechanics for example seems to owe little to kinaesthesia. Yet, it is a paradigmatic case where meaning is the result of active consciousness, beginning with the preparation of the experiment or of the technical context for insight: we are conscious of a quantum object as a constituted phenomenon. Our lived body is just expanded by instruments which, in turn, result from a theoretical commitment: this is the richest form of interaction in the sense mentioned above, with no meaningful object without active knowing subject. A (conscious) intentional process is at the origin of this form of knowledge.

As Tieszen explains, consciousness, for Husserl, is consciousness of something. It could be an ideal object of mathematics. The later being the result of a formation of sense founded on underlying acts and contents, which make possible the ideal construction. We would like to exemplify by considering Euclid's action on space by ruler and compass. This action organises figures in space by rotations, translations and reflections, to put it in modern terms. A dialogue with the Gods for sure, but also active measure of ground. But, how to define and measure surfaces, a technique that, in its mathematical generality, is the key Greek invention? In order to conceive *exact* metric *surfaces* one has to conceive lines *without thickness*; there is no way to give a mathematical sound notion of surface, without first proposing, with Euclid's clarity, lines with 0 thickenss. Then, as a consequence of intersecting lines, one *obtains* dimensionless points, as Euclid defines them (a remark by Wittgenstein). These extraordinarily abstract concepts, point, line etc. are the idealized result of a praxis of measure of surfaces and access to the world by translating and rotating ruler and compass, far away, but grounded on sensory experience. Rotations, translations and symmetries are "principles of (geometric) constructions", a notion to be often used in the sequel. In Euclid's geometry they are used in proofs and they define the geometric objects as given by *invariant* properties w. r. to these transformations.

Tieszen stresses several times the relevance of *invariants* in Husserl's foundational approach. "Mathematical objects are invariants that persist across acts" carried out in different contexts. The practical constructions of mathematics, in human space and time, are also stabilized by language and, then, by writing, says Husserl: their constituted ideal nature is primarily the result of their invariance, as conceptual constructions, with respect to suitable transformations of context. And this is extremely modern: invariant structures and transformations were first the foundational core of Riemann's geometry, in Klein's approach, then of Category Theory. Husserl seems to precede the underlying philosophy of Category Theory by his analysis of mathematical knowledge. These invariants are then the *essences* and, thus, provide the only possible ontology for mathematics; they are the result of different "fulfilled mathematical intentions", as constructions. And these constructions have a horizon, the space of the historical praxis which leaves as trace the most stable invariants of all our mental practices, the structures and objects of mathematics. Then, underlines Tieszen, "truth is within this horizon" as there is no, for Husserl, absolute mathematical truth nor evidence. Yet, mathematical theories are not arbitrary creations (consider the example of Greek geometry above), they are no conventional games of signs: we do not solve open problems by convention, as they are the result of a meaningful and motivated construction. The genetic analysis gives evidence for the objectivity of the constitution of mathematics, in the interface between us and the world.

One of the challenges of this approach, that stresses the exactness and stability of mathematical structures, is the correlation with the inexact (i.e. morphological) nature of everyday sensory objects. Even more so, it is the challenge of the relation of mathematics to the sciences of life, where the stability may be global but is not due to exactness. We will go back to this at the end.

Tieszen dedicates a section to the "Origin of Geometry". Even if the significance of that essay is very often present in Tieszens book, we think that it is not sufficiently stressed. This mature text of Husserl's is a splendid progress and a revision for his overall philosophy of mathematics. It radically departs from the partial proximity one can find in Husserl's early books with Frege's logicism and even Hilbert's formalism. However, in this short section, Tieszen presents very clearly Husserl's view on the role of bodily action, by the kinaesthetic and orientation systems, in organising space (the distance of an object is the evaluation of the movement needed to reach it, says Poincaré). Abstraction is then seen as "limitation of attention" and reflection as "adoption of a theoretical attitude" (one can see here the path towards the abstract and reflected nature of our science, which begins with Euclid's Geometry, the original mathematics). Measure, in order to be exact, requires ideal shapes, as given by the notions of point, line etc we mentioned above. These are invariants that found knowledge, from Euclid up to Einstein's Invariantentheorie (the early name of Relativity Theory). Riemann's Geometry, which was born to understand gravitation, underlies this late developments, where, jointly to Category Theory, one sees "invariance and objectivity go hand in hand", since "invariance is a cornerstone of rationality and science". The challenge is to propose, at the same time, the right transformations that preserve the invariants, *i.e.* to give the right Category (or metric phase-space, to put it in physical terms).

Part I.2

The second part of Tieszen's book is largely devoted to an effort to find some Husserl in Gödel's philosophy. Gödel, in the last part of his life, became, apparently, a close reader of Husserl. But, Gödel's reading seems more concentrate on some of the writings by Husserl of the beginning of the century and does not seem to span the mature reflection of Husserl, typically to the "Origin of Geometry", in landmark of Husserl's foundational reflection on Mathematics. In particular, Gödel stresses the possible ontological understanding of Husserl's phenomenology far away from the theory of invariance we sketched above. He proposes in particular an identification of physical and mathematical objects, made not in the ground of a similar *construction* of objectivity, but in reference to a similar objective and *autonomous* existence. Of course, the foundation of physical knowledge is strictly related to the mathematical one, but this common foundation may be understood, in phenomenological terms, by reversing bottom-up Gödel's realism: their simultaneous constitution is the result of a common praxis, not of a similar transcendent reality. Quantum physics is the highest example for this: an electron is a solution of Dirac's equation and nothing else; but Dirac's equation is given on the grounds of robust empirical evidence, *prepared* by a theoretically oriented acting subject. Here is the virtuous circle of knowledge, which is hypothetical - principles driven, where these principles are grounded on a friction, on a "reality", whatever it may mean, which opposes *frictions* or *resists* and *canalises* our endeavour towards knowledge construction.

So, if one reads Gödel the other way round (both mathematical and physical objectivity - and objects, in a sense - are constituted, including the book and the table mentioned by Gödel), then this is (modern) phenomenology, otherwise one stays with the current interpretation of Gödel as an ordinary Platonist in mathematics. It is not by chance that this interpretation is prevailing, as most published writings of Gödel largely favour this understanding. Of course, the table and the book in Gödel's working place, pre-exist the knowing subject, as they are the result, both the object and the concept (of table, of book), of previous human activities. But also the mountain, which is out there, for sure, is delimited and given a name, isolated from the context, by our historical endeavour towards organising the world (where is it its lower bound, exactly? *We* draw the mountain's contours). For Gödel, instead, concepts are abstract objects, which exist independently of our perception of them, "perceived" by some physical "ad hoc" organ, which allows intuiting the essence. The lesson instead we learn from (the late) Husserl gives even to intuition the structure of a constituted: intuition in not an absolute, it is the result of our historical praxis, beginning with our phylogenetic history. More closely to us, the mathematician's intuition of the continuum, for example, 150 years after the splendid construction of Cantor's, is deeply indebted to the Cantorian real line: we *see* the continuum in that way and it is very hard to appreciate different continua (Leibniz's insight, for example.)

In short, Gödel's philosophy of mathematics contains transcendence and very little transcendental constitution, in spite of Tieszen's claim. And Tieszen rightfully recalls Merleau-Ponty's stress on the change towards a theory of existence as preceding essence in the late Husserl, in "Krisis" in particular: existence as invariance and as a result of a constitution. It is a process of free variation that gives the conceptual stability of mathematical (or more generally, conceptual) rules and structures. And this process is similar in mathematics and in physics; while for Gödel it is static perception that is similar in mathematics and physics, as if Physical knowledge (years after Quantum Mechanics) were based on sense perception. Perception, both everyday, common perception, and organised one, is far from static. Even vision requires an activity (saccades at least): perception is the result of an action, which, by its variations, *singles out* stabilities and invariance. Late unpublished (or recently published) reflections by Gödel show an increasing appreciation of the dynamic of thought in Husserl and the role of transcendental constitution in his philosophy. But this unknown writings had little or no relevance in the major influence that Gödel has had in the formation of the platonist approach to foundations. And a philosophy matters also for the effect that it has.

As a matter of fact, the modern perspective in phenomenology even more radically departs from Gödel's blend of realism and idealism: by, perhaps, forcing slightly Husserl (and surely beyond Tieszen's approach), we can even say that *any constitution is contingent*, as it is the result of a history and a praxis (a praxis in this world, with its frictions to our action).

Whatever is the true teaching by Husserl, our strong stand here is the opposite of the Fregean absolutes that still invade Gödel's views. Of course, though, this constitution has a pre-human history: we share with many animals basic counting praxes, appreciation of borders and trajectories. These are invariants of pre-linguistic activities and, thus, partly precede conscious intentionality, in Husserl's sense.

More generally, mathematics is practiced in "our space of humanity", by our "historical communicating community", as Husserl says, not by an individual subject: this adds to it its further, conscious and intersubjective stability. At once, then, one also understands the effectiveness of mathematics, which suddenly becomes reasonable: it follows from this contingency. There is no real line in the world, nor imaginary number i, but we organise the world by these construed concepts, which are strictly conceived by us, against Gödel's view, and are conceived by *acting* in *this* world, this is why they are not arbitrary. Mathematics is the result of this contingent friction between us and the world by a complex praxis of "action - abstraction of the action". The memory of a prey's trajectory, say, is already "abstract": it is the retention of a protension, of the inexistent line, preceding the prey to be caught and that the predator traces in advance, by saccades. And later, the concept of line is the result of our symbolic-linguistic culture, by language, drawing and writing, a further stabilization by intersubjective practices. Here lies its objectivity. Clearly, the resulting invariants, both in mathematics and, perhaps, in general conceptual constructions, may be surely transcendental and, once constituted, non-contingent. By this, we mean that they may be invariant w.r. to transformations of reference system, in physics or mathematics, or of humanly possible forms of life. It is the *constitution* of these transcendental invariants that is contingent.

Similarly, as humans, we constructed this chair and this table, which were not already there, nor was the concept of chair, of table, but resulted, both the object and the concept, from our constructive action and linguistic exchanges. Once constructed, the table, the chair and their concepts stay there, go cross generations and history, through changes, and transcend each individual life. And we also invented language and the alphabet, which are not in the world; yet, by their constitutive history, they are extremely (yet, reasonably) effective in understanding each other and the world, no less than mathematics, in their domains of application. They are at least as effective as this chair, which seats us so well, it fits us and the physical world! And we understand each other by talking and by (re-)producing language in writing: where is the miracle? It is no more than in the panglossian nose and spectacles, which surprisingly fit each other so well. And we can even use words with certain prevailing meanings to express very different situations, by metaphors say. That is, we may transfer linguistic expressions and conceptual structures, by formal or semantic drifts, similarly as complex numbers can be used to express conjugate values in Quantum Physics, three centuries after their invention.

In conclusion and against Logicism and Platonism metaphysical necessities, which need miracles and/or inspire an unreasonable effectiveness when transferred to the world, we claim that mathematics is objective and effective, *exactly because* it is the *constructed contingency of maximally stable invariants* of our action in space-time and of reasoning, by gestures and language, along our phylogenetic and human history. Or, also, we *define* mathematics as the set of conceptual practices that is maximally stable and independent from the contingency of their constitutive path (they are invariant under transformations of humanly possible frames, as we said). And this, since Euclid's triangles, which, by *definition*, do not depend on the thickness of the traits, or the colour of the ink. This maximal stability and independence, which is part of their construction, makes the mathematical concepts more stable than a chair and a table or their concepts; they better go through history.

Finally, let's observe that mathematical objects are *limit* constructions, obtained by a conceptual "critical" transition (see part 2), where the constitutive contingency is lost at the limit. Euclid's line with no thickness or the "transcendental" number π is the result of a geometric construction, pushed to the limit. But, in the end, their objectivity does not depend on the specific-contingent and more or less abstract reference to actual traits or sequence of rational numbers, needed to conceive or present them: at the limit, the transition to infinity provides us with a perfectly stable conceptual object (a mathematical ideality, some like to say). Their value and their sense do not depend on the specific construction or contingent converging sequence of rationals, yet needed to conceive or define them. And the conceptual transition is irreversible: no way to go from π to a given specific converging sequence; there are infinitely many of them and π is the invariant under transformation (change) of equiconvergent sequence. In the second part we will return at length on this understanding of the constitution of mathematical idealities as a conceptual (we will say "critical") transition, in analogy to the physics of critical phase change.

Let's go back to Tieszen's book. In a very scholarly fashion, Tieszen's chapters on Gödel's allow to grasp the differences between the mature Husserl and Gödel's philosophy, as well as the ontological shift of Gödel's which may at most be referred to Husserl's work at the beginning of the century. Yet, we would continue further to dig into the non-phenomenological reference to absolutes in Gödel's thinking. Tieszen explains how Gödel believed Hilbert to be correct in supposing the decidability of all number-theoretic questions, in spite of his own undecidability (incompleteness) theorem: rationality needs just to be extended by new laws and procedures. This rationalistic optimism, a belief in absolute-ideal essences and their possible perfect knowledge, is paradigmatic of how both Formalism and Platonism detached the foundation of mathematics from physics and allows us to understand the opposite views of mathematicians more deeply concerned with the latter (philosophically at least, as Hilbert greatly contributed to Mathematical physics): Poincaré, Weyl, Enriques. How could a leading Nominalist and a leading Ontologist (in an broad semi-Husserlian sense), Hilbert and Gödel respectively, both possibly say that there could be no undecidable assertion? And this when Poincaré had shown that, given a formal system of a few equations (the Newton-Laplace system for three bodies in their gravitational fields, 1890), one could state a (formalizable) proposition, parametrized on a sufficiently distant time, that could be provably shown to be undecidable, under the intended (and best) physical approximation of initial data. This was the first great result on the mathematical unpredictability of non-linear deterministic systems. No added (physically possible) mathematical knowledge can solve this. This is why Poincaré cries out against Hilbert's "sausage machine's view of mathematics" (an aggressive stand recalled also by Tieszen): he has an entirely different *Philosophy of Knowledge*. Let's analyse this point more closely.

Preliminary reflections on incompleteness, in mathematics and in physics

In Longo [31], it was suggested that Poincaré's three-body theorem (see [5] for an introduction and an historical account) is an epistemological predecessor of Gödel's undecidability result. Of course, Hilbert and Gödel were speaking about *purely mathematical 'yes or no'* questions, while unpredictability shows up, at finite time, in the relation between a physical system and a mathematical set of equations (or evolution functions). That is, in order to give unpredictability, Poincaré's Negative Result, as he called his proof of the non-analyticity of the equations for three-body system, needs a reference to physical measure. Measure is always, in classical (and relativistic) physics, an interval, that is an approximation, by which *non-observable* initial fluctuations may give observable thus unpredictable evolutions, in presence of non-linearity of the mathematical determination - a set of equations or an evolution function (main reasons: the initial interval expands exponentially, by the so-called Lyapounov exponents, and it is "mixed", [18], [19]). Yet, one can reformulate the problem in terms of a formal trajectory "reaching or keeping away" from a given target neighbourhood: if the deterministic system is chaotic, the mathematical question cannot be decided (see [26], [27] for surveys). However, this is slightly unsatisfactory as it only shows undecidability properties of chaotic dynamics (and not the converse) and it partly relies on abstract properties of real computable numbers, instead of intervals, that do not need to fully express the mathematics of physical systems. In particular, an effective measure theory (along the lines of Lebesgue's measure, the locus for dynamic randomness) needs to be introduced. The problem then is to have a sound and purely mathematical treatment of the epistemological issue (and obtain a convincing mathematical correspondence between unpredictability and undecidability).

Now, unpredictability of deterministic systems is randomness in classical physics (see [3]) and it may also be expressed as a limit or asymptotic notion. Under this form, it may be soundly turned into a purely mathematical issue. That is, randomness, as a mathematical limit property, lives in formal systems of equations or evolution functions, with no need to refer to physical processes and their approximated measure.

On these purely mathematical grounds, the second author conjectured that a rigorous formal link could be shown, between Poincaré's unpredictability and Gödel's undecidability, by passing through Martin-Löf number-theoretic randomness. This is a "gödelian" notion of randomness, as it is based on Recursion Theory and yields a strong form of (strong) undecidability for infinite 0-1 sequences (an infinite sequence is random if it passes all *effective statistical tests*, see [37][2]). On the side of physical dynamics, its mathematical counterpart can be found in reference to Birkhoff's notion of ergodicity, which refers to infinite trajectories, a purely mathematical approach at the infinite limit, with no need to refer to the interval of physical measure to engeder randomness. That is, mathematical dynamical systems, in their (Lesbesgue) measurable spaces, allow to define generic points and *infinite* random trajectories, in the ergodic sense [41].

Recently, M. Hoyrup and C. Rojas, under Galatolo's and the second author's supervision, proved that dynamic randomness (à la Poincaré, thus, but asymptotically, following Birkhoff), in suitable effectively given measurable dynamical systems, is equivalent to (a generalization of) Martin-Löf randomness. This is a non-obvious result, spreading across two doctoral dissertations (available by summer 2008, see aknowledgements) and gives an indirect, but relevant, we believe, technical link between Gödel's incompleteness as undecidability and Poincaré's unpredictability.[3]

On more philosophical grounds, Poincaré, in several writings, also tried to relate the foundations of mathematics to that of physics, passing by cognition and action in space. Similarly for Weyl and Enriques, who insisted, during their entire life, on a parallel foundation of scientific knowledge in these exact Sciences, see [2]. We insisted here on the connections between the issues of

---

[2] Martin-Löf's randomness, inspired by Kolmogorof's ideas, has been developed by many, Chaitin most remarkably, see [13] for a classic.

[3] Other links betwen physical and algorithmic randomness may be found in the litterature, yet the connection via ergodic theory seems new. In particular, [14] relates algorithmic randomness to quantum indetermination, a different issue; in Part 2 we will discuss the difference between incompleteness and indetermination in Quantum Mechanics.

decidability and predictability, at the core of mathematical and physical theoretizing. Of course, unpredictability in not in the world, it is not an ontological matter, it concerns the *relation* of our forms of (mathematical) knowledge *to* the world and this by the role of physical measure (our form of access to the world). However, as we hinted, mathematics brings it within a purely theoretical realm, by pushing to the asymptotic limit the mathematical treatment of the intended physical processes in space-time continua, following Birkhoff ergodicity. And this relates, as we said, to algorithmic randomness.

Observe now that, since the beginning of the XXth century, physicists have been discussing the immense philosophical challenge of the intrinsic indetermination of Quantum Physics (again, an issue related to measure) and that, in 1935, the possible "incompleteness" of Quantum Mechanics was proposed as a key theoretical issue, as we shall see in Part II. These internal limitations to knowledge (physico-mathematical unpredictability, intrinsic indetermination *vs* quantum incompleteness), whose understanding opened the way to two major scientific domains, in classical and Quantum physics, were simply out of the scope of the discussion of the two metaphysical rationalisms of Hilbert and Gödel, in foundations.

Back to Gödel's philosophy

Gödel's philosophy, in our opinion, seems far away from any internal debate in physics such as the one we will discuss in Part II (an example is given by the mathematically remarkable, but physically unsound paper on the circularity of relativistic time, see [23]). It escapes though trivial physicalism, as for Theory of Mind. As Tieszen explains, Gödel argues against Turing's claim that mental procedures cannot carry any further than mechanical procedure, as both mind (in the brain) and machines are both "finite state devices". To this, Gödel observes that mind is not static, but constantly developing. A very modern view, as it is now clear that there are no "states" in brain/mind, but only processes: Turing's Machines "instantaneous descriptions" simply do not make sense in cognition. In particular, Gödel remarks that abstract meanings or concepts may be the result of limit procedures (to put it short: after a few iterations, we "look at the horizon" and we construct and understand, say, irrational numbers, projective limits, transfinite ordinals.)

In chapters 6 and 7, Tieszen further develops his close analysis of Gödel's path towards phenomenology. This is a rather unaccomplished path, as we explained above and as one may deduce also from Tieszen's many references and comparisons. Yet, this path is very interesting, as it brings Gödel to a clarification of the role of meaning in Arithmetic. With the help of notions from phenomenology, Gödel derives from his incompleteness results the need for a reference to meaning as "categorial intuition" in (arithmetic) proofs. Thought structures or thought contents, in Gödel's words, are the result of insights which go beyond the combinatorial properties of symbols and require a reflection upon the *meanings* involved. This is what we call the meaningful construction principles, as rooted in our cognitive history (from evolution to human history); the well-ordering of integer numbers is an example (see [32]). It is in this sense that formal proof principles are incomplete w. r. to construction principles, see also below. As we recalled, for the late Husserl as well, these meanings are the result of a "formation of sense" which is grounded in our "historical spaces of humanity".

While missing this point about the *constitution* of meaning and principles, Gödel takes up another fundamental theme of phenomenology: intentionality. In Tieszen's interpretation, Gödel uses and understands it as *directedness towards invariants*. This fundamental structure of thought allows us to understand even sense perception as a non passive but active and constitutive first step of knowledge building: the early, pre-conceptual, singling out of invariants. Of course, from this perspective, Benacerraf's dilemma or alike, to which Tieszen refers in a highly critical way, are just part of the new scholastic which F. Enriques had forecasted long before ("the Philosophy of Mathematics is heading towards a new scholastic: the Scylla of Ontologism and the Charybdis of Nominalism", he observed in 1937). As a matter of fact and well beyond Quine's Platonism, in particular in Set Theory (meaning as truth is "out there"), Gödel proposes an understanding of meaning as content, that is as a result of intentional processes. Husserl's intentionality again allows understanding categorization, from perception to thought, and thus content as part of meaning formation.

As we are talking of more or less naive forms of Platonism, which regained relevance by, helas, too common interpretations of incompleteness, a reference should be made here to Penelope Maddy's book. Tieszen begins by some high praises of it and continues with . . . an extremely severe critique of any relevant idea in that book. Tieszen explains the misunderstanding of Gödel's refined Platonism which is brought back to the usual flat ontology, in contrast to the phenomenological components that Tieszen showed to contain. He insists on the wrong "bon sens" (our words) attitude of considering that "(some) sets are part of the physical world" and perceived by us as such, without any of the constructed conceptual stability and invariance that is proper to mathematics. This brings Maddy to an unavoidable relativism, which is rather alien, says Tieszen, to that science. There are three eggs, three atoms and so forth and, ho miracle, we have the mathematical "set of three" and the *concept* of "three"! But mathematics is a science of structures: there is no mathematics without structures. Set Theory originated in the *logical* foundations of mathematics and passed aside the foundation of modern physics exactly because this refers to geometric invariants or geodetic principles (which are symmetry invariants: they reflect a structure). Mathematics actively organises the world and our forms of scientific knowledge, *it shapes them* simultaneously, since it is grounded on invariant structures, beginning with perception: the line is not a set of points (let alone a set of eggs), but a *gestalt*. It may be logically found or reconstructed, *a posteriori*, as a set of points (Cantor), or even without points, in some toposes (Lawvere, Bell, see [6]).

Tieszen analysis continues by scholarly and stepwise demolishing Maddy's book (in a very gentle and motivated way, of course, though some polemic tension may be appreciated in between the lines). One further issue should be mentioned here; it concerns the naive attitude of confusing "bon sens" (which is different from "common sense") with "formation of sense": there is a huge gap between the two and this is called the constructed objectivity of science. Science always goes against "bon sens". Greek mathematics did not begin by looking at sets of eggs, but by daring to propose non-existing lines with no thickness and nondimensional points. Modern science began against the "bon sens" evidence of the Sun turning around the Hearth. But even the concept of number is a constructed and complex *invariant* w.r. to ordering, organizing and "small" counting, as ancient (pre-human?) practices ([17] is a classic about this). It is complex, as it is the result of many active experiences of transformations and their invariants.

But then, why not to propose that the unity of consciousness (and mathematical meaning) is due to Quantum entanglement as global effects in the brain? This is the question discussed in chapter 10. Personally, we are not against occasional audacious speculations as the one proposed by Penrose in several bestsellers and analyzed by Tieszen. However, as it is a matter of finding physical phenomena in a material structure (the brain) it would be better, at least in principle, to start from some empirical evidence. Of course, there is none of this and Penrose's starting point is Gödel's incompleteness theorem. And a few assumptions.

First, *awareness is a physical action* and any (classical) physical action can be simulated computationally (in Turing's sense); then, *truth is an absolute matter*, to which we have access by awareness. So, again, in order to escape the Scylla and Charybdis of Nominalism and Ontologism, Penrose suggests a shortcut from microphysics to consciousness.

As for the role of Gödel's theorem, unfortunately this is based on a misunderstanding of the proof of it: Tieszen quotes a fine analysis of Penrose's mistakes made by Feferman. The result of this misunderstanding is the belief that one can deduce from it the absolute and transcendent nature of truth. Tieszen criticizes this point not only on the technical grounds of Feferman's remarks but also along the lines of his phenomenological analysis of meanings as intentions. In short, infinity, as meaningful thought structure, constituted along history, steps in the proof of consistency of Arithmetic. This may be more closely understood by an analysis of recent "concrete" incompleteness results, carried on in [32] and, more informally, in [33].

There is no place here to criticize further Penrose's physicalism. It takes explicitly for granted that current physical *theories* are complete w.r. to the world (this was also Aristotle's opinion): so, if a phenomenon is not classical (relativistic) - thus computable (really? see [36]), it must be describable by some Quantum Theory. If the founding fathers of Quantum physics had had the same attitude, they would have searched in existing theories, Relativity Theory, say, or in variants of (thermo-)Dynamics the solution to strange phenomena such as the energy spectrum of the hydrogen or the three bodies' problem of the helium. They proposed instead a radically new theory.

It is interesting to see how often physicists, who in general defend non trivial philosophies of knowledge within their discipline, when looking at life or cognitive phenomena, just claim: this must be understood in terms of (reduced to) one of our theories (usually: the one I know best). Typically, when discussing of General Relativity and Quantum Mechanics they talk of "unification" not of reduction, as many do when referring to biology. Unification, in contrast to reduction, means that one must be ready to invent a new theory that changes radically both pertinent notions of field or even the intended objects (String Theory) or the structure of space-time (Non Commutative Geometry). The point is that in order to "unify" one must have *two* robust field theories and, in biology, we miss exactly an autonomous and proper notion of "biological (causal) field" (this is discussed at length in [2]). There is no use to analyse any further Penrose's claim that all classical physical processes are computational, since we know, for example, that even the evolution of our planetary system is provably non computable. As a matter of fact, recent results of unpredictability in deterministic systems [30] prove that there is no way to compute the relative positions of all planets and the Sun in more than ten million years. That is, the system will have a position in the continuum of space that no *digital* machine can compute; this a "concrete" and difficult version of Poincaré's theorem on non-linear Dynamical Systems and their unpredictability, as uncomputability. It is a "concrete" result, similarly as the famous combinatorial results by Paris and Harrington, or Friedman version of Kruskal's theorem, which provide concrete combinatorial example, that is (interesting) propositions about interger numbers, that are provably unprovable in Formal Arithmetic, see [39], [24] (and [32] for a discussion).

Of course, by this we do not pretend to exhaust the discussion on mind, as developed also in Tieszen's chapter on Penrose: even a sound biological theory of brain would still be far from our symbolic culture and the phylogenetic and historical formation of sense in our "communicating community", which is the place where meaning, consciousness and intentionality are formed. That is, a purely biological theory of brain would be incomplete w.r. to cognition, as this should be embedded in our social and historical "forms of life": brain signs have meaning only within a context. A contingent distinction of *theories* does not imply dualism, as much as the distinction of quantum field from relativistic one is not dualism, but a distinction of phenomenal levels. And unification is a difficult matter. Yet, Gödel's attitude on the matter, as explained by Tieszen, is a traditional ontological dualism Mind *vs* Brain: the first has access to (infinitary) truth and meaning, by a metaphysical ahistorical path, the second should be less complex than our planetary system as, according to Gödel, it is fully computational. Unfortunately, Penrose's answer to this ordinary dualism is highly insufficient.

Part I.3

This part of Tieszen's book is largely dedicated to Weyl and Poincaré's philosophies, with a final reflection returning to Frege and Husserl. First, though, a very interesting understanding of Intuitionism is presented. Dummett's approach, in particular, is surveyed, with a clear presentation of his Wittgensteinian constructivism: "meaning is use". Unfortunately, Dummett seems to focus only on linguistic sense of use, in a clear contraposition with the Brouwerian orthodoxy (for Brouwer, mathematics is languageless!). In addition to language, use as action, as gestures, as (imagined) figures and drawings as forms of presence in space and time or of human interaction contribute to the transcendental constitution and to single out invariants by our intentional attitude. In our phenomenogical perspective, meaning goes well beyond the linguistic truth-conditions and brings us back again to intentionality, broadly construed (here we would like to pause and insist, with Merleau-Ponty, that, well before consciousness, "le mouvement et l'action sont l'intentionalite' originaire"). Tieszen stresses another crucial point which is not captured by the linguistic turn: the intentional grasp of meaning doesn't need to be fully determinate nor clear and exact. This allows inserting the phenomenological theory of meaning in human contexts where communication is enriched (and made possible) by polysemy, ambiguity, cross references .... Mathematics, in our approach, singles out and is determined as the locus of maximal stability and invariance among our conceptual constructions: no polysemy is allowed, in principle, no ambiguity: we force a - relative - stability of meaning; this is mathematics.

Tieszen's analysis of Weyl's constructivism is unfortunately limited to his views on the foundations of mathematics. The point though is that both Weyl's and Poincaré's views should not be detached from their Philosophies of Nature, though difficult it may be to spell them out, in particular from the overloaded and unorganised writings by Poincaré. Thus, while it is true that in "Das Kontinuum" Weyl spends several pages in sketching a (mathematically remarkable) predicative theory of reals, however, his short "flirt" with brouwerian intuitionism is motivated by his broader constructivist perspective that always tries to relate the foundations of mathematics to our general "human endeavour towards knowledge". Thus, like Brouwer, but well beyond Brouwer's psychological time, Weyl reflects to the continuum as space-time structure. He distinguishes between space continua and the phenomenal time continuum, the time we experience in consciousness. This cannot be reduced to analytic representations by points, in his view. Tieszen gives a very clear account of Weyl's understanding of time (based on a "specious or extended present") as well as of his short lasting predicativism; yet, the presentation makes an insufficient effort in connecting "Das Kontinuum" and the contemporary work in "Raum-Zeit-Materie". Moreover, later on, Weyl invented gauge theory, in physics, as an analysis of invariance, with a peculiar role of symmetries. Considering the constitutive role of both these notions for the foundations of mathematics *and* physics, one should find also there the rich perspective of Weyl's.

An informal survey of this physico-mathematical unity may be found in "Symmetry", Weyl's last book. Observe also that, while stressing the interest of Brouwer's free choice sequence as an approach to the continuum, Weyl, in several places, radically disagrees with Hilbert's formalist project as "the idea of a potential mechanization trivializes mathematics" (in "Das Kontinumm", where in an early section Weyl hesitantly conjectures the incompleteness of Arithmetic, in 1918! This should be noted more often).

Weyl's reflections are of a rare depth in the XXth century Philosophy of mathematics, scattered in several writings, from "Das Kontinuum" (1918) to Weyl's simple, but deep masterpiece that we mentioned, "Symmetry" (1952). One can find there the key role of symmetries in the constructed objectivity of our physico-mathematical knowledge. There is no ontological miracle nor miracle whatsoever, but the role of mathematics as a science structuring the world, by its very definition. One sees, in that book, the classification of planar symmetries (by rotation, translation, reflection: the Greek insight) as the pre-requisite for understanding that of finite groups. Other classifications, from Platonic solids to crystals' symmetries are then understood as conceptual, if not technical, consequences of these regularities, which make space-time intelligible and objective. Gauge invariance, a result of rotation/translation symmetries, is also seen as a foundation of Relativity and Quantum Theories (more on symmetries in the foundations of physics and mathematics may be found in [21], [2]). In short, Weyl shows that we *singled out from* and *imposed on* the world, also by our own bodily symmetries, a few regularities as tools for knowledge (for understanding, organizing), of which symmetries are the core part, and we called it "mathematics". In some cases, this is exhaustive of our spatio-temporal and linguistic representations: it covers them completely (by classifications). In no way does it follow from this, however, that mathematics can be detached from our own existence in this world and its concrete, active representations; on the contrary, it roots mathematics in them, starting with these resonance of symmetries, between us and the world. [4]

The passage to Poincaré's philosophy is motivated by Tieszen in the best possible way, in our opinion, that is by the call for an epistemological or *cognitive* dimension of proof. Poincaré refers to proofs by their meaning and geometric organization (against the flat arithmetic coding by Hilbert), and stresses the cognitive grounding of mathematical structures. Proof is the realization of a mathematical expectation, it is grounded on possibly new concepts, it requires a conceptual investigation, as in the contentual approach by

---

[4] "But perhaps this question can be answered by pointing toward the essentially historical nature of that life of the mind of which my own existence is an integral but not autonomous part. It is ... contingency and necessity, bondage and freedom, and it cannot be expected that a symbolic construction of the world in some final form can ever be detached from it." [48, p. 62]

Martin-Löf quoted by Tieszen. The geometric intuition of the real line precedes, in Poincaré's view, its logical foundation. In general, the analysis of the construction of mathematical concepts and structures (the epistemology of mathematics), which is extraneous to the formalist and logicist (Platonist) approaches, is a matter of a "living process in which the mind remains active": an analysis of human cognition, thus, in our sense. And here Tieszen moves to Husserl's "Krisis", a lesson for today even more than for his time: by "the construction and mastery of formal systems, Science becomes nothing but technique". It loses meaning and "forgets its historical origin". It loses also students, a dramatic process in Western world, as, once that making physical experiments, for example, is transformed into implementing computer simulations (isn't any classical-relativistic physical process computational?), why should one study physics? And many claim that also understanding biology or Cognition is only a matter of good computer models and compilers. Thus, financing increasingly goes only to short term technical and "competitive" industrially oriented projects, possibly producing quick computational models of whatever natural process. This is gradually killing theoretical long-term collaborative construction of knowledge, often grounded on "negative results" such as Gödel's or Poincaré's. In their time, these results were the opposite of positive modelization. It happened though that they opened the way to major scientific areas, the geometry of dynamical systems and computability, whose practical fall-out are under everybody's eyes.

The epistemological relevance of the major negative results we have been talking about goes toghether with their technical interest and actually motivated them. Tieszen's final chapter goes back to Husserl's philosophy in order to single out more closely, from his early writings, the role of meaning and intentionality as constitutive of the epistemological analysis and, in our views, of the analysis of proofs as well (in view of the incompleteness of formalisms). Husserl's reflection may thus be mentioned both for the need to preserve "meaning and sense" to the general scientific enterprise, beyond its technical developements, and for opposing Frege's project of eliminating intuition from proofs (meaning would be "obtruding", according to Frege!).

Conclusion on mathematical incompleteness

We tried to better understand incompleteness, which is usually seen just as an incomplete covering of semantics by syntax, as the gap between proof principles (on which formal deductions are based) and construction principles (the locus of meaning, in the constitutive relation between mathematics and our life world). Both principles are the result of the contingent constitution we mentioned above, but the former are the late commers of this process and, by principle, they forget the constitutive path. The latter instead may allow a reconstruction of the cognitive and historical gestures that lead to them and yield meaning.[5]

---

[5] More on this may be found in [32], [2] and [33].

This view of principles and incompleteness further specifies Husserl's approach in mathematics and it is even more remote from Frege's view, as he actually believed in Peano Arithmetic's "categoricity" (to put it in modern terms: there is only one model, or syntax coincide wit semantics), a much stronger property than Hilbert's completeness. As we shall see next, the discussion on "meaning" and "interpretation", also in relation to "completeness", is at the core of theoretical and foundational analysis also in physics.

Part II
Incompleteness and uncertainty: differences and similarities between physics and mathematics

In the view of discussing some themes around the notion of incompleteness, such as it appeared with force during the 30s, a period which saw the flourishing of Gödel's great logical theorems, and such as it started to foster debates and arouse new perplexities in physics during the same era, we will delve here into some concepts of quantum physics.

Firstly, and somewhat trivially, we know that in physics the accumulation of empirical proof does not suffice to account for the totality of the theoretical construction which represents phenomena. This is how, in physics and at a very first level, is manifested the incompleteness of proof principles, as grounded on empirical evidence, relative to construction principles. The latter are principles of conceptual construction and are, often, limit principles. No empirical evidence showed to Galileo that bodies never stop. Yet he dared to propose the principle of inertia at the non-existing limit of absence of friction, which is the only general and pertinent one (cf. [2] for more on this; chapter 4 for example, shows the constructive role of the geodesic principle).

However, this incompleteness has been thought way beyond this first and simple level, both in classical and in quantum frames.

Completeness/incompleteness in classical theories

We first return to a conceptual comparison between the positions of Laplace and Hilbert on the one hand and of Poincaré and Gödel on the other. The first two refer to a sort of strict completeness of theories that are physical in one case, and mathematical in the other: they would be complete in the sense that any statement concerning the future would be decidable (this is the Laplacian predictability of systems determined by a finite set of equations) or, regarding logico-formal derivations, the completeness or decidability of arithmetics (or of any sufficiently expressive axiomatic theory: this is Hilbert's completeness and decidability conjecture).

The other two demonstrate incompleteness: Poincaré, with his theorem on the three-body problem, showed unpredictability of interesting non-linear systems, which we may understand as undecidability of future states. Their dynamics will be said to be sensitive to the initial conditions, and in that, deterministic, yet unpredictable: minor variations, possibly below the level of observability, could cause major changes in the evolution of the system. Gödel proved at once the unprovability of coherence and the intrinsic undecidability and incompleteness of arithmetics (and of all its formal extensions), by constructing an undecidable statement, equivalent to the formal assertion of consistency (which is thus also unprovable within the system). We hinted to the epistemological and the recent technical link between the physico-mathematical problem and the purely mathematical one. It can then be interesting to try to characterize the main types of physical theories in terms of this relationship to completeness.

If relativistic theories may indeed proclaim theoretical completeness in the sense defined above, it very well appears that theories of classical dynamics on the one hand and quantum theories on the other hand, may present two distinct manners by which they manifest incompleteness.

In the case of chaotic dynamic systems, as we observed, unpredictability is associated to the sensitivity to the initial conditions joint to the non-linearity, typically, of the (formal) determination. It may however be observed that a (theoretically) infinite precision regarding the initial data is meant to generate a perfectly defined evolution (deterministic aspect of the system), or that a reinitialization of the dynamic system with rigorously *identical* values leads to reproducible results. Of course, this in principle and from the mathematical viewpoint, because physically speaking, we are still within the context of an approximation and the result of a measurement in classical and relativistic is always, in fact, an interval and not a unique point, in spite of the supposed mathematically continuous (space-time)background. Hence, we may notice that an essential *conceptual transition* appears between that which pertains to a finite level of approximation and that which constitutes a singularity, at the "actual" infinite limit of precision (or, one could say, between the unpredictable and the theoretically reproducible).

In quantum theories, contrastingly, as we shall see, be it an issue of relationships of indetermination or of non-separability, unpredictability is intrinsic to the system, it is inherent to it. In this case, the degree of approximation matters little: there is no conceptual *break* between finite and infinite, since, as for measurement, there is no supposed continuous space-time background (conitnuity may be found in Hilbert spaces, before measurement, but this is a different issue). Another way to observe this consists in noticing that this time, regardless of the rigor of the reinitialization of the system, the results of the *measurement* will not necessarily be individually reproducible (probabilistic character of quantum measurement), even if the law of probability of these results can be perfectly well known.

It is randomness then which is at the center of these theories, and this by subtle differences from classical frameworks. This point is delicate and we refer the reader to [3].

Incompleteness in quantum physics

In the debate on the completeness of quantum physics, Einstein, Podolsky and Rosen (EPR) highlighted three characteristics of the physical object which they believed to be fundamental in order to be able to speak of a *complete* theory (see [20], [10]):

1. the reference to that which they called elements of *reality* (as existing objects "beyond ourselves", independently of measurement or access, say);
2. the capacity to identify a principle of *causality* (including in the relativist sense);
3. the property of *locality* (or of *separability*) of physical objects.

Bell inequalities, [7], and their experimental verifications, notably by Aspect and his team, [1], have shown that the third EPR postulate was not corroborated: experience shows that two quantum objects having interacted remain for certain measurements a single object, consequently *non separable*, regardless of their distance in space. In other words, for two quantons having interacted, even if they are afterwards causally separated in space, any measurement of a value on the one would *instantaneously* determine the value of the other, against fundamental principles of Relativity Theory.

Presented this way, the eventual *incompleteness* (or, conversely, *completeness*) of quantum physics seems to have nothing to do with what is meant by completeness or incompleteness in logic, which have been addressed above.[6] However, a deeper examination reveals that what appears at first glance as a lexical telescoping may not be completely fortuitous.

To each of these characteristics "required" by EPR in physics, one may, indeed, without distorting the significations too much, associate characteristics "required" by mathematics (recall that principles of proof correspond to formal deductions in mathematics and to empirical "evidence" in physics):

1. elements of reality would be put into correspondence with proofs that construct existence, that is, the effectiveness of mathematical constructions (which, axiomatically or not, we have seen the difference, cause mathematical structures to exist - similarly as we construct, isolate or "point-out" objects in physics, see 1 above);
2. the principle of causality would be put into correspondence with the effectiveness of the administration of proof (which presents and works upon the rational sequences of demonstrations, be they stemming from formalism as such or not - the deductive chain is here placed in correspondence with the causal one in physics, see 2 above);
3. the property of locality (or of separability) would be put into correspondence with the autonomy of mathematical theories and structures inasmuch as they would be "locally" decidable (or that within a formal theory,

---

[6] By accepting, for this discussion, to use the framework of arithmetics or of set theory (ZFC type) and its models, a framework within which the questions relative to logical completeness were first raised.

any statement or its negation would be demonstrable, see next for a relation to 3 above).

Now it is precisely this local autonomy of theories, this "locality" in terms of decidability, which seems to be contradicted by the theorems of incompleteness in mathematics. The latter indeed refer to a sort of globality of mathematical theories in that one may need to use stronger principles to prove a statment in a given theory. Techinically, recent results by Friedman show that, relatively simple combinatorial statements of Peano Arithmetic may require increasingly large ordinals or cardinals to be proved.[7] In a sense, the "global" structure of orders, even of the entire mathematics (if one believes that ordinals and cardinals express the proof theoretic power in mathematics) seems to step in the proof of local properties of the first order arithmetic. So, the adjunction of (finetely or recursively many) axioms to a theory doesn't render "decidable", at most "more expressive" (or capable of deciding previously undecidable statements), but at the cost of generating a new theory which requires the same treatment itself, because, remaining formal, it would still be incomplete. In other words, there is no way to isolate arithmetic nor any other sufficiently expressive mathematics and deduce within it "completely": tools from any other branch of mathematics may be need in a proof of a statement of the given (apparently simple) theory.

But we can probably push the analysis further than suggested by these conceptual analogies.

In mathematics, if we refer to the interpretation of Gödelian incompleteness in terms of discrepancy between construction principles (structural and significant) and proof principles (formal) that is, in terms of incomplete covering, between *semantics* and *syntax* (all achievable propositions are not formally derivable, or, more traditionally: semantics exceeds syntax), then a closer relationship may be established. This relationship concerns, among other things, the introduction and the plurivocity of the term *interpretation*, according to whether it is used in a context of model theory or if it is taken in its common physical sense. In model theory, the excess of semantics (construction principles) with regard to syntax (proof principles) is first manifested in distinct interpretations (existence of non-isomorphic models, that is, non-categoricity) of a same syntax (for example, the non-standard model of Peano's arithmetic, cf. above). Gödelian incompleteness furthermore demonstrates that some of these models realize different properties (technically, they are elementarely non-equivalent).

In physics, if we accept to see the equivalent of syntax in the mathematical structure of quantum mechanics and the equivalent of semantics in their proposed conceptual-theoretical interpretations (hidden-variable theories, for example, see below), then the "semantics" would also exceed "syntax" and, consequently, a certain form of incompleteness (in this sense) would be man-

---

[7] See http://www.math.ohio-state.edu/users/friedman/

ifest.[8] But more profoundly, in the case of quantum physics, it is the excess of the "possible", the quantum states in a Hilbert space, over the "actual", once states are measured, which best illustrates this type of parallelism and of comparison. In other words, the result of a measurement corresponds to a plurality of potential states leading to it, each with its well defined probability. A sort of non-categoricity of the states (non-isomorphic, even if they belong to the same system) relatively to the well defined result of the measurement. The latter operates here like a sort of "axiomatic" constraint in that it stems, as we have seen, from the physical principles of proof, which are empirical (empirical proofs are "constraint" by physical measure).

The concept of incompleteness was then understood by EPR in the sense that quantum physics should be deterministic in its core and that its probabilistic manifestations would only be due to the lack of knowledge of "hidden variables" and of their behaviors. This actually amounts to saying "there are hidden causal relationships between particles that are not described by the theory". Now, the EPR argument is experimentally contradicted by the violation of the Bell inequalities and the fact that the property of non-separability is indeed inherent to quantum physics. In this sense quantum theory has been shown to be complete (there are no hidden variables).

Can we nevertheless speak of incompleteness in a sense different than that of EPR without however it being totally extraneous? In other words, would it be possible to formulate a proposition that is undecidable in the sense that it would be true according to one model and false according to another? Let's consider the crucial statement which can be attributed to EPR: "there are hidden variables". As we have just seen, this statement is false according to the usual model (standard interpretation) of quantum mechanics. Yet a theory presenting the same properties as quantum mechanics can be constructed, in which this statement is true *on condition* that an adjective is added: "there are *non-local* hidden variables".[9]

---

[8] If we want to continue with the analogy and in parallel with Logic, we will also notice that this search for hidden variables, which prove to be non-local, evokes in a way the method of *forcing*, which enabled Cohen to demonstrate the independence in ZFC of the Continuum Hypothesis and of the Axiom of Choice (by constructing a model not realizing them, whereas Gödel had constructed a model that does, see [28]). The hidden variables in question are indeed "forcing" for the physical model they nevertheless continue to respect. This is somewhat analogous, conceptually speaking, to the logical situation where forcing propositions are compatibly integrated with the original axiomatic construction: one adds or forces extending variables or properties, previously "hidden" or not assumed.

[9] "Local variables" is an expression which is also equivalent to "variables attached to particles" (they depend only on properties specific to a given particle each: this is locality). To speak then of non-local variables is to express the fact that the value of a variable which governs the behavior of a particle may not only depend on this particle, but may also depend on (remote) other particles. This is also another way to consider the non-separability we have just mentioned. As a matter of fact, the distinction (separability) between two particles having

Incompleteness *vs* indetermination

Sometimes, confusion is set in between the concept of incompleteness (as presented by EPR) and that of indetermination (as highlighted by Heisenberg). So let's try to further explain how are the relationships between incompleteness and indetermination, which do not cover the same conceptual constructions.

The issue of incompleteness such as raised by EPR leads, we have seen, to the search for "hidden variables" which would "explain" the counter-intuitive behavior of quantum phenomena. As we observed, it is possible to elaborate hidden-variable theories, but these variables are themselves non-local, therefore simply postponing the intuitive difficulties. In this regard, it is better to preserve the canonical version of quantum physics, the Bell inequalities and the experiments by Aspect which highlight the property of quantum *non-separability* (that for two separated quantons, which have previously interacted, measurement on one would determine the value of the other). By highlighting the fact that one of the origins of this situation stems from the use of *complex* numbers (in state vectors or wave function) as additive quantities (principle of superposition), whereas that which is measured is a *real* number and refers to the squares of the modules of these quantities. We will return to this in a moment.

Quantum indetermination ("the uncertainty principle"), for its part, mobilizes somewhat different concepts: it consists in the treatment of explicit variables (non-hidden), such as positions and momentums, of which the associated operators will present a character of *noncommutativity* (measuring first one observable, then the other does not commute: a fundamental property which will lead moreover to the development of non-commutative geometry, the current insight into the space of microphysics). It is in fact an issue of the constraints which weigh the Plank constant (small but non-null) upon physical measurement, precluding simultaneous measurement with an "infinite" precision of two conjugated magnitudes such as positions and the corresponding momentums, which we have just mentioned.

If we wanted to roughly distinguish the two types of conceptual ambiguity introduced by these quantum properties with regard to habitual intuition, we would say that incompleteness refers rather to an *ambiguity of object* (is the object local or global? How is it that according to the nature of the experi-

---

interacted is a representation that stems directly from classical physics, be it relativistic. For its part, quantum physics proves to be fundamentally non-local, that is to contain entangled (non-separable) quanta. Thus, a type of valid proposition in an interpretation (no hidden variables and non-separability) can be false in another (existence of hidden variables, but by specifying *non-local* variables). In fact, more broadly, the controversy first initiated by L. de Broglie continues among some physicists, currently a small minority, with regards to the character of causal determination, which would be classical but hidden, undescribed by the theory: an incompleteness of quantum physics.

ment it appears to manifest either as a particle or a wave?[10]). Indetermination instead leads to an *ambiguity of the "state"* of the system: a quantum object of which we know the precise position would be affected by an imprecise velocity; conversely, the precise knowledge of a velocity would entail an "indeterminacy" regarding the position occupied. In a purely mathematical way, a quantum object which we would manage to "stop" would occupy all space.

In fact, both versions of these quantum specificities refer to a difficulty of describing quantum phenomena "classically" within time and space of classical or relativistic theories, while their description within their own "abstract" spaces (a Hilbert space, for instance) is perfectly clear.[11]

Regarding space, we will note several traits which make of our usual intuition of space (and even of the Riemannian manifold of general relativity) an instrument which is unadapted to properly represent quantum phenomenality:[12]

i) Firstly, as we have indicated above, quantum quantities are defined at the onset on the field of complex numbers C, as opposed to classical and relativistic quantities which are defined at the onset of the field of real numbers R. It stems from this that in quantum physics, what is added (principle of superposition) is not what is measured (complex amplitudes are added - as vectors, their square norms are measured - as real numbers). At the same time and for the same reason, quantum objects so defined (wave function or state vector, for example) are no longer endowed with the "natural" order structure associated to real numbers (a total order).

ii) Then, as we have seen, the definition of the observables makes it so that some among them (corresponding to the conjugated magnitudes) are not commutable, as opposed to the classical case. This, in the context of the geometrization of this physics, necessitates the introduction of a geometry that is itself non-commutative, thus breaking, as we said, with all previous traditions [16].

iii) Finally, the enquiry may lead even further, with regard to the relationships between quantum phenomenality and the nature of our usual geometric

---

[10] Situations with regard to which Bohr was lead very early on to introduce the concept of *complementarity* (in the sense of a complementarity specific to the quantum object, which could manifest itself, according to the type of measurement performed, either as particle or as a wave), which was the object of many controversies.

[11] These differences between the notions of incompleteness, as meant by EPR, and of indetermination make our conceptual analysis concerning incompleteness, from Gödel to EPR, very different from the technical correlation, à la Chaitin, beween Gödel's theorem and quantum indetermination in [14].

[12] For example, with inseparability, everything seems to occur as if an event locally well defined in the state space of the definition of quantum magnitudes - an Hilbert space typically, was to *potentially* project itself upon two distinct points in our usual state space.

space. May the latter be Newtonian or Riemannian, it will admit a representation as a set of points and its continuum stems from an indefinite divisibility. Now given the existence of a scale of length (possibly minimal, cf. [38]) such as the Plank length, recent string theories lead to ask if in fact this space would not escape a description in terms of punctual elementarity (eventually to the benefit of another, in terms of interval elementarity,[13] or in terms of higher dimensionality such as "branes").

It should be clear that these specific issues lead to a conceptual revolution in our relating to physical space, at least the space of microphysics. The key idea is that geometry, as a human construction, as we stressed in part, is the consequence of the way we *access* to space, possibly by *measure*. So Euclid started by accessing, *measuring*, with rule and compass. Riemann analyzed, more generally, the *rigid body* (and characterized the spaces where this tool for measure is preserved: those of constant curbature). Finally, todays noncommutative geometry, in Connes' approach, begins by reconstructing space by quantum measure. In microphysics, this happens to be noncommutative (measure this and, later, that, is not equivalent to measuring that and, later, this). And this takes us very far from the space of senses or even classical/relativistic spaces. In conclusion, measure, by rule and compass, by the rigid body, or quantum measure is the form of access we have to space and events in it. This access may differ, yet it may provide a geometric way to a novel unity, by explaining first how to pass from one mathematical organization/understanding of space to another.

Constitutions of objectivity. Conceptual comparisons between mathematics and physics

We will now try to understand some aspects of the contingent constitution of mathematical idealities, in analogy to some physical processes, the phase transitions.

"Actual infinity", once conceptually constructed by constitutive intersubjectivity, is not only conducive to imagining the idea of God (and theology) or, like Giordano Bruno, the infinity of worlds. It is, on one side, a mathematical concept, and on the other it is involved in the process of constitution of the mathematical objects themselves (both finitary and infinitary). The idea is that mathematical objects are "substantially" (if we may say so) a concentrate of actual infinity brought into play by human beings, within their symbolic culture. Let's explain and argue.

The transcendental constitution of mathematical objectivities (from the finitary ones which are the triangle or the circle, to the structures of well

---

[13] We would then maybe pass from a Cantorian representation of continuity to a representation by interval interlockings such as proposed by Veronese or to the nil-potent infinitesimals, see [6].

order, or even to the categories of finite objects of which we have spoken) actually involves a very fundamental *change of level*: the process of abstraction of acts of experience and of the associated constructions (see Part I) leads to this *transition*, which constitutes the forms thus produced, into abstract structures as eidetic objects. This transition (which also leads to their conditions of possibility) presents all the characteristics we ascribe to "physical criticality" (the theory of "critical phase transitions", [8], [29]). Notably, this describes the passage from the *local* (such or such empirical form) to the *global* (the structure which is defined abstractly and which is to be found in *all* particular manifestations), as well as the passage from a certain ("subjectivizing") heteronomy to an autonomy (objectivity), and from a certain instability (circumstantial) to a stability (a-eventual). For example, even Euclid's passage from the empirical practice of lines, in measuring "geo", to the concept of thickless line or o dimensionless point, is a conceptual transition, which may be better specified, in our view, in analogy to the critical ones in physics. It is namely these constituted characteristics, resulting from a sort of passage to the effective infinite limit as process of constitution, which lead to mathematical platonic thinking. The latter though, as much as the formal axiomatic approach, forgets the constitutive process itself.

The examples we can use to try to account for this conceptual transition, which leads to the constitution of new objectivities, are varied. In mathematics, we recall the example of the sum of rational numbers ($1/n!$, for example) which gives, at the actual infinite limit, a transcendental number (e, in this case). This may be seen as a critical transition, or "space transition", which leads from one field (the rationals, $\mathbb{Q}$) into another one encompassing it (the reals, $\mathbb{R}$). In physics, we may find an equivalent in a change of phase, associated to the divergence of an intensive magnitude of the system (a susceptibility, for example, which formally is considered to go to infinity) and to the passage from local to global (divergence of the correlation length of interactions). In biology, it would be a case of a change in the level of functional integration and regulation (the organism in relation to its constituents, for example).

If mathematical structures are also the result of the search for the most stable invariants, as is conceptually characterized in the preceding, it is then probably due, at least in part, to this process of constitution mobilizing a form of actual infinity and leading to a sort of stable autonomy, at the limit. Let's continue with the physical metaphor of phase changes. A phase transition can be manifested for instance in a symmetry breaking of the system and a concomitant change (sudden or more progressive) of an order parameter (the total magnetic moment for a para-ferromagnetic transition, density for a liquid/solid transition). In fact, the phase transition is, in a way or another, a transition between disorder (relative) and order (also relative). If we keep these characteristics in mind and make them into a conceptual trait that is common to the transcendental constitution of mathematical objectivities, we will readily notice the disordered situation with regards to the often uncoordinated collection of "empirical" mathematical beings, and the ordered situation in mathematical objects and structures as such, as resulting from the process of abstraction and of constitution.

So it is easily conceivable why the axiomatization, or even the logicization of the statements characterizing these mathematical structures, are genetically and in some respects conceptually *second* relatively to the mathematical activity and to the process of constitution itself, as we emphasize in the first part of this paper. In our view, here are the roots of the mathematical incompleteness of formalisms. Indeed, the formal statements describe *order* consecutively to the "phase transition" we have just invoked. But mathematical thinking concerns as much the process of transition as the putting into form and description of its result. And to go even further, we could almost say that the evacuation of the "contingent disorder" accompanying transition-constitution corresponds to the evacuation of the "significations" associated to the structures over the course of their elaboration and to the "infinitary involvement" it presupposes. This is why the purely logico-formal foundations lead to pure syntaxes devoid of meaning and intrinsically incomplete.

It is probably also one of the reasons enabling to understand why formalism "works" when it is an issue of describing the order resulting from the transition in question (the constituted mathematical structures), but that it fails from the moment it is given the task of *also* describing the transition itself, that is, the process of constitution as such, from and in the terms of its result. In fact, one may consider that formalism fails to capture "actual infinity" that enabled the passage and which has become a major characteristic of the objectivities thus constructed. In this sense, the "semantic" aspect is the most deeply involved in the occurrence of this effective passage to the infinite limit, whereas the syntactic aspect is much more relative to the rigorous, possibly axiomatic, description of the once stabilized *results* of this passage. We insist that it is the non-coincidence of these two dimensions that is at the origin of the properties of incompleteness, that we saw as a discrepancy or gap between construction principles (conceptual) and proof principles (formal), of which we speak in part I. The results of incompleteness are a demonstration of these lackings.

So here ends, in our view, the proposed conceptual analogy with phase transitions in physics because we know that in physics, as we have recalled earlier, renormalization theory proves itself in a way to be able to address the critical transition itself. This difference in behavior relatively to the processes put into play is to be referred to the difference between the objects considered themselves, such as they are elaborated in physics and in mathematics: if the construction principles are similar, as we have shown in [2] and hinted here, the proof principles are completely different (empirical *vs* formal), and it is indeed regarding the status of the proof that the difference is manifest.

It is probably what transpires in this dichotomy introduced in [2] relatively to elementarity. As a matter of fact, we opposed the elementary and *simple* (related to the artificial processes of algorithmic calculus, to the concatenation of simple logical gates, or even to any artifact) to the elementary and *complex* (related to natural processes such as strings in quantum physics or cells in biology). Quantum physics and biology address natural phenomena that are confronted to elementarities that are rather complex and hence they seem irreducible to processes grounded on simple elementarities, in the sense of artificial computation and general artefacts. Besides the role of meaning in deduction (a role stressed by "concrete" incompleteness in particular, see [32]), this further prevents from reducing scientific judgment to a calculus, in this sense, without denying, of course, the interest of the complementary understanding provided by the formal and computational descriptions.

Conclusion

Let's conclude this paper by stressing the perspective that guided our work. In our views, the epistemological investigation of mathematics cannot be detached from a constitutive analysis of concepts and structures (and thus of the very object of knowledge) in other scientific disciplines, such as physics. This is the project that, in a very preliminary and modest fashion, but along the same phenomenological approach, we try to pursue in several papers and in [2], an extension, within a scientific project, of some of the ideas we hinted here. The analogies and differences in the "phenomenology of incompleteness" is a fundamental part of it. We believe that further work should lead to an analysis of this phenomenon in other disciplines (see [35] for some reflections on a form of causal incompleteness in biology).

Acknowledgements

We would like to thank the anonimous referees for their many comments and questions. Longo's papers are downloadable: http://www.di.ens.fr/users/longo.

References

1. A. Aspect, P. Grangier and G. Roger: Experimental Realization of the Einstein-Podolsky-Rosen-Bohm Gedankenexperiment: A New Violation of Bell's Inequalities. *Phys. Rev. Let.* **9** (1982) p 91
2. F. Bailly and G. Longo: *Mathématiques et sciences de la nature. La singularité physique du vivant* (Hermann, Paris 2006). (English introduction, downloadable)
3. F. Bailly and G. Longo: Randomness and Determination in the interplay between the Continuum and the Discrete. *Mathematical Structures in Computer Science* **17**, 2 (2007)
4. H. Barreau and J. Harthong (eds): *La mathmatique non standard* (Ed. CNRS 1989)
5. J. Barrow-Green: *Poincaré and the three body Problem. History of Mathematics* XI (AMS 1997)

6. J. L. Bell: *A Primer of Infinitesimal Analysis* (Cambridge Univ. Press, Cambridge 1998)
7. J. S. Bell: On the Einstein-Podolsky-Rosen Paradox. *Physics* **1** (1964) p 195
8. J. Binney, N. J. Dowrick, A. J. Fisher and M. E. J. Newman: *The Theory of Critical Phenomena: An Introduction to the Renormalization Group* (Oxford Univ. Press, Oxford 1992)
9. M. Bitbol: *L'aveuglante proximité du réel* (Flammarion, Paris 2000)
10. D. Bohm: The Paradox of Einstein, Rosen and Podolsky. *Quantum Th.* (1951) p 611
11. V. Brattka and G. Presser: Computability on subsets of metric spaces *Theoretical Computer Science*, **305** (2003)
12. V. Brattka and K. Weihrauch: Computability on subsets of Euclidean space I: Closed and compact subsets. *Theoretical Computer Science* **219** (1999)
13. C. Calude: *Information and Randomness: An Algorithmic Perspective* (Springer, Berlin 2002)
14. C. Calude and M. Stay: From Heisenberg to Gödel via Chaitin. *International Journal of Theoretical Physics* **44**, 7 (2005) pp 1053-1065
15. P. Collins: Continuity and computability of reachable sets. *Theoretical Computer Science* **341** (2005)
16. A. Connes: *A. Non-commutative Geometry* (Academic Press, New York 1994)
17. S. Dehaene: *The Number Sens* (Oxford Univ. Press, Oxford 1997)
18. R. L. Devaney: *An introduction to Chaotic Dynamical Systems* (Addison-Wesley, Reading 1989)
19. F. Diacu: *Singularities of the N-Body Problem* (Publications CRM, Montreal 1992)
20. A. Einstein, B. Podolsky and N. Rosen: Can Quantum-Mechanical Description of Physical Reality be Considered complete? *Phys. Rev.* **41** (1935) p 777
21. B. van Fraassen: *Lois et symetries* (Vrin, Paris 1994)
22. G. Frege: *The Foundations of Arithmetic* (1884) (Evanston 1980)
23. K. Gödel: Remark About the Relationship between Relativity Theory and Idealistic Philosophy. In: *Albert Einstein: Philosopher-Scientist*, ed by P. A. Schilpp (Open Court, LaSalle Ill.) pp 557-562. Reprinted with corrections and additions in *Collected Works*, ed by S. Feferman et. al., Vol. 2 (Oxford University Press, Oxford 1990)
24. L. Harrington et al. (eds): *H. Friedman's Research on the Foundations of Mathematics* (North-Holland, Amsterdam 1985)
25. D. Hilbert: *Les fondements de la géométrie* (1899) (Dunod, 1971)
26. M. Hoyrup: Dynamical Systems: Stability and Simulability. *Mathematical Structures in Computer Science* **17**, 2 (2007)
27. M. Hoyrup, A. Kolcak and G. Longo: Computability and the Morphological Complexity of some dynamics on Continuous Domains. Invited survey *TCS* (2008) To appear
28. T. Jech: *Set Theory* (Springer, Berlin 1997)
29. H. J. Jensen: *Self-Organized Criticality, Emergent Complex Behavior in Physical and Biological Systems* (Cambridge Lectures in Physics 1998)
30. J. Laskar: Large scale chaos in the Solar System. *Astron. Astrophysics* **287**, L9 L12, (1994)
31. G. Longo: *Laplace (A note on incompleteness)*. An item of J-Y. Girard's Logic Dictionnary, at the end of "Locus Solum", MSCS, vol. 11, n.3, (Cambridge Univ. Press, Cambridge 2001)

32. G. Longo: Reflections on Incompleteness or On the proofs of some formally unprovable propositions and Prototype Proofs in Type Theory (invited Lecture: *Types for Proofs and Programs* Durham UK, Dec. 2000). In: *Lecture Notes in Computer Science* **2277** ed by Callaghan et al. (Springer, Berlin 2002) pp 160-180
33. G. Longo: The Cognitive Foundations of Mathematics: human gestures in proofs and mathematical incompleteness of formalisms. In: *Images and Reasoning*, ed by M. Okada et al. (Keio University Press, Tokio 2005)
34. G. Longo: Laplace, Turing and the "imitation game" impossible geometry: randomness, determinism and programs in Turing's test. In: *The Turing Test Sourcebook*, ed by R. Epstein, G. Roberts and G. Beber (Kluwer, Dordrecht 2007)
35. G. Longo and P. E. Tendero: The causal incompleteness of Programming Theory in Molecular biology, *Foundations of Science* (to apprear). French version in: *Evolution des concepts fondateurs de la biologie du XXIe siècle*, ed by Miquel (DeBoeck, Paris 2007)
36. G. Longo and T. Paul: *The Mathematics of Computing between Logic and Physics*. In: *Computability in Context: Computation and Logic in the Real World*, ed by Cooper and Sorbi (Imperial College Press, World Scientific 2008) To appear
37. P. Martin-Löf: The definition of random sequences. *Information and Control* **9** (1966) pp. 602-619
38. L. Nottale: *La relativité dans tous ses états* (Hachette, Paris 1999)
39. J. Paris and L. Harrington: A mathematical incompleteness in Peano Arithmetic. In: *Handbook of Mathematical Logic*, ed by J. Barwise (North-Holland, Amsterdam 1978)
40. F. Patras: *La penseé mathématique contemporaine* (PUF 2001)
41. K. Petersen: *Ergodic Theory* (Cambridge Univ. Press, Cambridge 1990)
42. J-L. Petit: *Solipsisme et Intersubjectivité* (Cerf 1996)
43. S. Y. Pilyugin: *Shadowing in dynamical systems* (Springer, Berlin 1999)
44. M. B. Pour-El and J. I. Richards: Computability in analysis and physics. *Perspectives in mathematical logic* (Springer, Berlin 1989)
45. J-M. Salanskis: *L'hermneutique formelle* (Ed. CNRS 1991)
46. J. Tappenden: Geometry and generality in Frege's philosophy of Arithmetic. *Synthese* **102**, 3 (1995)
47. R. Tieszen: *Phenomenology, Logic, and the Philosophy of Mathematics* (Cambridge Univ. Press, Cambridge 2005)
48. H. Weyl: *Philosophy of Mathematics and of Natural Sciences* (Princeton University Press, Princeton 1949)
49. K. Weihrauch: Computable Analysis. In: *Texts in Theoretical Computer Science* (Springer, Berlin 2000)

Index

Abadi, M., 156
Aczel, P., 45, 47
Aho, A., 172
Al-Kindi, 196, 219
Alberti, L. B., 206, 219
Alexander, J., 61, 171
Ammon, K., 128–130, 139
Archimedes, 69, 70, 72, 73, 78, 79
Architas, 69
Argand, J., 199
Aristotle, 2, 31, 254
Arnold, V., 63
Artin, E., 165, 167, 171
Aschbacher, M., 33, 47
Aspect, A., 261, 264, 269
Asperti, A., 155
Atiyah, M., 63
Avigad, J., 14, 24, 31, 32, 47
Azzouni, J., 6, 7, 24

Bacon, R., 196, 219, 220
Baillot, P., 148, 151, 155
Bailly, F., 219, 269
Baire, R., 39, 228–230
Banach, S., 238
Barendregt, H., 42, 43, 47
Barreau, H., 269
Barrow-Green, J., 269
Basin, D., 139
Battaglia, D., 172
Baumslag, G., 172
Baxter, R. J., 62
Bell, J. L., 253, 270

Bell, J. S., 202, 204–206, 219, 261, 263, 264, 269, 270
Benacerraf, P., 253
Bendixon, I., 31
Berardi, S., 193
Bernays, P., 30, 34–36, 44, 47, 118, 126, 131
Beth, E., 182
Bianchi, L., 56
Bibel, W., 139
Biedenharn, L. C., 172
Binney, J., 270
Birkhoff, G., 251, 252
Birman, J., 63, 172
Bishop, E., 193
Bitbol, M., 270
Blum, L., 225, 226, 241
Böhm, C., 45
Bohm, D., 269, 270
Bohr, N., 265
Boolos, G., 139
Borel, E., 31, 230
Bourbaki, N., 13
Bourgain, J., 47
Boyer, R. S., 130
Brattka, V., 229, 230, 235, 241, 270
Brendle, T., 63
Bridges, D., 185, 193
Bridson, M., 172
Brouwer, L., 175–179, 183, 185, 186, 190, 192, 243
Brunelleschi, F., 206
Brüning, S., 129, 139

Bruno, G., 266
Bundy, A., 128, 130, 132, 139
Byrnes, J., 120, 139, 140

Calude, C., 219, 270
Cannon, J., 172, 173
Cantini, A., 47
Cantor, G., 25, 31, 39, 129, 165, 247
Cardano, G., 218
Carnap, R., 11, 12, 24
Carroll, L., 32, 47
Casari, E., 31, 47
Cassirer, E., 202, 217, 219
Cauchy, A., 73, 227, 228, 230, 231, 234, 236, 238, 239
Cayley, A., 163, 164
Cellucci, C., 2, 3, 24, 31, 47, 219
Cenzer, D., 237, 238, 242
Chaitin, G., 219, 224, 242, 251, 265, 270
Chern, S.-S., 49, 55, 59, 60, 62, 63, 170, 171
Chomsky, N., 160, 162, 172
Chuang, I., 173
Church, A., 32, 40, 41, 44, 89, 93, 130, 141, 153, 155, 200, 220, 232
Cipra, B., 78
Cittadini, S., 121, 140
Cohen, P., 31, 131, 139, 263
Collins, P., 270
Connes, A., 266, 270
Conway, J., 61, 172
Cook, S., 149, 155
Cooper, W., 16, 20, 24
Coppola, P., 155
Coquand, T., 185
Cornaros, C., 47
Corsi, G., 116
Crosilla, L., 45, 47
Crossley, J., 150, 155
Crutchfield, J., 173
Cucker, F., 225, 226, 241
Curry, H., 30, 40, 86, 93, 141, 142, 144, 155
Curtis, R., 172
Cutland, N., 155

Dal Lago, U., 155, 156
Daniel, M., 63
Danos, V., 156

Dawson, J., 14, 25
de Broglie, L., 200, 264
de Swart, H., 182
de Vries, F., 240, 241
Dedekind, R., 45, 46, 118, 119, 131, 139, 224
Dehaene, S., 270
Dehornoy, P., 165, 172, 173
Dennis, E., 173
Descartes, R., 23–25, 27, 206, 207, 209, 219
Detlefsen, M., 78, 88, 89, 91, 93
Deutsch, D., 220
Devaney, R., 270
Devlin, K., 16, 25
Diacu, F., 270
Diels, H., 25
Dijksterhuis, E., 78
Dimitracopoulos, C., 47
Dirac, P., 200, 219, 247
Donnelly, K., 47
Dowrick, N., 270
Dummett, M., 176, 177, 179–183, 185, 188–190, 193, 256

Earman, J., 220
Eguchi, T., 63
Einstein, A., 53, 56, 243, 246, 261, 269, 270
Ekert, A., 220
Enriques, F., 250, 251, 253
Epstein, D., 173
Eratosthenes, 70
Erdös, P., 37
Euclid, 5, 20, 23, 72, 245, 246, 249, 266, 267
Eudoxus, 69, 72
Euler, L., 56, 65–69, 73, 78
Eutocius, 76

Facchi, P., 220
Fano, G., 220
Fearnley-Sander, D., 130, 139
Feferman, S., 38, 46, 47, 118, 140, 220, 254
Fermat, P., 3, 4, 26, 68, 191
Feynman, R., 52, 58, 173, 220, 225, 242
Feys, R., 93, 155
Fisher, A., 270

Fitch, F., 124, 140
Flament, D., 220
Fleuriot, J. D., 9, 26
Floyd, W., 172
Fraenkel, A., 118
Freedman, M., 173
Frege, G., 1, 6–8, 13, 25, 220, 246, 256, 258, 259, 270, 271
Friedgut, E., 47
Friedman, B., 255, 262
Friedman, H., 37, 38, 191, 270

Gaboardi, M., 156
Galilei, G., 21, 25, 259
Gandy, R., 201, 220
Garnerone, S., 63, 173
Gay, W., 240, 242
Gentzen, G., 1, 2, 6, 7, 9, 13, 25, 29, 84–86, 93, 95, 97, 129, 141, 156
Gersten, S., 172
Gherardi, G., 241, 242
Gilkey, P., 63
Gilman, R., 173
Ginsburg, S., 173
Girard, J.-Y., 145, 146, 148, 151, 156, 270
Giunchiglia, F., 130, 139
Giusti, E., 31, 47
Gödel, K., 4, 10, 11, 14, 24, 25, 31, 39, 117, 119–121, 123–125, 127–133, 135, 137, 139, 140, 183, 202, 220, 243, 246–255, 259, 260, 263, 270
Goerdt, A., 47
Goldbach, C., 31, 191
Gonseth, F., 35, 47
Gonthier, G., 156
Goodman, N., 173
Gouvea, F., 78
Govindarajan, T., 63
Grangier, P., 269
Gray, D., 47
Gray, R., 22, 26
Grosholz, E., 5, 9, 25
Gross, D., 59, 63
Grosseteste, R., 195, 196, 220
Guadagnini, E., 63
Guénon, R., 224, 242
Guerrini, S., 155
Gustavsson, J., 154, 156

Hadamard, J., 207, 220
Hafner, J., 47
Hales, T., 78
Hamming, R., 5, 25
Hanson, A., 63
Hardy, G., 14, 25, 47
Harrington, L., 255, 270, 271
Harrison, M., 173
Hart, W., 20, 25
Harthong, J., 269
Hausdorff, F., 31, 51
Heath, T., 78
Hegel, G., 46
Heidegger, M., 18, 25
Heil, M., 19, 25
Heisenberg, W., 219, 264, 270
Hellman, G., 193
Hersh, R., 15, 25, 65, 78
Heyting, A., 93, 177, 179
Hilbert, D., 1, 2, 6, 7, 9, 10, 13, 25, 26, 31, 32, 35, 40–42, 118, 126, 131, 170, 190, 201, 203–205, 209, 210, 213, 218, 220, 246, 249, 250, 252, 257, 259, 260, 263, 265, 270
Hintikka, J., 6, 26
Hippocrates, 3, 4, 76
Hofmann, M., 150, 156
Holt, D., 173
Hopcroft, J., 173
Hosoi, T., 95, 116
Howard, W., 30, 86, 93, 141, 142, 144, 156
Hoyrup, M., 251, 270
Hughes, R., 210, 220
Hume, D., 24, 83, 89, 92
Hunt, G., 22, 26
Husserl, E., 243–249, 253, 256, 258, 259
Hutter, D., 139

Intrigila, B., 45, 48
Ireland, A., 139
Ishihara, H., 46, 47

Jackiw, R., 63
Jaeger, F., 63
Jaffe, A., 49–51, 63
Jankov, V., 95, 100, 116
Jech, T., 270
Jensen, H. J., 270

Jones, V., 60–63, 170, 171, 173

Kac, M., 78
Kahle, R., 40, 48
Kalmar, L., 142, 144, 146
Kant, I., 17, 26, 27
Kauffman, L., 63
Kaul, R., 63
Kemp, M., 220
Kepler, J., 63, 207
Kitaev, A., 173
Kleene, S., 128, 149, 226
Klein, F., 213, 245
Knorr, W., 6, 23, 26
Kohlenbach, U., 37, 39, 48
Kolcak, A., 270
Kolmogorov, A., 86, 93, 177
König, D., 231
Korteweg, D., 240, 241
Koyré, A., 207, 209, 220
Kreisel, G., 32, 39, 48, 150, 156, 183
Kreitz, C., 228
Kripke, S., 8, 26, 182, 193
Krivine, J.-L., 193
Kruskal, M., 37, 255
Kwiat, P., 220

Lafont, Y., 156
Lagrange, J., 56
Lakatos, I., 65, 73, 75, 78
Landahl, A., 173
Landau, L., 52, 146
Laplace, P., 220, 250, 259, 270, 271
Larsen, A., 173
Laskar, J., 270
Laurent, O., 156
Laurent, P., 61
Lawvere, F., 253
Lebesgue, H., 251
Lee, G., 37, 48
Leibniz, G., 247
Leighton, R., 220
Leonardo da Vinci, 207–210, 218, 220
Levinson, I., 174
Lévy, J.-J., 156
Lévy, S., 173
Lickorish, W., 63
Lie, S., 56, 60
Lindberg, D., 220

Lindstrom, I., 140
Lindstrom, S., 140
Liouville, J., 62, 63
Lipton, J., 193
Löb, M., 124, 127–129, 132, 133, 135, 140
Löbl, M., 38
Lolli, G., 2, 26, 48, 220
Longhi, R., 220
Longo, G., 219, 220, 250, 269–271
Lorentz, H., 57
Louck, J., 172
Lubarski, R., 46, 48
Luker, M., 78, 88, 89, 91, 93

Mac Lane, S., 34, 48
MacFarlane, J., 194
Mach, E., 21, 26
Mach, L., 198, 199, 214
Macintyre, A., 1, 26
Mackie, I., 156
Maddy, P., 253, 254
Mäenp, P., 6, 26
Mancosu, P., 31, 47, 48
Marion, J.-Y., 156
Markov, A., 43, 227
Martin-Löf, P., 45, 86, 93, 185, 187, 194, 251, 258, 271
Martini, S., 156
Marzuoli, A., 63, 173
Mathai, G., 155
Matousek, J., 38
Maxwell, J., 56, 59
McCarthy, J., 131
McCarty, D., 194
McKenzie, R., 173
McLaughlin, D., 124
McNaughton, R., 173
Meikle, L., 9, 26
Menaechmus, 3, 4
Merleau-Ponty, M., 247, 256
Merli, R., 220
Mills, R., 55, 59, 63
Minkowski, H., 56, 57
Missiroli, G., 220
Moore, C., 173
Moore, J., 130
Moran, A., 154, 156
Motwani, R., 173

Murawski, A., 151, 156
Myhill, J., 45, 235, 236

Newman, M., 270
Newton, I., 23, 24, 26, 63, 207, 220, 250
Nicomedes, 75, 76
Nielsen, M., 173
Norton, S., 172
Nottale, L., 271

Oliva, P., 48
Ong, C.-H., 151, 156

Pachos, J., 173
Pacioli, L., 207, 218
Padoa, A., 2
Panofsky, E., 196, 208, 209, 221
Pappus, 6, 26, 75, 76
Paris, J., 255, 271
Parker, R., 172
Parry, W., 172
Pascal, B., 2
Pascazio, S., 220
Patras, F., 271
Patterson, M., 173
Paul, T., 271
Peano, G., 4, 11, 37, 118, 119, 230, 231, 237, 259, 262, 271
Pedicini, M., 148, 155
Pennings, T., 16, 26
Penrose, R., 174, 221, 254, 255
Perelman, G., 50, 51, 64
Petersen, K., 271
Petit, J-L., 271
Pieri, M., 2
Piero della Francesca, 206, 218, 221
Pilyugin, S., 271
Pinto, J., 156
Planck, M., 58
Plato, 2, 69, 70, 76, 77
Plotkin, G. D., 44, 156
Plutarch, 75
Podolsky, B., 261, 269, 270
Poincaré, H., 46, 50, 51, 57, 243, 246, 250, 251, 255–259, 269
Poisson, S. D., 63
Pólya, G., 12, 26, 31, 46, 48, 65, 66, 69, 78
Pour-El, M., 230–237, 240–242, 271

Pozzi, G., 220
Prawitz, D., 29, 48, 93, 190, 194
Prawitz, H., 93
Preskill, J., 173
Presser, G., 270
Pringsheim, A., 38
Proclus, 26
Putnam, H., 65, 69, 78

Quaife, A., 128–130, 137, 140
Quine, W., 174, 253
Quinn, F., 49–51, 63

Racah, G., 170
Raff, P., 47
Ramadevi, P., 63
Ramus, P., 17
Rasetti, M., 63, 172–174
Rathjen, M., 45–48
Rav, Y., 7, 26, 30, 32, 34, 48
Regnier, L., 156
Reidemeister, K., 167
Remes, U., 6, 26
Remmel, J., 237, 238, 242
Rényi, A., 37
Ribet, K., 3, 4, 26
Ricci Curbastro, G., 64
Richards, J., 230–237, 240–242, 271
Richman, F., 185, 191, 194
Riemann, B., 245, 246, 266
Robbins, H., 66
Roger, G., 269
Rojas, C., 251
Ronchi, S., 156
Rosen, N., 261, 269, 270
Rosser, J., 41, 130
Rota, G., 13, 22, 26, 29, 48
Rufini, E., 72, 79
Russell, B., 8, 17, 26, 30, 35, 36, 46, 48

Salam, A., 59
Salanskis, J-M., 271
Sambin, G., 185, 187, 194
Sand, M., 220
Sands, D., 154, 156
Scarani, V., 221
Schnirelmann, L., 31
Schönfinkel, M., 40
Schrödinger, E., 205, 238–241

Schub, M., 225, 226, 241
Schuster, P., 46
Schütte, K., 38, 46
Schützenberger, M., 172
Schwarz, J., 59, 63
Scott, D., 42
Seely, R., 155
Selberg, A., 32
Shankar, N., 128, 130, 131, 140
Shapiro, M., 172
Shimura, G., 3, 4, 13, 21, 26
Short, H., 172
Sieg, W., 120, 140, 221
Silva Bueno, J., 19, 25
Simons, J., 49, 55, 59, 60, 62, 63, 170, 171
Simpson, S., 37, 48, 231, 242
Smale, S., 225, 226, 241
Sobolev, S., 240, 241
Stålmarck, G., 82
Statman, R., 45, 48, 144, 156
Stay, M., 219, 270
Steiner, M., 65, 69, 79
Suslin, M., 31
Swart, E., 79

't Hooft, G., 59
Taniyama, Y., 3, 4, 13, 21, 26
Tappenden, J., 271
Tarski, A., 10, 26
Tassi, G., 116
Taylor, R., 4
Teller, P., 79, 88–90, 94
Tendero, P., 271
Terui, K., 155, 157
Thielscher, M., 139
Thompson, R., 165, 168, 171, 173, 174
Thurston, W., 51, 64, 173
Tieszen, R., 244–250, 252–258, 271
Timmermans, B., 27
Tortora de Falco, L., 156
Troelstra, A., 157, 194
Turing, A., 32, 141–143, 145, 153, 160, 200–202, 204, 218, 221, 223, 225–229, 232, 236–238, 241, 252, 254, 271
Turing, A. M., 160
Tutte, W., 63
Tymoczko, T., 66, 79, 87, 88, 90, 94

Ulam, S. M., 78
Ullman, J. D., 173
Urquhart, A., 149, 155

Vaihinger, H., 21, 27
Valentini, S., 193
van Benthem, J., 20, 27
van Dalen, D., 194
van Emde Boas, P., 157
van Fraassen, B., 221, 270
Vanagas, V., 174
Vassiliev, V., 165
Veronese, G., 266
Vertigan, D., 63
Viallet, C., 63
Villafiorita, A., 130, 139
Voghera, N., 93
von Neumann, J., 141, 217, 221

Wadsworth, C., 153, 157
Walsh, T., 130, 139
Wang, Z., 173
Weiermann, A., 37, 38, 48
Weierstrass, K., 38
Weihrauch, K., 223, 225, 228, 231–235, 238–242, 270, 271
Weil, A., 3
Weinberg, S., 59
Weinfurter, H., 220
Weldman, D., 182
Welsh, D., 63
Westerståhl, D., 85, 194
Weyl, H., 46, 204, 209, 211, 212, 218, 221, 243, 250, 251, 256, 257, 271
Wheeler, J., 205, 221
Wigner, E., 170
Wilczek, F., 59
Wiles, A., 4
Wilson, R., 58, 60, 62, 170–172
Witten, E., 60, 64, 171, 174
Wittgenstein, L., 179, 190, 245
Wootters, W., 221

Xenophanes, 9

Yang, C., 55, 59, 62, 63
Yutsis, A., 174

Zanardi, P., 173, 174
Zehnder, L., 198, 199, 214
Zeilinger, A., 220

Zeno, 215, 217, 220
Zermelo, E., 118
Zhang, B., 240, 242

Zhong, N., 223, 231–235, 238–242
Zureck, W., 221

Printed in July 2008